AF576061

Legumes as Food Ingredient

Legumes as Food Ingredient: Characterization, Processing, and Applications

Editors

Alfonso Clemente
Jose C. Jimenez-Lopez

MDPI • Basel • Beijing • Wuhan • Barcelona • Belgrade • Manchester • Tokyo • Cluj • Tianjin

Editors
Alfonso Clemente
Spanish National Research Council (CSIC)
Spain

Jose C. Jimenez-Lopez
Spanish National Research Council (CSIC)
Spain

Editorial Office
MDPI
St. Alban-Anlage 66
4052 Basel, Switzerland

This is a reprint of articles from the Special Issue published online in the open access journal *Foods* (ISSN 2304-8158) (available at: https://www.mdpi.com/journal/foods/special_issues/Legumes_Food_Ingredient_Characterization_Processing_Applications).

For citation purposes, cite each article independently as indicated on the article page online and as indicated below:

LastName, A.A.; LastName, B.B.; LastName, C.C. Article Title. *Journal Name* **Year**, *Volume Number*, Page Range.

ISBN 978-3-0365-0614-2 (Hbk)
ISBN 978-3-0365-0615-9 (PDF)

Cover image courtesy of Jose R. Fernandez.

Contents

About the Editors

Alfonso Clemente He is a staff scientist at the Spanish National Research Council, working at the Estación Experimental del Zaidín (Granada, Spain). He has been working in legume seeds for the last 20 years, being involved in several national and international related projects. Alfonso Clemente joined different labs (Institute of Food Research, 1999–2000; John Innes Centre, 2000–2002; Sainsbury Laboratory, 2003–2004) in the UK to broaden his laboratory skills and scientific knowledge. Currently, he is the President of the Spanish Legume Association (Asociación Española de Leguminosas, www.leguminosas.es) having strong interaction with a relevant network of scientists and agricultural associations and agri-food companies in the field. He is author of more than 120 scientific manuscripts and an editorial board member of the World Journal of Gastroenterology and Frontiers in Bioscience, among others.

Jose C. Jimenez-Lopez Bs. in Biochemistry and Molecular Biology (1998) and Bs. in Biological Sciences (2001), Ms. in Agricultural Sciences (2004), University of Granada, Spain and PhD degree in Plant Cell Biology (2008) at the Spanish National Research Council (CSIC). He developed a Full-time Postdoctoral research associate at Purdue University, USA (2008-2011). Marie Curie Research Fellow (FP7-PEOPLE- 2011-IOF) (2012–2015) at the University of Western Australia and CSIC working in human health benefits of legume seed proteins, their allergy molecular aspects and cross allergenicity. He is a Senior Research Fellow (Ramon y Cajal research program), currently working in the functionality, health benefits, and allergy implications of proteins from reproductive tissues (pollen and seeds) in crop species of agro-industrial interest (mainly legumes). He is an Author of more than 60 peer-review journal articles, 25 book chapters. His work has been presented in more than 130 international congresses. He is an Active member of different Scientific Societies: Spanish and International Legume Society; Spanish and EU Microscopy societies. He is Editor of multiple books, Reviewer for more than 35 Peer-Review Journals of editorial as Elsevier, Springer, Wiley, Frontiers, etc, and international expert panels member for funding grants evaluation.

Preface to "Legumes as Food Ingredient: Characterization, Processing, and Applications"

Grain legumes like lupines, chickpeas, lentils, peas, and beans, among others, are well-recognized as outstanding sources of protein and other health promoting compounds as starch, fibre, vitamins, and minerals for the human diet. Legumes are essential and economically affordable food crops for people around the globe, being major ingredients in the Mediterranean diet, and playing a vital role particularly for developing countries.

Recent advances have been made in unravelling the beneficial effects of legumes beyond meeting basic nutrient requirements. Thus, there is a growing body of scientific evidences regarding the health benefits of some of their bioactive components in inflammatory-related diseases, cardiovascular diseases, obesity and weight management, type-2 diabetes, cancer, among others. Anti-nutritional properties as food allergies is a key subject that is also being analysed at molecular and clinical level for safety, while increasing knowledge are being used to help developing molecular tools for diagnosis and therapy.

Therefore, efforts are being also made in studies focused in the sensory and techno-functional properties of food with particular relevance as effective method for the delivery of improved nutritional and nutraceutical compounds for food industry applications.

This special issue brings together a range of scholarly review and research articles focused on legume crops, key components of healthy diets, sensory properties and seed compounds allergy molecular features. Here we summarize some of the highlights derived from the nine research articles and a review published in this special issue.

Marin-Manzano et al. reported the prebiotic properties of non-fructosylated α-galactooligosaccharides from pea seeds, with strong bifidogenic activity with potential use in infant formula.

Lima-Cabello et al. developed a molecular highly specific method for detection, identification and quantification of the major lupines allergen b-conglutin in natural in processed food, contributing to a more efficient management of allergens by the food industry, the regulatory agencies and clinicians, thus helping to keep the health safety of the consumers.

Matsuo et al. analysed the levels of various major allergens from genetically modified (GM) soybean using antigen-immobilized ELISA and immunoblotting. No significant differences either in these levels non in the serum IgE-reactive protein profiles of the patients analysed, concluding that, in general, GM soybeans are not more allergenic than non-GM soybeans. Khazaei et al. wrote a review describing lentil seeds as outstanding source of plant-based proteins and a viable alternative animal feed and food processing formulations for human. Authors focuses on the current knowledge of seed protein, extraction and isolation methods, bioactive peptides, and food applications of lentil protein.

Shelat et al. focused in wattle seed as primary food within indigenous communities in Australia. This research encompasses the nutritional and sensory properties of this underutilized but abundant Acacia specie, showing that the species analysed had good nutritional value for human

Schlegel et al. investigated the influence of fermentation process using different microorganisms on the sensory profile, techno-functional properties and protein integrity of lupin protein isolates. This study concludes that fermentation allows the development of food ingredients with good functional properties in foam formation and emulsifying capacity, with a well-balanced aroma and taste profile.

El Youssef et al. investigated how to reduce off-flavours in plant-derived food mainly associated with the presence of aldehydes, ketones, furans, and alcohols, by using fermentation processes with lactic acid bacteria (LAB) and yeasts. This study conclude that these fermentations may be a powerful tool for the improvement of the sensory perception of a pea protein-based product.

Khrisanapant et al. characterized the volatile and fatty acid profiles across different legume seeds using solid-phase microextraction gas chromatography–mass spectrometry and GC coupled with a flame ionisation detector (GC-FID), respectively. Such information may contribute to develop a database of legume-based ingredients with specific volatile characteristics aiming to reduce undesirable odours.

Kim et al. investigated the use of defatted soybean flour in food as a potential source of dietary fiber, without changing the physical and sensory properties in the making of high-quality food such as tofu.

Sozer et al. described the use of lactic acid fermentation as a potential modification tool for faba bean flour implementation in improving the nutritional profile of gluten-free breads, while preserving the crumbliness, evenness of pore size and springiness of breadcrumb.

Alfonso Clemente, Jose C. Jimenez-Lopez
Editors

Editorial

Introduction to the Special Issue: Legumes as Food Ingredient: Characterization, Processing, and Applications

Alfonso Clemente * and Jose C. Jimenez-Lopez

Estación Experimental del Zaidín (CSIC), Consejo Superior de Investigaciones Científicas (CSIC), 18008 Granada, Spain; josecarlos.jimenez@eez.csic.es
* Correspondence: alfonso.clemente@eez.csic.es; Tel.: +34-9-5857-2757

Received: 29 September 2020; Accepted: 20 October 2020; Published: 23 October 2020

Abstract: Legumes are major ingredients in the Mediterranean diet, playing an essential role in developing countries. Grain legumes, such as lentil, chickpea, pea, lupin and beans, among others, are recognized as good sources of proteins, starch, fiber, vitamins and minerals for human nutrition, being an essential food crop for people worldwide. Due to their nutritional and techno-functional properties, legumes are widely used by the food industry as ingredients in a wide range of products for general and specific groups of the population, including vegetarians, diabetics or celiac patients. The Special Issue "Legumes as Food Ingredients: Characterization, Processing, and Applications" covers key aspects regarding the nutritional quality of legume flours and their derived products, as well as the health benefits of some of their bioactive components. The amounts of antinutritional components, such as certain allergens that might pose risks to sensitized consumers, are reported to be reduced by processing. Several pretreatments, including fermentation with lactic bacteria and yeasts, are used to improve the nutritional and sensory profile of the legume-derived products, increasing their acceptance by consumers.

Keywords: legumes; pulses; nutritional properties; health benefits; processing; microbiota; allergens; fermentation; sensory properties

---

Grain legumes and their derived products are increasingly used in food industry due to their nutritional composition and technological properties. The presence of bioactive compounds in legume seeds has been reported to exert beneficial effects in terms of tackling chronic disorders, including diabetes, cardiovascular diseases, and inflammatory and carcinogenic processes. As a result, Governmental and Health Agencies recommend as a healthy habit the regular consumption of legumes. These are largely used as ingredients by the food industry, being employed in a wide range of processes, including heat treatment, roasting, milling, canning, germination and fermentation, among others, that might influence their nutritional and sensory profile. On this issue, nine research papers and one review are published in this Special Issue.

Marin-Manzano et al. [1] reported the prebiotic properties of non-fructosylated α-galactooligosaccharides from pea seeds in infants. These oligosaccharides were fully fermented by fecal microbiota, lactic acid and short chain fatty acids were generated, and strong bifidogenic activity was observed. As a result, its potential use has been suggested as a prebiotic in infant formula.

The presence of allergen traces in foodstuffs might elicit allergic reactions in sensitized consumers. Regarding this, a sensitive and accurate ELISA methodology to detect, identify and quantify the lupin major allergen β-conglutin (Lup an 1) in raw and processed (roasted, fermented, boiled, cooked, pickled, toasted, pasteurized) lupin-derived foods was developed [2]. This assay might be considered as a reliable method to be used for the detection of lupin in a range of food matrices. In a comparative study, the levels of eight major soybean allergens in genetically and non-genetically modified soybeans were evaluated using ELISA and immunoblotting techniques; the overall data indicated that genetically modified soybeans did not show significant differences in the levels of the soybean allergens compared to non-genetically modified soybeans [3]. Lentil seeds are a source of high-quality proteins, being increasingly used in food industry as ingredients in a wide range of food formulations. Khazaei et al. [4] reviewed the current knowledge on lentil proteins and their bioactive peptides, the methods used for their extraction and isolation, and their potential applications in the food industry.

The nutritional and sensory profiles of four underutilized *Acacia* species have been reported [5]. Their nutritional value suggests their potential to be included in the human diet or used in food formulations. Khrisanapant et al. [6] reported the characterization of volatile (aldehydes, alcohols, ketones, esters, terpens and hydrocarbons) and fatty acid profiles for up to 11 legume species, identifying some compounds that are specific for certain legume species. This information could be used by plant breeders and the food industry for the elaboration of legume products with specific volatile profiles.

The fermentation of legume flours is carried out by the food industry for the improvement of the nutritional, functional or sensory properties of legume products. Thus, the lactic acid fermentation of faba bean flours has been demonstrated as a potential pretreatment for improvement of the nutritional quality of gluten-free faba breads, without affecting their sensory properties [7]. Schlegel et al. [8] reported the fermentation of lupin protein isolated with up to eight different microorganisms, and the effects on the sensory and techno-functional (protein solubility, foaming and emulsifying capacity) properties, and the protein's resistance to the fermentative process, were both evaluated. The fermentation of a pea protein solution with a starter culture of lactic bacteria and different yeasts significantly reduced the presence of green-off compounds; such results are of interest for the food industry in terms of increasing the acceptability of plant matrices to consumers [9].

Finally, Kim et al. [10] reported the use of superfine defatted soybean flour obtained by jet milling for the preparation of fiber-enriched tofu, without affecting the physical and sensory properties of the product.

Funding: This research received no external funding.

Conflicts of Interest: The authors declare no conflict of interest.

References

1. Marin-Manzano, M.C.; Hernandez-Hernadez, O.; Diez-Municio, M.; Delgado-Andrade, C.; Moreno, F.J.; Clemente, A. Prebiotic Properties of Non-Fructosylated α-Galactooligosaccharides from Pea (*Pisum sativum* L.) Using Infant Fecal Slurries. *Foods* **2020**, *9*, 921. [CrossRef] [PubMed]
2. Lima-Cabello, E.; Alché, J.D.; Jimenez-Lopez, J.C. Narrow-Leafed Lupin Main Allergen β-Conglutin (Lup an 1) Detection and Quantification Assessment in Natural and Processed Foods. *Foods* **2019**, *8*, 513. [CrossRef] [PubMed]
3. Matsuo, A.; Matsuchita, K.; Fukyzumi, A.; Tokumasu, N.; Yano, E.; Zaima, N.; Moriyama, T. Comparison of Various Soybean Allergen Levels in Genetically and Non-Genetically Modified Soybeans. *Foods* **2020**, *9*, 522. [CrossRef] [PubMed]
4. Khazaei, H.; Subedi, M.; Nickerson, M.; Martinez-Villaluenga, C.; Frías, J.; Vandenberg, A. Seed Protein of Lentils: Current Status, Progress, and Food Applications. *Foods* **2019**, *8*, 391. [CrossRef] [PubMed]
5. Shelat, K.J.; Adiamo, O.Q.; Olarte Mantilla, S.M.; Smyth, H.; Tinggi, U.; Hickey, S.; Rühmann, B.; Sieber, V.; Sultanbawa, Y. Overall Nutritional and Sensory Profile of Different Species of Australian Wattle Seeds (*Acacia* spp.): Potential Food Sources in the Arid and Semi-Arid Regions. *Foods* **2019**, *8*, 482. [CrossRef] [PubMed]
6. Khrisanapant, P.; Kedebe, B.; Leong, S.Y.; Oey, I. A Comprehensive Characterization of Volatile and Fatty Acid Profiles of Legume Seeds. *Foods* **2019**, *8*, 651. [CrossRef] [PubMed]
7. Sozer, N.; Melama, L.; Silbir, S.; Rizzello, C.G.; Fander, L.; Poutanen, K. Lactic Acid Fermentation as a Pre-Treatment Process for Faba Bean Flour and Its Effect on Textural, Structural and Nutritional properties of protein-Enriched Gluten-Free Faba Bean Breads. *Foods* **2019**, *8*, 431. [CrossRef] [PubMed]

8. Schlegel, K.; Leidigkeit, A.; Eisner, P.; Schweiggert-Weisz, U. Technological Properties of Fermented Lupin Protein Isolates. *Foods* **2019**, *8*, 678. [CrossRef] [PubMed]
9. El Youssef, C.; Bonnarme, P.; Fraud, S.; Peron, A.C.; Helinck, S.; Landaud, S. Sensory Improvement of a Pea Protein-Based Product Using Microbial Co-Cultures of Lactic Acid Bacteria and Yeasts. *Foods* **2020**, *9*, 349. [CrossRef] [PubMed]
10. Kim, Y.N.; Muttakin, S.; Jung, Y.M.; Heo, T.Y.; Lee, D.U. Tailoring Physical and Sensory Properties of Tofu by the Addition of Jet-Milled, Superfine, defatted Soybean Flour. *Foods* **2019**, *8*, 617. [CrossRef] [PubMed]

Article

Prebiotic Properties of Non-Fructosylated α-Galactooligosaccharides from PEA (*Pisum sativum* L.) Using Infant Fecal Slurries

María del Carmen Marín-Manzano [1], Oswaldo Hernandez-Hernandez [2], Marina Diez-Municio [2], Cristina Delgado-Andrade [1], Francisco Javier Moreno [2] and Alfonso Clemente [1,*]

[1] Estación Experimental del Zaidín (CSIC), Consejo Superior de Investigaciones Científicas (CSIC), 18008 Granada, Spain; mamen.marin.manzano@gmail.com (M.d.C.M.-M.); cristina.delgado@eez.csic.es (C.D.-A.)

[2] Institute of Food Science Research (CIAL, CSIC-UAM), 28049 Madrid, Spain; o.hernandez@csic.es (O.H.-H.); m.dmunicio@gmail.com (M.D.-M.); javier.moreno@csic.es (F.J.M.)

* Correspondence: alfonso.clemente@eez.csic.es; Tel.: +34-9-5857-2757

Received: 26 April 2020; Accepted: 8 June 2020; Published: 13 July 2020

Abstract: The interest for naturally-occurring oligosaccharides from plant origin having prebiotic properties is growing, with special focus being paid to supplemented products for infants. Currently, non-fructosylated α-galactooligosaccharides (α-GOS) from peas have peaked interest as a result of their prebiotic activity in adults and their mitigated side-effects on gas production from colonic bacterial fermentation. In this study, commercially available non-fructosylated α-GOS from peas and β-galactooligosaccharides (β-GOS) derived from lactose were fermented using fecal slurries from children aged 11 to 24 months old during 6 and 24 h. The modulatory effect of both GOS on different bacterial groups and bifidobacteria species was assessed; non-fructosylated α-GOS consumption was monitored throughout the fermentation process and the amounts of lactic acid and short-chain fatty acids (SCFA) generated were analyzed. Non-fructosylated α-GOS, composed mainly of manninotriose and verbascotetraose and small amounts of melibiose, were fully metabolized and presented remarkable bifidogenic activity, similar to that obtained with β-GOS. Furthermore, non-fructosylated α-GOS selectively caused an increase on the population of *Bifidobacterium longum* subsp. *longum* and *Bifidobacterium catenulatum/pseudo-catenulatum*. In conclusion, non-fructosylated α-GOS could be used as potential ingredient in infant formula supplemented with prebiotic oligosaccharides.

Keywords: galactooligosaccharides; GOS; gut microbiota; pea; prebiotic; raffinose oligosaccharides; short-chain fatty acids (SCFA)

1. Introduction

The colonisation of the gastrointestinal tract (GIT) by microorganisms is an essential process in our life cycle and starts during the gestation period [1,2]. Evidence shows that the GIT colonisation rapidly increases after birth by aerobic microorganisms that decrease the concentration of oxygen. This lowering of redox potentially allows the colonisation of anaerobic bacteria [3]. The continued succession of microorganisms during the first years of life has an important effect on the long-term maturity of the GIT microbiota, which is affected by diet, mode of delivery, antibiotic treatments, genetics, intestinal mucin glycosylation and other factors [4].

Non-digestible dietary ingredients are capable of modulating composition and metabolic function of gut microbiota. In this context, the prebiotic concept was recently defined by the International Scientific Association for Probiotics and Prebiotics (ISAPP) as "a substrate that is selectively utilized by host microorganisms conferring a health benefit" [5]. It is now recognized that the prebiotic effect extends beyond bifidobacteria and lactobacilli and includes other genera such as *Akkermansia* and *Faecalibacterium*. These bacterial groups, among others, are involved in the production of lactic acid and short-chain fatty acids (SCFA), which perform crucial physiological functions as metabolic regulators and immunomodulators, decreasing the growth of potential intestinal pathogens. The SCFA are responsible for the decrease of pH in the intestinal lumen, an increase of calcium absorption and shortening of gastrointestinal transit time. Besides, SCFA play an important role in cross-feeding processes and are major sources of energy for colonic epithelial cells and other cell types located in peripheral tissues [4,6].

The human-milk oligosaccharides (HMOs) are one of the most important dietary ingredients modulating the infant microbiome. Only certain bifidobacterial species, such as *Bifidobacterium longum* subsp. *infantis*, are able to utilize HMOs as metabolic substrates [7] with specific preferences depending on the *Bifidobacterium* strain and the HMOs structure [8,9]. Due to different clinical and social conditions, human milk is not a realistic choice for some infants and, in such cases, infant formulas are the most suitable alternative. However, access to HMOs for infant formula is extremely limited [10]. As a result, different oligosaccharides have been utilised to mimic the function of HMOs, such as fructooligosaccharides (FOS) and β-galactooligosaccharides (β-GOS). Other natural sources have also been utilised to obtain carbohydrates for infant formulas with well reported prebiotic activities such as inulin [11], pectin-oligomers [6], raffinose [12] and resistant starch [13]. For infant formulas, research has mainly focused on inulin, fructooligosaccharides (FOS) and β-galactooligosaccharides (GOS); indeed, there is still a wide range of oligosaccharides that could mimic, at least to some extent, the functional properties of HMOs.

Plant-based GOS, with α-galactosidic linkages instead of β-linkages, are water soluble carbohydrates and are particularly abundant in legume seeds as soybean, chickpea, lentil, faba bean and pea. They are commonly named raffinose family oligosaccharides (RFOS) and show a terminal sucrose unit linked by the glucose monomer to galactoses via α-(1→6) linkages. Like β-GOS, α-GOS are not hydrolysed in the upper gastrointestinal tract due to the absence of the enzyme α-galactosidase expressed by somatic cells in the mammalian GIT [14]. As a result, they reach the large intestine where are fermented by gut microbiota and can exert prebiotic properties [15]. However, it is well-known that RFOS cause discomfort, bloating and flatulence to consumers due to the presence of the ending fructose monomer [16,17].

Non-fructosylated α-GOS include melibiose (CAS 585-99-9), manninotriose (CAS 13382-86-0) and verbascotetraose (CAS 1111-08-6), which are essentially raffinose, stachyose and verbascose without the ending fructose units. Non-fructosylated α-GOS from peas is produced by the activity of β-fructosidases which are able to split sucrose from the α-GOS chain into glucose and fructose, being commercially available as AlphaGOS®. This particular α-GOS mixture has claimed to reduce the post-prandial glycaemic responses, being recently approved by the European Food Safety Agency (EFSA) [18], and exert prebiotic activity in adult faecal inoculum [19]. Available scientific data suggest that the administration of prebiotic oligosaccharides to healthy infants does not raise safety concerns with regards to adverse effects. In this sense, non-fructosylated α-GOS has been proven safe in infant formula in a concentration up to 8 mg/mL after preclinical evaluation in neonatal piglets [20].

Despite the various scientific evidence available regarding non-fructosylated α-GOS derived from peas, no reports have been published so far regarding the effect of these plant oligosaccharides on infant microbiota either in vitro or in vivo. Taking into account the published data reporting the bifidogenic properties of non-fructosylated α-GOS in adults, and its safety in infant formula, we hypothesized that a similar effect could be also found in infants, despite the differences between GIT microbiota in adults and in infants. Such increase in the number of bifidobacteria might be considered a major shift in the gut microbiota towards a potentially healthier composition. Important functions have been attributed to bifidobacteria including a protective role against pathogens, promote gut epithelium integrity and modulate the host immune system. Therefore, the main aim of the study was to evaluate the effect of non-fructosylated α-GOS in the microbiota present in infant faecal samples with particular attention paid to their bifidogenic properties.

2. Material and Methods

2.1. Chemicals

MTBSTFA (N-terbutil-dimetil-silil-N-metil-trifluoroacetamid) was purchased by Fluka Chemie AG (Buchs, Switzerland). Short chain fatty acids (SCFA) and organic acids were obtained from Sigma-Aldrich. All other chemicals were of analytical grade.

2.2. Non-Fructosylated α-GOS and β-Galactooligosacharides (GOS) Characterization

Non-fructosylated α-galactooligosaccharides (α-GOS, commercial brand AlphaGOS®) derived from pea seeds were kindly provided by Olygose (Venette, France). The chemical composition of α-GOS is detailed in Table 1 and shows as major components mannitriose and verbascotetraose with degree of polymerization of three (DP3) and four (DP4), respectively.

Table 1. Chemical composition of AlphaGOS®.

Parameter	% (Dry Matter)
Dry matter [a]	96.3
Crude proteins [a]	<0.2
Crude ash [a]	<0.2
DP1 + Maltose + Sucrose [b]	<0.2
DP2: Melibiose [b]	3.9
DP3: Manninotriose [b]	49.2
DP4: Verbascotetraose [b]	43.0

[a] Provided by the supplier; [b] quantified by GC-FID (see Section 2.10 for details); DP: degree of polymerization.

An industrially available galacto-oligosaccharide mixture derived from lactose (β-GOS) was used in this study for comparative purposes as prebiotic reference. This mixture was initially composed by galactose (1%), glucose (21%), DP2 (37%), DP3 (22%), DP4 (11%), >DP4 (8%). Gel filtration chromatography was carried out for the removal of mono- and disaccharides due to the presence of high levels of digestible carbohydrates including lactose; the DP of collected fractions was determined by electrospray ionization mass spectrometry (ESI-MS). The purified β-GOS mixture consisted in a complex mixture of oligosaccharides from DP3 to DP6 and fully free of mono- and disaccharides. The trisaccharide fraction of β-GOS (35.2% of total carbohydrates) contained mainly 4′-galactosyl-lactose (15.8%), 6′-galactosyl-lactose (5.3%), as well as other minor galactobioses linked to the reducing glucose unit by β-(1→2) and β-(1→6) glycosidic linkages (Supplementary Materials Table S1) [21].

2.3. Infant Faecal Samples

Parents/guardians of children of the Los Angeles Nursery (Granada, Spain) were invited to attend a meeting where the nature of the study was explained and those who agreed to participate signed the informed consent. Parents/guardians were given a small survey to learn about the overall health and eating habits of donors. The participants fed on solid food and none of them had taken probiotics, fermented foods, antibiotics or received medical treatment that could affect the intestinal microbiota at least a month before the stool samples were collected. The study protocol was conducted in accordance with the ethical recommendations of the Declaration of Helsinki.

Stool samples were collected from eight children in the age range between 11–24 months. They were taken by our laboratory staff after deposition in nappies, avoiding the edges and the area in contact with the nappy and also avoiding urine that was absorbed in the cellulose. Immediately after deposition, fecal samples were housed in closed jars containing Anaerocult A tablets (Anaerocult®, Darmstad,

Germany) in order to maintain anaerobic conditions during the transport to the laboratory. The fecal samples (~5 g), within 2 h after collection, were homogenized in a stomacher by dilution 1/10 (*w*/*v*) with anaerobic 0.1 M sodium phosphate buffer, pH 6.8, by using sterile plastic filter bags, and filtrated through Miracloth (Calbiochem, Darmstad, Germany) to remove solid residues.

2.4. In Vitro Faecal Fermentation

Faecal homogenates were transferred into sterile Hungate tubes containing a specific basal medium for cultivation of infant faeces (BMIF) by dilution 1/10 (*v*/*v*) [22] and were stabilized overnight under anaerobic conditions at 37 °C. After that, 0.3% (*w*/*v*) of specific substrate (non-fructosylated α-GOS and β-GOS) was added to stabilized faecal mixtures. Fermentation experiment for the fecal sample of each subject was carried out once for each GOS-type. An additional tube was kept without inoculum and without GOS as a negative control and the positive control contained inoculum but not GOS. In addition, the medium was supplemented with 1/1000 (*w*:*v*) of resazurine to control the strict anaerobiosis conditions. Fermentations were carried out in anaerobic chambers at 37 °C and samples were taken at time 0, 6 and 24 h. One mL of culture was centrifuged (12,000× *g* for 15 min). The pellet was frozen at −80 °C and freeze-dried for the study of bacterial populations; the supernatant was used for direct determination of pH at the different collection times (Crison Instruments S.A., Barcelona, Spain) and in order to quantify lactic acid and SCFA.

2.5. DNA Extraction

DNA was extracted from pellets of fecal batch cultures using FavorPrep™ Stool DNA Isolation Mini Kit (Favorgene Biotech Corp, Shuttleworthstraße, Vienna, Austria). A NanoDrop ND-100 spectrophotometer (NanoDrop Technologies, Wilmington, DE, USA) was used for DNA quantification; purified DNA samples were stored at −80 °C.

2.6. Microbial Faecal Populations Determined by Quantitative PCR (qPCR) Analysis

Quantitative PCR was used to evaluate the effect of non fructosylated α-GOS and β-GOS on microbial composition in fecal homogenates after 6 and 24 h of in vitro fermentation in comparison with the control group (absence of oligosaccharides). Different microbial groups including total bacteria, Bacteroides, lactobacilli, bifidobacteria, *Eubacterium rectale*/*Clostridium coccoides*, *Clostridium leptum*, enterobacteria and *Faecalibacterium prausnitzii* were distinguished and quantified using qPCR. The 16S rRNA gene-targeted group-specific primers used in this study are listed in Supplementary Materials Table S2. qPCR conditions used in this study were as previously reported [23]. For quantitative analysis of the different bifidobacteria species (*B. adolescentis*, *B. bifidum*, *B. catenulatum*/*pseudo-catenulatum*, *B. breve*, *B. longum* subsp. *infantis* and *B. longum* subsp. *longum*), the PCR conditions were one cycle of 94 °C for 5 min, then 40 cycles of 94 °C for 20 s, 55 °C for 20 s, and 72 °C for 1 min. The species-specific primer pair for bifidobacteria is listed in Supplementary Materials Table S3. In the case of *Eubacterium rectale*/*Clostridium coccoides* and *Clostridium leptum* groups, PCR conditions were an initial denaturation step at 94 °C for 5 min followed by 40 cycles at 94 °C for 20 s, 50 °C for 20 s and 72 °C for 1 min for primer annealing and product elongation [23]. The fluorescent product was detected in the last step of each cycle. Following amplification, melting temperature analysis of PCR products was performed to determine the specificity of the PCR. The melting curves were obtained by slow heating at 0.5 °C increments from 55 to 95 °C, with continuous fluorescence collection. A plasmid standard containing the target region was generated for each specific primer set using DNA extracted from fecal homogenates. The amplified products were cloned using the TOPO TA cloning kit for Sequencing (Invitrogen, Barcelona, Spain) and transformed into *E. coli* One Shot Top 10 cells (Invitrogen). Sequences were submitted to the ribosomal RNA database to confirm the specificity of the primers. For quantification of target DNA copy number, standard curves were generated using serial 10-fold dilutions of the extracted products by using at least six non-zero standard concentrations per assay. The bacterial concentration in each sample was measured as $\log_{10}$ copy number by the interpolation of the C_t values obtained by the fecal homogenates samples and the standard calibration curves. Each plate included triplicate reactions per DNA sample and the appropriate set of standards.

2.7. Short-Chain Fatty Acids (SCFA) and Lactic Acid Analysis

SCFA and lactic acid were analyzed as described previously by Tabasco et al. [24], with some modifications. For quantitative analysis, 55 μL of a 100 mM 2-ethylbutyric acid in distilled water was added as internal standard to 550 μL of bacterial supernatant after in vitro fermentation. SCFA and organic acids were extracted by the addition of 275 μL concentrated HCl and 1 mL diethyl ether followed by vortexing 1 min. Then, samples were centrifuged at 3000× *g* for 10 min and 50 μL of the ether layer was transferred to a GC-microvial to minimize ether evaporation, adding 10 μL of the derivatization agent MTBSTFA. Hermetically stoppered microvials were heated at 80 °C for 20 min, afterwards the reaction mixture was kept for 24 or 48 h at room temperature to ensure total derivatization. An external calibration curve was built using a standard solution mixture made from pure compounds (formate 10 mM, acetate 60 mM, propionate 20 mM, butyrate 20 mM, iso-butyrate 5 mM, valerate 5 mM, iso-valerate 5 mM, lactate 17 mM, succinate 17 mM) and derivatized as described for samples. A gas chromatograph Thermo Scientific (Trace GC Ultra, Rodena, Italy) equipment was used, with a mass detector ion trap (Thermo Scientific ITQ 900, Rodena, Italy) and with an automatic sampling system (Triplus). The chromatograph was equipped with a Thermo Scientific column (TR-5MS 30 m × 0.25 mm × 0.25 μm) and a 2.5 m × 1 μm 0.32 ID pre-column (Teknokroma, Guard Column TR30005, San Cugat del Vallés, Spain). The samples were introduced via split (ratio 50:1) injection with the port heated to 275 °C. Helium was used as the carrier gas at a flow rate of 1.0 mL/min. The oven temperature was initially held at 63 °C for 3 min, increased to 190 °C with a 20 °C per min rate, held at 190 °C for 6 min, then raised to 230 °C with a 40 °C per min rate, where it was held for 1 min. The mass spectrometer interface temperature was set to 250 °C. For monitoring and confirmation analysis, the electronic impact (EI, 70 eV) mode was used. For the MS/MS experiments, the adequate ion precursor and a collision energy was selected for each compound, in order to perform the quantification. Identification and quantification of SCFA and organic acids was based on the use of the relative response factors calculated for the target compounds in the standard solutions at different concentrations against the IS. Measurements were conducted at least twice.

2.8. PCR-Denaturing Gradient Gel Electrophoresis (DGGE) Analysis of Bifidobacteria

The 16S rRNA genes were amplified by PCR from extracted DNA of pellets from fecal batch cultures of infants. The *Bifidobacterium* genus-specific primers (Bif164-F and Bif662-GC-R) and PCR amplification conditions were as previously reported [23]. PCR fragments were separated by DGGE by using a denaturing gradient of 40 to 65%. The gels were visualized by silver-staining, dried at 37 °C and scanned. The total number and dendogram of similarity cluster analysis of DGGE bands were determined by Quantity One analysis software (BioRad). DGGE profiles were determined by cluster analysis using the Dice similarity coefficient and the unweighted-pair group method by means of arithmetic average clustering algorithm (UPGMA). The richness (S) of the bifidobacterial community was established from the number of electrophoretic bands in individual samples. Shannon index (H), was calculated as reported by Magurran [25] as: $H = -\sum (pi \cdot \ln pi)$, where *pi* is the abundance of every species. The evenness (E) of the bacterial community was further estimated as $E = H/\ln S$ [26].

2.9. Excision and Sequencing of DGGE Bands

The DGGE bands were excised and eluted by using sterilised distilled water at 4 °C. The primers Bif164-F and Bif662-R were used to amplify the bands of interest as previously described [23]. PCR products were purified and then cloned using the TOPO TA Cloning kit for Sequencing (Invitrogen). Plasmids DNA were isolated from selected transformants with the GenElute Plasmid Miniprep kit (Sigma-Aldrich, St. Louis, USA), being inserts sequenced. Searches for sequence similarity were carried out using the BLAST algorithm [27] of the GenBank database Release 232.0 (www.ncbi.nlm.nih.gov)

for identification of the nearest relatives of the partial 16S rRNA sequences. A sequence similarity ≥98% of the 16S rRNA gene was used as the criterion for identification of bifidobacterial species.

2.10. Carbohydrate Quantification by Gas Chromatography Coupled to Flame-Ionization Detector (GC-FID)

The carbohydrate concentration before and after in vitro fecal fermentation was analysed by GC-FID. The carbohydrate fraction was derivatized to their corresponding trimethylsilyl oximes (TMSO) following the method of Brobst and Lott [28]. The samples were dried and the oximes were formed by adding 350 μL of hydroxylamine chloride in pyridine (2.5% *w/v*) at 70 °C for 30 min. The resulting oximes were silylated with hexamethyldisilazane (350 μL) and trifluoroacetic acid (35 μL) at 50 °C for 30 min. The mixtures were centrifuged at 7000× *g* for 4 min. The supernatants were stored at a temperature of 4 °C prior to analysis. The chromatography separation was carried out in an Agilent Technologies gas chromatograph (Mod 7890A, Santa Clara, USA) with a fused silica capillary column DB-5HT (5%-phenyl-methylpolysiloxane; 30 m × 0.25 mm × 0.10 μm) (Agilent). The oven temperature was set to 150 °C and then increased to 380 °C at a rate of 3 °C/min. The injector and detector temperatures were set to 280 °C and 385 °C, respectively. One mL/min of nitrogen was used as carrier gas and the injections were performed in split mode (1:20). Data acquisition and integration were performed using the Agilent ChemStation software.

2.11. Statistical Analysis

The effect of non-fructosylated α-GOS and β-GOS on microbiota composition and SCFA amounts of fecal contents was analyzed using a linear mixed model: repeat measure (SPSS Statistics version 22.0, Madrid, Spain). The Bonferroni method with a p value ≤ 0.05 was used for adjustments for major effects and the time 0 h was used like a covariate.

3. Results and Discussion

3.1. Microbiota Composition after In Vitro Fermentation of Non-Fructosylated α-GOS and β-GOS with Infant Fecal Samples

Using infant fecal slurries, all bacteria and up to seven different bacterial groups were analysed before and after 6 and 24 h of the in vitro fermentation of non-fructosylated α-GOS and β-GOS. Table 2 shows the population of these bacterial groups, which is the average of data resulting from eight infant donors. During treatment, the control samples (in the absence of oligosaccharides) did not show significant differences, with the exception of bifidobacteria and enterobacteria that slightly increased after 6 and 24 h. *Bifidobacterium* spp. increased significantly in non-fructosylated α-GOS and β-GOS groups but no differences were observed between 6 and 24 h treatment. Indeed, the bifidogenic effect in non-fructosylated α-GOS and β-GOS groups was significantly greater than that found in control samples. To the best of our knowledge, no data has been previously reported regarding the effect of non-fructosylated α-GOS in the modulation of infants' microbiota, using either in vivo or in vitro systems. On the contrary, the bifidogenic properties of β-GOS in infants' microbiota is well documented. Thus, the bifidogenic activity of β-GOS using infant fecal slurries in a three-stage in vitro culture method that mimics the proximal, transversal and distal colon has been reported [29]. A strong bifidogenic activity in an in vivo intervention study in infants by using a formula-fed containing β-GOS was also reported [30]. A significant decrease in enterobacteria was observed in non-fructosylated α-GOS and β-GOS samples, while a slight increase was found in the control group. *C. coccoides/E. rectale* group, Lactobacillus spp., *C. leptum* group, *F. praustnizii, Bacteroides* and total bacteria did not show significant differences in both non fructosylated α-GOS and β-GOS treatments (Table 2).

Table 2. Microbiota population from in vitro-fermented infant faecal samples with non-fructosylated α-GOS and commercial β-GOS derived from lactose.

Log10 Copy Number/ Gr Dry Faeces	Control			α-GOS			β-GOS			*p*-Values				Pooled SEM
	Bs	6 h	24 h	Bs	6 h	24 h	Bs	6 h	24 h	T	Prebiotics	T * Prebiotics	Bs	
All bacteria	9.78	9.83	9.75	9.77	9.79	9.79	9.89	9.77	9.83	0.871	0.316	0.600	<0.001	0.027
Bifidobacteria spp.	7.49	7.59 [a]	7.68 [a]	8.01	8.59 [b]	8.62 [b]	7.61	8.18 [b]	8.35 [b]	0.285	<0.001	0.823	<0.001	0.043
Clostridium coccoides/ Eubacterium rectale group	8.54	8.60	8.54	8.63	8.61	8.66	8.61	8.53	8.55	0.994	0.144	0.606	<0.001	0.020
Lactobacilli spp.	5.01	4.78	4.40	5.02	4.86	5.02	5.02	4.69	4.80	0.788	0.150	0.208	<0.001	0.068
Clostridium leptum subgroup	7.57	7.67	7.73	7.72	7.68	7.69	8.00	7.93	7.88	0.900	0.049	0.719	<0.001	0.026
Enterobacteria spp.	7.81	7.85 [b]	7.88 [b]	8.24	8.12 [a, b]	8.12 [a, b]	7.93	7.69 [a]	7.70 [a]	0817	<0.001	0.967	<0.001	0.028
F. praustnizii	8.13	8.14	8.12	8.20	8.09	8.10	8.13	8.04	7.99	0.716	0.100	0.893	<0.001	0.023
Bacteroides	9.63	9.67	9.61	9.67	9.59	9.60	9.63	9.51	9.54	0.855	0.065	0.707	<0,001	0.021

Outcomes were analyzed using a linear mixed model: repeat measure. A Bonferroni method was used for adjustments for major effects (time of treatment and type of GOS) [a,b] Mean value with different letters means that differences among treatments was significantly different at p-value < 0.05. Unadjusted mean values $n = 8$. Bs: Baseline like covariate. * Interaction among major effects.

Analysis of different bacterial groups after in vitro fermentation of non-fructosylated α-GOS and β-GOS suggest that these oligosaccharides are preferentially metabolized by bifidobacteria. Although this study is mainly focused in the bifidogenic properties of both GOS types, prebiotic targets extend beyond stimulation of bifidobacterial and lactobacilli, and recognizes that health benefits can derive from effects on other beneficial taxa. Such is the case of *F. praustnizzii* and the *Clostridium coccoides/Eubacterium rectale* bacterial group that includes species that are known butyrate-producers, thereby contributing to important processes linked to colonic health, including the protection against inflammatory bowel diseases. As reported in Table 2, in vitro fermentation of non-fructosylated α-GOS and β-GOS did not affect the growth of these butyrate-producers whilst enterobacteria numbers significantly decreased. In this sense, further studies are planned to determine the preventive role of non-fructosylated α-GOS in the colonization of pathogenic enterobacteria.

The modulation of gut microbiota by RFOS has been previously evaluated using in vitro fermentation batch cultures, inoculated with a standardised human adult faecal sample; a higher increase of *Bifidobacterium* spp. using RFOS when compared with FOS and β-GOS was observed, being the bifidogenic effect dose-dependent [19]. However, to the best of our knowledge, there is not specific information regarding the potential prebiotic properties of non-fructosylated α-GOS. Recently, the prebiotic potential of melibiose-derived gluco-oligosaccharides linked by α-(1→3) and α-(1→6) glycosidic bonds as unique carbon source has been evaluated by testing the growth of bifidobacteria (*B. breve*, *B. longum* subsp. *longum*, *B. animalis*) and lactobacilli (*L. reutieri*, *L. plantarum* and *L. rhamnosus*) [31]. In agreement with our data, lactobacilli were unable to metabolize melibiose-derived oligosaccharides whereas bifidobacteria were able to utilize them.

Although no data regarding the modulatory effect of non-fructosylated α–GOS in infant fecal microbiota has been previously reported, it has been suggested they could be safely used in infant formulas. In a study carried out with neonatal piglets to mimic infancy conditions, an intervention during three weeks having a daily consumption of 3.6–3.9 g non-fructosylated GOS/kg body weight demonstrated to be well tolerated with no adverse effects in terms of clinical signs, body weight, feed consumption, haematology, organ weight and histopathology [20].

3.2. *Effect of Non-Fructosylated α-GOS on Bifidobacterial Composition*

As previously pointed out, non-fructosylated α-GOS and β-GOS had a strong bifidogenic effect in infant inoculum after 6 h of fermentation and was kept at 24 h of treatment. To obtain a more comprehensive assessment of the impact of both GOS on the population structure of bifidobacteria in children's faecal samples, 16S rRNA gene profiles were generated by means of genus-specific primers PCR-DGGE. Control samples reflected an inter-individual variation in faecal bifidobacterial community being grouped in two cladograms with electrophoretic bands differing from 3 up to 11 (Supplementary Materials Figure S1). Dendrogram analysis indicated a grouping of the samples by children but they were not clustered after non-fructosylated α-GOS and β-GOS treatment (Figure 1). The electrophoretic bands were excised for sequencing and bifidobacterial species identified. Sequencing allowed us to identify *B. catenulatum/pseudocatelunatum*, *B. longum* subsp. *longum and B. longum* subsp. *infantis* as the most frequently present bifidobacterial species in infants. Whereas *B. longum* subsp. *infantis* was clearly identified, *B. longum* subsp. *longum* did not achieved such level of recognition at subspecies level and might be an experimental limitation that needs to be considered. The presence of *B. adolescentis and B. bifidum* was identified in only one of the infants. These species have been reported in many studies as the dominant bifidobacteria in infant faecal samples. Significant differences in bifidobacteria communities display differential metabolic features and have strong implications in infant physiology state and health [32–34]. Delivery mode (vaginal birth or assisted delivery) and feeding modes (breastfeeding, milk formula and mix-fed) have influences on bifidobacterial composition at species level [32]. Significant changes in the gut microbiota of infants occur when cessation of breastfeeding and introduction of solid foods cause an important shift in gut microbiota composition and begins to resemble a stable adult-like microbiome [35]. Under these circumstances, the supplementation of infant formulae with prebiotic oligosaccharides (GOS/FOS) results in significantly higher stool colony counts of bifidobacterial; it has also been associated to clinical effects such as increased stool frequency and stool softening [36].

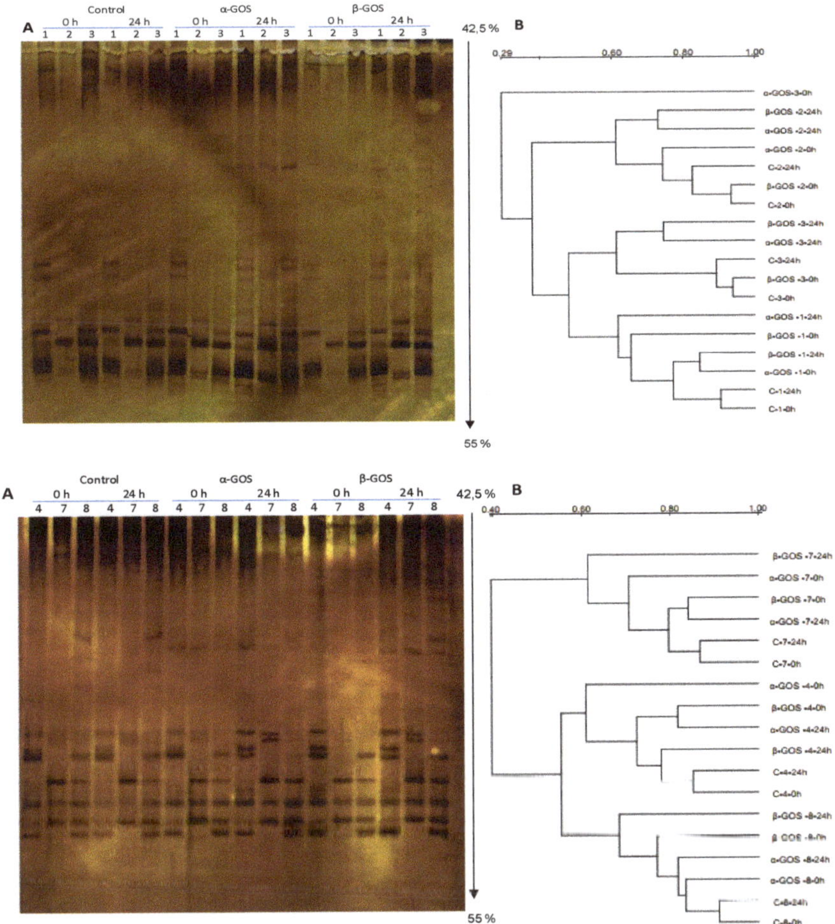

Figure 1. (**A**) DGGE profiles and (**B**) dendrogram of the Bifidobacterium population from faecal samples of six infants and three experimental groups (control, non-fructosylated α-galactooligosaccharides (α-GOS) and β-galactooligosaccharides (β-GOS)) at time 0 and 24 h. Samples were grouped in two gels corresponding to 1, 2, 3 and 4, 7, 8 infants, respectively. Cluster analysis of Denaturing Gradient Gel Electrophoresis (DGGE) pattern profiles was performed using the Dice similarity coefficient and the unweighted-pair group method by means of arithmetic average clustering algorithm (UPGMA). The vertical arrow shows the direction and concentration of the denaturing gradient.

To provide an interpretation of the DGGE pattern, three diversity indices of DGGE profiles were calculated (Table 3). The number of bands or richness was significantly higher ($p < 0.05$) in β-GOS after 24 h of fermentation compared to control and non-fructosylated α-GOS that only demonstrated a trend. Shannon and Evenness indices followed the same pattern for each GOS treatment.

Table 3. Diversity indices of bifidobacterial samples after non-fructosylated α-GOS and β-GOS 24 h treatment of infant faecal samples.

	Time	Enrichment		Shannon Index		Evenness Index	
		Mean	**SD**	**Mean**	**SD**	**Mean**	**SD**
Control	0 h	8.33 [a]	4.03	2.04 [a]	0.42	0.65 [a]	0.13
	24 h	8.33 [a]	4.03	2.04 [a]	0.42	0.65 [a]	0.13
α-GOS	0 h	10.83 [a]	3.97	2.33 [a]	0.36	0.74 [a]	0.11
	24 h	12.50 [a]	4.68	2.47 [a]	0.36	0.78 [a]	0.10
β-GOS	0 h	7.50 [a]	2.07	1.99 [a]	0.26	0.63 [a]	0.08
	24 h	12.00 [b]	3.16	2.46 [b]	0.24	0.78 [b]	0.07

[a, b], within each row, mean values bearing a different superscript letter differ significantly ($n = 6$; $p < 0.05$). Significance main effects were determined by GLM REP (General Linear Model by one-way ANOVA).

3.3. Modulation in Bifidobacteria Species Population by Non-Fructosylated α-GOS and β-GOS

In this study, five different bifidobacterial species were quantified at 0 and 24 h after fermentation of non-fructosylated α-GOS and β-GOS with fecal slurries of eight infants (Table 4). The highest level of bifidobacterial species in fecal samples corresponded to *B. longum subsp- longum* and *B. catenulatum/pseudocatenulatum*. It is well-known that bifidobacterial species show differential ability for breakdown, uptake and utilization of carbohydrates. In particular, non-fructosylated α-GOS and β-GOS significantly stimulated the selective growth of *B. longum* subsp. *longum* after 24 h fermentation whereas *B. adolescentis* and *B. longum* subsp. *infantis* were not affected. A significant ($p < 0.05$) growth of *B. catenulatum/pseudocatenulatum* after 24 h fermentation of non-fructosylated α-GOS was observed. Although *B. breve* is a common intestinal bacteria in children's microbiota, mainly in those fed breast milk, attempts to quantify *B. breve* by qPCR was unsuccessful, likely due to their low levels in fecal samples of children. No match for *B. breve* was found when the faecal bifidobacterial community of children was analyzed by DGGE. In both GOS groups, a significant ($p < 0.05$) decrease in *B. bifidum* population compared to controls was shown. The requirement of prebiotic specificity has recently been questioned by several authors [35,36]. Such controversy is likely to affect to more complex dietary carbohydrates that need bacteria consortium for their degradation instead of GOS compounds that have demonstrated to be selectively bifidogenic [37,38]. Sequencing analysis of fecal samples from healthy human volunteers consuming GOS revealed an increase of bifidobacterial populations [39]. Access to genome sequences of several bifidobacterial strains from infant fecal isolates such as *B. bifidum* PRL2010, *B. breve* UCC2003 and *B. longum* subsp. *infantis* ATCC15697 has increased our knowledge of the metabolic machinery within the genus *Bifidobacterium*. Bifidobacteria contains a high number of glycosyl hydrolases-encoding genes with ability to hydrolyze a wide range of complex carbohydrates (e.g., amylose, amylopectin, maltodextrin) as well as stachyose, raffinose and melibiose, which represent carbohydrates widely present in the human diet [40,41]. Further studies regarding the bifidobacterial enzymes involved in the selective degradation of non-fructosylated α-GOS will allow us to understand their potential as prebiotic ingredient in infant formula.

Table 4. Effect of non-fructosylated α-GOS and β-GOS on bifidobacterial species of infant faecal samples.

	Control		α-GOS		β-GOS		p-Values		Pooled SEM
Log10 Copy Number/Gr De Contenido Fecal	**0 h**	**24 h**	**0 h**	**24 h**	**0 h**	**24 h**	**Pb**	**Bs**	
Bifidobacterium adolescentis	4.45	4.68	4.65	5.11	4.52	4.52	NS	<0.001	0.193
Bifidobacterium bifidum	4.55	5.09 [b]	4.70	4.04 [a]	4.57	4.27 [a]	0.045	<0.001	0.157
Bifidobacterium catenulatum/pseudo-catenulatum	7.47	7.41 [a]	7.25	7.65 [b]	7.44	7.74 [a, b]	0.068	<0.001	0.094
Bifidobacterium longum subsp. *infantis*	4.96	5.12	5.22	5.40	5.17	5.48	NS	<0.001	0.117
Bifidobacterium longum subsp. *longum*	8.25	8.26 [a]	8.03	8.47 [b]	8.36	8.71 [b]	0.010	<0.001	0.065

Outcomes were analyzed using a linear mixed model: repeat measure. A Bonferroni method was used for adjustments for major effects (time of treatment and type of GOS). [a, b] Mean value with different letters means the differences among treatments was significantly different at p-value < 0.05. Unadjusted mean values $n = 6$. Pb: Prebiotics. Bs: Baseline like covariate.

3.4. Lactate and Short-Chain Fatty Acid (SCFA) Concentration

Under anaerobic conditions, non-digestible dietary carbohydrates including GOS are fermented by gut microbiota and thereby producing SCFA that contributes to metabolic regulation and control of energetic metabolism of the host [42]. In our study, lactate and SCFAs were quantified after fermentation of non-fructosylated α-GOS and β-GOS (Table 5). Acetate was the most abundant SCFA, showing a significant ($p < 0.05$) increase after both GOS treatments. In the case of non-fructosylated α-GOS, the levels of acetate increased up to four times compared to baseline. The highest amount of lactate in infant fecal cultures was reported after 6 h in both GOS fermentations, being observed a decrease at 24 h likely due that normally lactate behave as intermediate in microbial metabolism because of onward conversion [43]. Similar to acetate, lactate concentration was higher in non-fructosylated α-GOS compared to β-GOS. Lactate and acetate are main fermentation products of bifidobacteria and are used by other bacteria species to produce butyrate and propionate. In our study, the amounts of propionate and butyrate were not affected after GOS treatment. In agreement with previous studies, GOS are highly selective for *Bifidobacterium* spp. and stimulated acetate production but not the production of propionate and butyrate [44]. Formate amounts were increased due to GOS fermentation, being formate levels after non-fructosylated α-GOS fermentation much higher compared to β-GOS, tripling its value after 24 h (Table 5). Other metabolic indicators such as iso-butyrate, valerate, iso-valerate and succinate were also quantified in this study but no significant changes were observed. The production of SCFA caused a significant drop of pH (0.8 and 1.0 units after 24 h treatment of β-GOS and non-fructosylated α-GOS, respectively) in fecal cultures. Such decrease in luminal pH has been associated to beneficial effects including a decrease in the growth of potential intestinal pathogens sensitive to pH, like some enterobacteria, as well as an increase in the mineral absorption of calcium, magnesium and likely iron [45].

Table 5. Concentration (mM) of lactic acid and short-chain fatty acids (SCFA) infant faecal samples after non-fructosylated α-GOS and β-GOS fermentation.

SCFA (mM)	Control			α-GOS			β-GOS			*p*-Values				Pooled SEM
	Bs	**6 h**	**24 h**	**Bs**	**6 h**	**24 h**	**Bs**	**6 h**	**24 h**	**T**	**Prebiotics**	**T * Prebiotics**	**Bs**	
Acetate	14.669	21.640 [a]	18.908 [a]	10.808	44.356 [b]	45.752 [b]	16.587	33.463 [b]	40.743 [b]	0.671	0.001	0.677	0.084	2.313
Propionate	3.163	6.491	6.116	2.915	9.130	8.522	3.971	8.196	10.072	0.839	0276	0.746	0.060	0.729
Butyrate	3.408	6.664	4.900	6.727	6.970	6.772	7.512	7.719	8.112	0.728	0714	0.831	0.044	0.746
Iso-butyrate	0.307	0.649	0.764	0.253	0.700	0.791	0.339	0.582	0.624	0.540	0.605	0.975	0.412	0.067
Valerate	1.174	4.228	4.788	1.139	2.216	0.607	1.197	2.102	2.249	0.779	0.059	0.678	0.000	0.531
Iso-valerate	0.317	0.635	0.827	0.303	0.775	0.719	0.236	0.602	0.657	0.682	0.897	0.806	0.052	0.077
Formate	0.611	2.400 [a]	2.447 [a]	0.523	3.651 [b]	8.954 [b]	0.785	4.042 [a, b]	3.115 [a, b]	0.258	0.046	0.117	0.390	0.643
Succinate	0.265	1.022	1.022	0.428	1.065	0.975	0.354	1.018	0.969	0.899	0.998	0.995	0.967	0.182
Lactate	0.248	0.782 [a]	0.500 [a]	0.211	3.908 [b]	1.554 [b]	0.343	1.353 [a, b]	0.559 [a, b]	0.028	0.009	0.225	0.015	0.250

Outcomes were analyzed using a linear mixed model: repeat measure. A Bonferroni method was used for adjustments for major effects (time of treatment and type of GOS). [a, b] Mean value with different letters means that differences among treatments was significantly different at p-value < 0.05. Unadjusted mean values $n = 8$. Bs: Baseline like covariate. * Interaction among major effects.

3.5. Consumption of Non-Fructosylated α-GOS Per Donor and Degree of Polymerization

The genome sequences of bifidobacteria display a remarkable enrichment in genes involved in breakdown, uptake, and utilization of a wide variety of resistant carbohydrates to digestion, playing an important role in the healthy infant gut. In this study, manninotriose (DP3) and verbascotetraose (DP4), accounting for more than 92% of total carbohydrates, were completely and readily fermented by the studied donors (Figure 2). This trend was not observed for the disaccharide melibiose, where the fermentation rate ranged between 67% and 95%. Regarding consumption of total non-fructosylated α-GOS, the observed values ranged from 97.6% (donor 1) to 99.9% (donor 5) after 24 h fermentation.

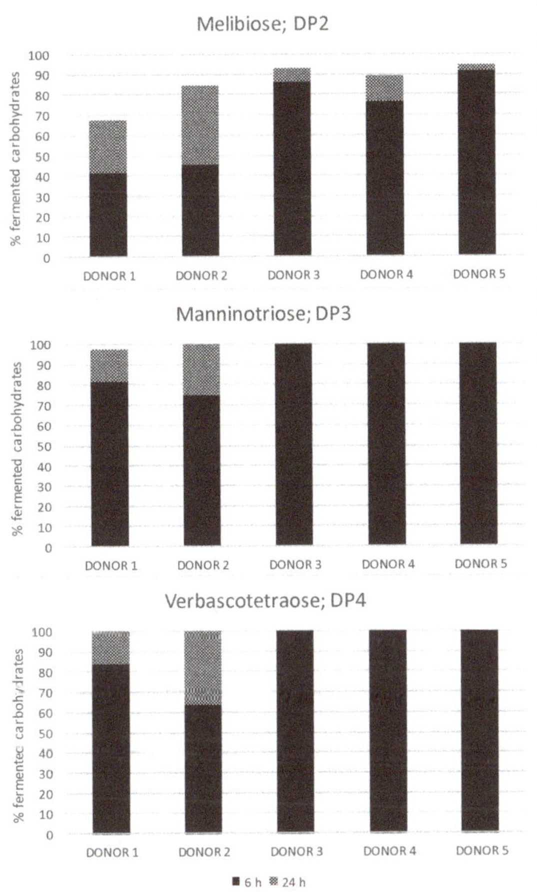

Figure 2. Relative percentage (%) of fermented melibiose (DP2), manninotriose (DP3) and verbascotetraose (DP4) following fermentation of non-fructosylated α-GOS with infant faecal samples ($n = 5$). Darker and lighter colours on the bars indicate the percentage of metabolized carbohydrate after 6 and 24 h of incubation, respectively.

Bifidobacteria can utilize a diverse range of carbohydrates that escape degradation in the upper part of the GIT, as evidenced by the clear enrichment for genes encoding carbohydrate-active enzymes [41]. Bifidobacteria have different specialized oligosaccharide transport systems, since most of their saccharolytic enzymes are intracellular [43]. The fermentation of non-fructosylated α-GOS, as shown in Table 4, significantly increased the growth of *B. longum* subsp. *longum* and *B. catenulatum/pseudocatenulatum* and decreased that of *B. bifidum*. The uptake of α-(1→6)-linked dietary oligosaccharides (ie., raffinose and a mixture of isomaltooligosaccharides) by several *Bifidobacterium* strains, that included *B. longum* subsp. *longum* ICM1217 and *B. longum* subsp. *infantis* ATCC15697 among others, was reported to be mediated by the solute binding protein (B1G16BP) associated with an ATP binding cassette (ABC) transporter [46]. Interestingly, *B. bifidum*, which lacks the genes encoding the B1G16BP ABC transporter, was the only one that did not grow on raffinose and isomaltooligosaccharides, which is in line with our findings. The ABC transporter uses energy produced by the ATP hydrolysis to carry out the oligosaccharides transportation to the intracellular compartment. Once internalized, these oligosaccharides are metabolized by glycosyl hydrolases (GHs) which hydrolyze the glycosidic bond between two or more carbohydrates. Melibiose and raffinose has been reported to be hydrolyzed by *B. longum* subsp. *longum* NCC2705 [40]. According to our results, the lower consumption of melibiose over manninotriose and verbascotetraose could be due to the higher preference of α-galactosides for Gal-(1→6)-Gal-linkages over Gal-(1→6)-Glc, and likely to potential preference for higher DPs. In addition, ligand preference of B1G16BP was ~35 fold lower for melibiose compared with raffinose, which might be explained by its high binding affinity for the non-reducing α-(1→6)-diglycoside, an structural motif present in manninotriose and verbascotetraose but not in melibiose [46]. Although it is clear major metabolic differences might occur at species level, such metabolic abilities might differ considerably between bifidobacterial strains. Thus, it has been reported that β-GOS fermentation is strain-dependent based on their DP and glycosidic linkages [47].

Outcomes were analyzed using a linear mixed model: repeat measure. A Bonferroni method was used for adjustments for major effects (time of treatment and type of GOS). [a, b] Mean value with different letters means that differences among treatments was significantly different at p-value < 0.05. Unadjusted mean values $n = 8$. Bs: Baseline like covariate. *Interaction among major effects.

4. Conclusions

The in vitro fermentation profile of non-fructosylated α-GOS by using infant fecal slurries has been investigated. Remarkable bifidogenic activity of non-fructosylated α-GOS was observed, similar to that obtained using commercial β-GOS, as well as a significant decrease of enterobacteria. A selective and significant increase in *B. longum* subsp. *longum* and *B. catenulatum/psudo-catelunatum* was observed. Therefore, we conclude that these plant-derived GOSs could be used as a prebiotic in infant formula.

Supplementary Materials: The following are available online at http://www.mdpi.com/2304-8158/9/7/921/s1, Figure S1: (A) DGGE profiles and (B) dendogram of the *Bifidobacterium* population from faecal samples of infants, Table S1: β-GOS composition after purification by size-exclusion chromatography, Table S2: PCR primers based on 16S rRNA sequences used for quantitative PCR for bacterial groups, Table S3: PCR primers based on 16S rRNA sequences used for quantitative PCR for Bifidobacteria species.

Author Contributions: Conceptualization, A.C., F.J.M.; methodology, A.C., F.J.M., M.d.C.M.-M., M.D.-M.; software, M.d.C.M.-M.; formal analysis, M.d.C.M.-M., M.D.-M., C.D.-A.; investigation, A.C., F.J.M.; resources, A.C.; writing—original draft preparation, A.C., O.H.-H.; writing-review and editing, A.C., O.H.-H., M.d.C.M.-M., F.J.M.; supervision, A.C., F.J.M.; project administration, A.C.; funding acquisition, A.C., F.J.M., O.H.-H. All authors have read and agreed to the published version of the manuscript.

Funding: This research was funded by the Spanish Ministry of Economy, Industry and Competitiveness, grant numbers AGL2017-83772-R and AGL2017-84614-C2-1-R (AEI/FEDER,UE); the Spanish Ministry of Science, Innovation and Universities grant number RTI2018-101273-J-I00 (JIN Program) and AGR2011-7626 from Junta de Andalucía.

Conflicts of Interest: The authors declare no conflict of interest.

References

1. Collado, M.C.; Derrien, M.; Isolauri, E.; De Vos, W.M.; Salminen, S. Intestinal integrity and *Akkermansia muciniphila*, a mucin-degrading member of the intestinal microbiota present in infants, adults, and the elderly. *Appl. Environ. Microbiol.* **2007**, *73*, 7767–7770. [CrossRef] [PubMed]

2. Rodriguez, M.A.; Murphy, K.; Stanton, C.; Ross, R.P.; Kober, O.I.; Juge, N.; Avershina, E.; Rudi, K.; Narbad, A.; Jenmalm, M.C.; et al. The composition of the gut microbiota throught life, with emphasis on early life. *Microb. Ecol. Health Dis.* **2015**, *26*, 26050. [PubMed]
3. Houghteling, P.D.; Walker, W.A. Mucosal why is initial bacterial colonization of the intestine important to the infant's and child's health? *J. Pediatr. Gastroenterol. Nutr.* **2008**, *6*, 2166–2171.
4. Verkhnyatskaya, S.; Ferrari, M.; De Vos, P.; Walvoort, M.T.C. Shaping the infant microbiome with non-digestible carbohydrates. *Front. Microbiol. Res. Top.* **2019**, *10*, 1–8. [CrossRef]
5. Gibson, G.R.; Hutkins, R.; Sanders, M.E.; Prescott, S.L.; Salminen, S.J.; Scott, K.; Stanton, C.; Cani, P.D.; Verbeke, K.; Reid, G. The International Scientific Association for Probiotics and Prebiotics (ISAPP) consensus statement on the definition and scope of prebiotics. *Nat. Rev. Gastroenterol. Hepatol.* **2017**, *14*, 491–502. [CrossRef]
6. Akkerman, R.; Faas, M.M.; de Vos, P. Non-digestible carbohydrates in infant formula as substitution for human milk oligosaccharide functions: Effects on microbiota and gut maturation. *Crit. Rev. Food Sci. Nutr.* **2019**, *59*, 1486–1497. [CrossRef] [PubMed]
7. Marcobal, A.; Barboza, M.; Froehlich, J.W.; Block, D.E.; German, J.B.; Lebrilla, C.B.; Mills, D.A. Consumption of human milk oligosaccharides by gut-related microbes. *J. Agric. Food Chem.* **2010**, *58*, 5334–5340. [CrossRef] [PubMed]
8. Ruiz-Moyano, S.; Totten, S.M.; Garrido, D.A.; Smilowitz, J.T.; Bruce German, J.; Lebrilla, C.B.; Mills, D.A. Variation in consumption of human milk oligosaccharides by infant gut-associated strains of *Bifidobacterium breve*. *Appl. Environ. Microbiol.* **2013**, *79*, 6040–6049. [CrossRef] [PubMed]
9. Wang, Y.; Jiang, K.; Ma, H.; Zeng, W.; Wang, P.G.; Yao, N.; Han, W.; Cheng, J.; Wang, W. Enzymatic production of HMO mimics by the sialylation of galacto-oligosaccharides. *Food Chem.* **2015**, *181*, 51–56. [CrossRef] [PubMed]
10. Barile, D.; Rastall, R.A. Human milk and related oligosaccharides as prebiotics. *Curr. Opin. Biotechnol.* **2013**, *24*, 214–219. [CrossRef] [PubMed]
11. Oswari, H.; Widodo, A.D.; Handayani, F.; Juffrie, M.; Sundjaya, T.; Bindels, J.; Hegar, B. Dosage-related prebiotic effects of inulin in formula-fed infants. *Pediatr. Gastroenterol. Hepatol. Nutr.* **2019**, *22*, 63–71. [CrossRef] [PubMed]
12. Ehara, T.; Izumi, H.; Tsuda, M.; Nakazato, Y.; Iwamoto, H.; Namba, K.; Takeda, Y. Combinational effects of prebiotic oligosaccharides on bifidobacterial growth and host gene expression in a simplified mixed culture model and neonatal mice. *Br. J. Nutr.* **2016**, *116*, 270–278. [CrossRef] [PubMed]
13. Gopalsamy, G.; Mortimer, E.; Greenfield, P.; Bird, A.R.; Young, G.P.; Christophersen, C.T. Resistant starch is actively fermented by infant faecal microbiota and increases microbial diversity. *Nutrients* **2019**, *11*, 1–16. [CrossRef] [PubMed]
14. Hernandez-Hernandez, O.; Olano, A.; Rastall, R.; Moreno, F.J. In Vitro digestibility of dietary carbohydrates: Toward a standardized methodology beyond amylolytic and microbial enzymes. *Front. Nutr.* **2019**, *6*, 61. [CrossRef]
15. Van den Ende, W. Multifunctional fructans and raffinose family oligosaccharides. *Front. Plant. Sci.* **2013**, *4*, 1–11.
16. Myhara, R.M.; Nilsson, K.; Skura, B.J.; Bowmer, E.J.; Cruickshank, P.K. Gas production from melibiose, raffinose and white bean extracts by bacteria of human fecal origin. *Can. Inst. Food Sci. Technol.* **1988**, *21*, 245–250. [CrossRef]
17. Martins, G.N.; Ureta, M.M.; Tymczyszyn, E.E.; Castilho, P.C.; Gomez-Zavaglia, A. Technological aspects of the production of fructo and galacto-oligosaccharides. Enzymatic synthesis and hydrolysis. *Front. Nutr.* **2019**, *6*, 78. [CrossRef] [PubMed]
18. EFSA NDA Panel (EFSA Panel on Dietetics Products, Nutrition and Allergies). Scientific Opinion on the substantiation of a health claim related to AlphaGOS® and a reduction of post-prandial glycaemic responses pursuant to Article 13(5) of Regulation (EC) N° 1924/2006. *EFSA J.* **2014**, *12*. [CrossRef]
19. Fehlbaum, S.; Prudence, K.; Kieboom, J.; Heerikhuisen, M.; van den Broek, T.; Schuren, F.H.J.; Steinert, R.E.; Raederstorff, D. In Vitro fermentation of selected prebiotics and their effects on the composition and activity of the adult gut microbiota. *Int. J. Mol. Sci.* **2018**, *19*, 3097. [CrossRef]

20. Kruger, C.; Zhou, Y.; Thorsrud, B.A.; Morel-Despeisse, F.; Chappuis, E. Safety evaluation of α-galacto-oligosaccharides for use in infant formulas investigated in neonatal piglets. *Toxicol. Res. Appl.* **2017**, *1*, 1–10. [CrossRef]
21. Hernandez-Hernandez, O.; Marín-Manzano, M.C.; Rubio, L.A.; Moreno, F.J.; Sanz, M.L.; Clemente, A. Monomer and linkage type of galacto-oligosaccharides affect their resistance to ileal digestion and prebiotic properties in rats. *J. Nutr.* **2012**, *142*, 1232–1239. [CrossRef]
22. Arboleya, S.; Binetti, A.; Salazar, N.; Fernández, N.; Solis, G.; Hernández-Barranco, A.; Margolles, A.; de los Reyes-Gavilán, C.G.; Gueimonde, M. Establisment and development of intestinal microbiota in preterm neonates. *FEMS Microbiol. Ecol.* **2012**, *79*, 763–772. [CrossRef]
23. Marín-Manzano, M.C.; Abecia, L.; Hernández-Hernández, O.; Sanz, M.L.; Montilla, A.; Olano, A.; Rubio, L.A.; Moreno, F.J.; Clemente, A. Galacto-oligosaccharides derived from lactulose exert a selective stimulation on the growth of *Bifidobacterium animalis* in the large intestine of growing rats. *J. Agric. Food Chem.* **2013**, *61*, 7560–7567. [CrossRef]
24. Tabasco, R.; Fernández de Palencia, P.; Fontecha, J.; Peláez, C.; Requena, T. Competition mechanisms of lactic acid bacteria and bifidobacteria: Fermentative metabolism and colonization. *Food Sci. Technol.* **2014**, *55*, 680–684. [CrossRef]
25. Magurran, A.E. *Ecological Diversity and its Measurements*; Princenton University Press: Princenton, NJ, USA, 1988; p. 179.
26. Pielou, E.C. *Ecological Diversity*; John Wiley & Sons, Inc.: New York, NY, USA, 1975; p. 165.
27. Altschul, S.F.; Gish, W.; Myers, E.W.; Lipman, D.J. Basic local alignment search tool. *J. Biol. Mol.* **1990**, *215*, 403–410. [CrossRef]
28. Brobst, K.; Lott, C.E. Determination of some components in corn syrup by gas-liquid chromatography of trimethylsilyl derivatives. *Cereal Chem.* **1966**, *43*, 35–43.
29. Grimaldi, R.; Cela, D.; Swann, J.R.; Vulevic, J.; Gibson, G.R.; Tzortzis, G.; Costabile, A. In Vitro fermentation of β-GOS: Impact of faecal bacterial populations and metabolic activity in autistic and non-autistic children. *FEMS Microbiol. Ecol.* **2017**, *93*, 1–10. [CrossRef] [PubMed]
30. Borewicz, K.; Suarez-Diez, M.; Hechler, C.; Beijers, R.; de Weerth, C.; Arts, I.; Penders, J.; Thijs, C.; Nauta, A.; Lindner, C.; et al. The effect of prebiotic fortified infant formulas on microbiota composition and dynamics in early life. *Sci. Rep.* **2019**, *9*, 1–13. [CrossRef]
31. Ispirli, I.; Kaya, Y.; Dertli, E. Bifidogenic effect and In Vitro immunomodulatory roles of melibiose-derived oligosaccharides produced by the acceptor reaction of glucansucrase E81. *Proc. Biochem.* **2020**, *91*, 126–131. [CrossRef]
32. Matsuki, T.; Watanabe, K.; Tanaka, R. Genus- and species-specific PCR primers for the detection and identification of bifidobacterial. *Curr. Issues Intest. Microbiol.* **2003**, *4*, 61–69.
33. Turroni, F.; Peano, C.; Pass, D.A.; Foroni, E.; Severgnini, M.; Claesson, M.J.; Kerr, C.; Hourihane, J.; Murray, D.; Fuligni, F. Diversity of Bifidobacteria Within the Infant Gut Microbiota. *PLoS ONE* **2012**, *7*, e36957. [CrossRef] [PubMed]
34. Yang, B.; Chen, Y.; Stanton, C.; Ross, R.P.; Lee, Y.K.; Zhao, J.; Zhang, H.; Chen, W. *Bifidobacterium* and *Lactobacillus* composition at species level and gut microbiota diversity in infants before 6 weeks. *Int. J. Mol. Sci.* **2019**, *20*, 3306. [CrossRef]
35. Berström, A.; Skoy, T.H.; Bahl, M.I.; Roager, H.M.; Christensen, L.B.; Ejlerskov, K.T.; Molgaard, C.; Michaelsen, K.F.; Licht, T.R. Establishment of intestinal microbiota during early life: A longitudinal, explorative study of a large cohort of Danish infants. *App. Environm. Microbiol.* **2014**, *80*, 2889–2900. [CrossRef] [PubMed]
36. Braegger, C.; Chmielewska, A.; Decsi, T.; Kolacek, S.; Mihatsh, W.; Moreno, L.; Piescik, M.; Puntis, J.; Shamir, R.; Szajewska, H.; et al. Supplementation of infant formula with probiotics and/or prebiotics: A systematic review and comment by the ESPGHAN committee on nutrition. *J. Pediatr. Gastroenterol. Nutr.* **2011**, *52*, 238–250. [CrossRef]
37. Bindels, L.B.; Delzenne, N.M.; Cani, P.D.; Walter, J. Towards a more comprehensive concept for prebiotics. *Nat. Rev. Gastroenterol. Hepatol.* **2015**, *12*, 303–310. [CrossRef]
38. Hutkins, R.W.; Krumbeck, J.A.; Bindels, L.B.; Cani, P.D.; Fahey, G.; Goh, Y.J.; Hamaker, B.; Martens, E.C.; Mills, D.A.; Rastall, R.A.; et al. Prebiotics: Why definitions matter. *Curr. Opin. Biotechnol.* **2016**, *37*, 1–7. [CrossRef] [PubMed]

39. Davis, L.M.; Martinez, I.; Walter, J.; Goin, C.; Hutkins, R. Barcoded pyrosequencing reveals that consumption of galactooligosaccharides results in a highly specific bifidogenic response in humans. *PLoS ONE* **2011**, *6*, e25200. [CrossRef] [PubMed]
40. Pokusaeva, K.; Fitzgerald, G.F.; van Sinderen, D. Carbohydrate metabolism in Bifidobacteria. *Genes Nutr.* **2011**, *6*, 285–306. [CrossRef] [PubMed]
41. El Kaoutari, A.; Armougom, F.; Gordon, J.I.; Raoult, D.; Henrisat, B. The abundance and variety of carbohydrate-active enzymes in the human gut microbiota. *Nat. Rev. Microbiol.* **2013**, *11*, 497–504. [CrossRef]
42. Canfora, E.E.; Jocken, J.W.; Blaak, E.E. Short-chain fatty acids in control of body weight and insulin sensitivity. *Nat. Rev. Endocrinol.* **2015**, *11*, 577–591. [CrossRef]
43. Flint, H.J.; Scott, K.P.; Louis, P.; Duncan, S.H. The role of the gut microbiota in nutrition and health. *Nat. Rev. Gastroenterol. Hepatol.* **2012**, *9*, 577–589. [PubMed]
44. Ventura, M.; O'Connell-Motherway, M.; Leahy, S.; Moreno-Munoz, J.A.; Fitzgerald, G.F.; van Sinderen, D. From bacterial genome to functionality; case bifidobacteria. *Int. J. Food Microbiol.* **2007**, *120*, 2–12. [CrossRef] [PubMed]
45. Alexander, C.; Swanson, K.; Fahey, G.C.; Garleb, K.A. Physiological importance of short-chain fatty acids from undigestible carbohydrate fermentation. *Adv. Nutr.* **2019**, *10*, 576–589. [CrossRef] [PubMed]
46. Ejby, M.; Fredslund, F.; Andersen, J.M.; Žagar, A.V.; Henriksen, J.R.; Andersen, T.L.; Svensson, B.; Slotboom, D.J.; Hachem, M.A. An ATP binding cassette transporter mediates the uptake of α-(1,6)-linked dietary oligosaccharides in *Bifidobacterium* and correlates with competitive growth on these substrates. *J. Biol. Chem.* **2016**, *291*, 20220–20231. [CrossRef] [PubMed]
47. Thongaram, T.; Hoeflinger, J.L.; Chow, J.; Miller, M.J. Prebiotic galactooligosaccharides metabolism by probiotic lactobacilli and bifidobacterial. *J. Agric. Food Chem.* **2017**, *65*, 4184–4192. [CrossRef]

Article

Narrow-Leafed Lupin Main Allergen β-Conglutin (Lup an 1) Detection and Quantification Assessment in Natural and Processed Foods

Elena Lima-Cabello [1], Juan D. Alché [1] and Jose C. Jimenez-Lopez [1,2,*]

1 Department of Biochemistry, Cell & Molecular Biology of Plants, Estacion Experimental del Zaidin, Spanish National Research Council (CSIC), Profesor Albareda 1, E-18008 Granada, Spain; elena.lima@eez.csic.es (E.L.-C.); juandedios.alche@eez.csic.es (J.D.A.)

2 The UWA Institute of Agriculture and School of Agriculture and Environment, The University of Western Australia, Crawley, WA 6019, Australia

* Correspondence: josecarlos.jimenez@eez.csic.es; Tel.: +34-958-181-600

Received: 25 July 2019; Accepted: 4 October 2019; Published: 18 October 2019

Abstract: The increasing prevalence of lupin allergy as a consequence to the functional characteristics of a growing number of sweet lupin-derived foods consumption makes the imperious necessity to develop analytical tools for the detection of allergen proteins in foodstuffs. The current study developed a new highly specific, sensitive and accurate ELISA method to detect, identify and quantify the lupin main allergen β-conglutin (Lup an 1) protein in natural and processed food. The implementation of accurate standards made with recombinant conglutin β1, and an anti-Lup an 1 antibody made from a synthetic peptide commonly shared among β-conglutin isoforms from sweet lupin species was able to detect up to 8.1250 ± 0.1701 ng (0.0406 ± 0.0009 ppm) of Lup an 1. This identified even lupin traces present in food samples which might elicit allergic reactions in sensitized consumers, such as β-conglutin proteins detection and quantification in processed (roasted, fermented, boiled, cooked, pickled, toasted, pasteurized) food, while avoiding cross-reactivity (false positive) with other legumes as peanut, chickpea, lentils, faba bean, and cereals. This study demonstrated that this new ELISA method constitutes a highly sensitive and reliable molecular tool able to detect, identify and quantify Lup an 1. This contributes to a more efficient management of allergens by the food industry, the regulatory agencies and clinicians, thus helping to keep the health safety of the consumers.

Keywords: vicilin; 7S-globulins; food allergens; Lup an 1; sweet lupin species; food labelling; processed food

1. Introduction

Lupine is a legume that belongs to the genus *Lupinus* and is included in the *Leguminosae* family, which is of great interest to the food industry, similarly for chickpeas, beans, peanuts, soya bean, lentils, and peas. *Lupinus* comprises between 200–600 different species [1]. However, only the four known as the sweet lupin group have gained interest since they are used in human food production [2] for their low levels of alkaloids [3] contained in their seeds. These four species include *Lupinus albus* (white lupine), *Lupinus angustifolius* (blue lupin or narrow-leafed lupin, NLL), *Lupinus luteus* (yellow lupine) [4], and *Lupinus mutabilis* (pearl or Andean lupin) that are mostly cultivated and consumed in central and South America [5].

Current interest for lupin seeds as a new functional food is growing [6], and the seeds from the sweet lupin species are becoming a crucial and alternative source of proteins for human consumption with nutritional and nutraceutical properties [7–10]. Sweet lupin species are a promising source of innovative ingredients for functional food, particularly those from the vicilin or β-conglutin family, which are the most abundant proteins in NLL seeds [11]. Among the most frequently found, flour is one of the basic products and a common form to use lupin seeds as an ingredient in a wide range of food i.e., bread, cake, pasta, pizza, sausage, spices, cream cheese, tofu, jam [12]. In addition, lupin might often replace soya bean, egg white, and milk in vegan food [5,13], and is used as a functional ingredient in gluten-free food [12,14].

In the last five years and due to this increased nutraceutical knowledge, the number and range of commercially available lupin based products increased. In this regard, the rising lupin-derived products consumption, and the growing number of allergy reactions in sensitized persons have also increased. The routes of sensitization not only arise as primary sensitization to ingested lupin seed proteins, but also occur due to the cross-reactivity in atopic subjects sensitized to seed proteins from other legumes, particularly to soybean and peanut [15–17]. A third route appears to be the occupational allergy in people that works daily with lupin flour and lupin derived products [18].

Although the prevalence in the general population of lupin allergy is still unknown [19], it has been estimated to be in the range of 0.3% to 8%, and particularly in children being 5% [5,13,20]. It is important to identify the lupin seed proteins involved in allergy reactions [21,22], with storage proteins being the main lupin allergens [23], particularly from the vicilin family (β-conglutins), which were named as Lup an 1 in NLL (WHO/IUIS Allergen Nomenclature Subcommittee). Due to the significant increase of reported cases of lupin allergy, and in order to keep safety among the population, the seeds from this legume together with soya bean and peanut were included in the European Union regulations (Regulation (EU) No. 1169/2011) as foods prone to induce allergy reactions. There is also a mandatory declaration in the labelling of the pre-packaged food ingredients, and overall, it provides information to consumers of these potential sources of food allergens. Thus, this demands very accurate and sensitive, highly specific quantitative methods to identify the main allergen protein content in food in order to identify the cross-contamination that justifies the precautionary labelling because of the presence of lupin as an ingredient. Thus, the aim of the present study was to develop a new highly specific ELISA method for the detection, identification and quantification of the lupin main allergen β-conglutin (Lup an 1), and the assessment of this method in natural and processed food. Since lupin flour is being mainly used as an ingredient in bakery products, vegetarian and vegan based products, this analytical method was validated in natural (non-processed) food and processed (fermented, cooked, boiled, roasted) food such as flour, bread, and biscuits; as well as others lupin based alternative foods as spreading, sauce, cheese, pickles, drink, butter, meat. This study also includes data from these commercial foodstuffs differing in their labelling ("may contain lupin traces", "lupine contained", "lupine content not included on the food package").

2. Material and Methods

2.1. Bioinformatics Analysis of β-Conglutin Protein Sequences

The sequences of seven available isoforms of NLL β-conglutin protein were retrieved from the NCBI database (accession number F5B8V9, F5B8W0–F5B8W5 corresponding to β1 to β7). A multiple sequence alignment and subsequent analysis was performed using ClustalW software (https://www.ebi.ac.uk/Tools/msa/clustalw2), based on Blosum62 matrix (BLOck SUbstitution Matrix) [24], and viewed using the Jalview viewer 2.2 (http://www.jalview.org). The Bioedit v 7.0.5.3 (http://www.mbio.ncsu.edu/BioEdit/bioedit.html) software was used to calculate the sequence identity matrices.

2.2. Analysis of the Antigenicity of β-Conglutin Proteins

The prediction of antigenicity for these β-conglutin protein isoforms was performed. Antigenic epitopes were determined on the basis of the following parameters: hydrophobicity [25]; amino acid surface accessibility [26,27]; antigenicity methods such as Hopp and Woods hydropathy [28], Welling [29], Parker [30], BepiPred-2.0 (sequential β-cell epitope predictor, http://www.cbs.dtu.dk/services/BepiPred), and Kolaskar and Tongaonkar [31]. The results were confirmed using the antigenicity prediction of GenScript service (https://www.genscript.com/antigen-design.html).

2.3. β-Conglutin Proteins Structure Modelling

The conglutin β1 protein sequence (F5B8V9) was retrieved and used for searching the best structural templates in the Protein Data Bank (PDB) (http://www.rcsb.org). The suitable templates for this sequence were selected using BLAST server (http://ncbi.nlm.nih.gov). To improve the best final templates identification and selection, the software BioInfoBank Metaserver (http://meta.bioinfo.pl) specializing in fold recognition homology was used, as well as the Swiss-model server (swissmodel.expasy.org). The best four identified templates (1uij, 2phl, 3s7e, and 2eaa) were retrieved from the PDB database, and implemented for homology modelling. The conglutin β1 protein model was built by the implementation of SWISS-MODEL via the ExPASy web server (swissmodel.expasy.org) using these top PDB closely related structural templates.

The structural errors in the initial structural 3D models were identified by using ProSA (prosa.services.came.sbg.ac.at/prosa.php), obtaining the first overall quality estimation of each model using QMEAN4 (swissmodel.expasy.org/qmean/cgi/index.cgi). The final structure of β-conglutin proteins was subjected to energy minimization using GROMOS96, which was implemented in DeepView/Swiss-PDBViewer v3.7 (spdbv.vital-it.ch) improving the van der Waals contacts and correcting the stereochemistry. The quality of the final models was assessed by assessing the protein stereology with PROCHECK (www.ebi.ac.uk/thornton-srv/software/PROCHECK) and ProSA programs, as well as the protein energy using ANOLEA (protein.bio.puc.cl/cardex/servers/anolea). The suitability of the model on the basis of the number of protein residues in the favoured regions was assessed by Ramachandran plot statistics.

To define the potential functional and interacting areas/clusters in the protein, the electrostatic Poisson-Boltzmann (PB) potentials were calculated by using APBS (DeLano Scientific LLC) molecular modelling software implemented in PyMOL 0.99 (www.pymol.org). The electrostatic PB potential values are given in units of kT per unit charge (k Boltzmann's constant; T temperature).

2.4. Construction of the Expression Plasmid

The protein expression in bacteria was achieved using the pET28a(+) vector (Novagen) with some modifications, such as an N-terminal 6xHis Tag. Vector pUC57 was used for cloning a synthetic gene encoding β-conglutin protein (GenBank HQ670409, β1), connected by restriction enzyme linker sequences, NcoI and XhoI (GenScript). The genetic construct included the expression vector pET28a(+)-conglutin β1-6xHis-Tag, which was obtained through the digestion of the pUC57- conglutin β1 construct with NcoI and XhoI restriction enzymes, followed by ligation of the β1 fragment into the pET28a(+) vector.

2.5. Overexpression of Conglutin β1

The final genetic construct containing the expression vector with the conglutin β1 gene was transformed into Rosetta™ 2(DE3) pLysS Singles™ Competent Cells (Novagen) for β1 expression. The protein expression was accomplished by using an auto-induction method [32]. Briefly, a colony of *Escherichia coli* containing the construct was isolated and grown for 24 h in ZY-medium plus kanamycin (50 μg/mL) at 37 °C in constant shaking (190 rpm). The culture was diluted 1:175 in Studier medium to grow for 7 h until the cell density reached 0.6 OD at 600 nm. The overexpression induction was achieved by adjusting the temperature to 20 °C during 17 h. The bacterial cell pellet was collected by centrifugation at 5000× *g* at 4 °C. The bacterial pellet was washed three times with PBS, pH 7.5, and after removing the supernatant, the pellet cell was flash frozen using liquid nitrogen. The resulting pellet was stored at −80 °C until further use.

2.6. Purification of the Recombinant Conglutin β1 Protein

Overall, β1 purification was performed following the company's recommendations (Qiagen) for Histidine tagged proteins. Briefly, the main steps comprised the cells breaking, followed by affinity chromatography using nickel-NTA spin columns, and linking a 6xHis-Tag at the C-terminal end of the β1 protein. The elution of this protein from the column was performed with an increasing imidazole concentration gradient (25–350 mM), collecting 2 mL fractions. All the fractions containing the protein were analyzed using SDS-PAGE, and those containing a single band with the expected molecular weight were pooled and dialyzed against PBS during 2 days with buffer changes every 12 h. The resulting samples were then aliquoted and flash frozen in liquid nitrogen to be kept at −80 °C until further uses. The purity of the protein samples was >95%. The typical yields were ~25–55 mg/mL.

2.7. Antibody Production Against β-Conglutin

The performance and specificity of an antibody highly depends on the nature of the binding to its protein target. A polyclonal epitope-specific antibody was developed using a combination of a lineal epitope identification and characterization (antigenicity assessment) in the target protein sequence and an affinity capture approach involving synthesized peptides.

The sequences currently available of seven NLL β-conglutin genes were retrieved from UNIPROT database (https://www.uniprot.org). The alignment of these sequences was analyzed to find a commonly shared antigenic peptide among the seven NLL β-conglutin proteins variants following an analysis of antigenicity described in previous sections, finding the following sequence: Nt– VDEGEGNYELVGIR –Ct. A synthetic peptide with this sequence was produced (Agrisera, Vännäs, Sweden). The chosen sequence (up to 14 amino acids + terminal cysteine) was synthesized in a 15 mg immunograde purity scale without any further modifications. In order to elevate the antigenicity of the synthetic peptide, 2 mg of it was coupled to a carrier protein, KLH, which is the keyhole limpet hemocyanin copper-containing protein, with a molecular size >450 kDa.

The KLH coupled synthetic peptide was used for animal (rabbits) immunization. The typical amount of antigen for a standard immunization protocol is ~500 μg/animal, and a preferable concentration >1 mg/mL dissolved in PBS. Five animals were subjected to a program of three immunization rounds (105 days long) with the same antigen. After 80 days, the test samples were delivered to our laboratory for evaluation, where sera, after this period of immunization, was tested to check for the capability to detect its target protein and also to show the potential background signal from these different sera in Western blot assays using recombinant and lupin seed protein extracts. The animals' analyzed polyclonal sera having target-specific antibodies directed towards linear epitope and giving Western blot bands of correct molecular weight against β-conglutin proteins and low or no background was used for further immunization and obtaining the sera. After the third immunization, all serum samples were checked for the presence of antigen specific antibodies using ELISA. A further proof of high specificity and no background produced by the antibody was obtained by a screening of pre immune serum. Testing the pre-immune samples made sure that no background signals were detected in the molecular weight region of the protein under investigation.

Furthermore, the antibody capture and clean up (affinity-purified) from antiserum was performed from the final rabbit immune serum by a liquid chromatography system against the same synthetic peptides as the affinity column tag. This step contributed to obtaining a higher specificity of the antibody, but also reduced the amount of the available antibody in the final serum.

2.8. Food and Biological Samples Used for Lupin Allergen Detection and Quantification

The different commercially available foods containing variable quantity of lupin seed were purchased from different supermarkets. More detailed information of each product is described in Table 1.

Table 1. Summary of samples analyzed in this study. The information includes brand, the lupin content, and the food processing state.

Product Number	Product	Product Information Related to Lupin Content	Web
1	Toasted bread-Crostini	Lupin protein	https://www.schaer.com/en-int/p/crostini
2	Gluten Free Maxi Sorrisi Chocolate Biscuit	It may content traces of lupin	https://www.schaer.com/en-int/p/maxi-sorrisi
3	Commercial lupin flour	Sweet lupin flour	https://www.arche-naturkueche.de/de/produkte/europaeische-kueche/backen-binden/s%C3%BCsslupinenmehl
4	Peanut butter	Roasted peanuts (97%), palm oil, and sea salt.	https://shop.wholeearthfoods.com/collections/award-winning-nut-butters/products/whole-earth-dark-roasted-peanut-butter-340-g
5	Carob spread hazelnut Chocolate duo	Lupin flour 5%	https://greenest.ee/en/product/carobella-carobella-chocolate-duo-bio-350-g
6	Seeds (*Lupinus albus* L.)	Seeds	https://www.semillascantueso.com/
7	Seeds (*Lupinus luteus* L.)	Seeds	https://www.semillascantueso.com/
8	Seeds (*Lupinus angustifolius* L.)	Seeds	https://www.semillascantueso.com/
9	Lupinen BOLOGNESE SAUCE	Sweet lupin seeds cooked (8%)	https://www.veggie-shop24.com/food/ready-meals/sauces-and-dips/alberts-lupine-bolognese-sauce-organic-300g
10	Fresh spread cheese	Lupin protein (6.6%)	https://www.veggie-shop24.com/food/cheese-alternatives/cream-cheese/made-with-luve-frisch-cremiges-streichglueck-fresh-creamy-spread-bliss-herbs-150g
11	Wheat toasted bread	Non lupin content	http://www.bimbo.es/productos/tostados#asies
12	Lupinen—Tempeh LUPEH	Boiled sweet lupin seeds 99%	https://www.veggie-shop24.com/food/vegan-basics/tofu-and-tempeh/alberts-lupeh-lupine-tempeh-organic-170g
13	Lupinen BURGER—MEDITERRANEAN	Boiled sweet lupin seeds 15%	https://www.veggie-shop24.com/food/meat-alternatives/burger-and-grill/alberts-lupine-burger-gluten-free-organic-200g
14	Pickled lupine	Lupin (*L. albus*)	https://es.openfoodfacts.org/producto/8480000330987/altramuces-encurtidos-hacendado
15	Lupinen Drink	Lupin protein (2.3%)	https://www.alles-vegetarisch.de/lebensmittel/milchalternativen-und-desserts/milchersatz/made-with-luve-lupinen-drink-natur-1l
16	TOFU smoked	Made with soya bean. Non lupin content.	https://www.veggie-shop24.com/food/vegan-basics/tofu-and-tempeh/alberts-tofu-smoked-organic-1kg
17	Boiled lentils	Non lupin content	
18	Boiled chickpea	Non lupin content	https://www.deliberry.com/mercadonamadrid/alimentacion-general/legumbres-y-verduras-envasadas
19	Boiled faba bean	Non lupin content	https://www.deliberry.com/mercadonamadrid/alimentacion-general/legumbres-y-verduras-envasadas

2.9. Proteins Extraction

The total protein (including allergen proteins) extractions were performed from the different food and biological samples. The samples were homogeneously grinded with a polytron homogenizer (Kinematica Polytron™ PT 2500E, VWR), always keeping the samples in a water-ice cool bath. For these proteins, extractions were used as the extraction buffer containing 100 mM Tris, pH 7.4, 250 mM NaCl, 1 mM EGTA, 1 mM EDTA, 1% Triton X-100, 0.5% Sodium deoxycholate, Protease inhibitor cocktail and 1 mM PMSF. The homogenates were centrifuged at 10,000× g to remove all the gross material and to keep the supernatant. The high NaCl concentrations contained in the extraction buffer assures specific vicilin proteins extraction [33] as β-conglutin (Lup an 1).

For the protein extraction from tissues, the samples were dissected and kept at a low temperature on ice-water to prevent proteolysis. Then, 50 mL of complete extraction buffer was added to 1 g of the sample in a tube to be homogenized with the polytron homogenizer. Afterward, the blades of the homogenizer were rinsed with the extraction buffer, and the homogenized sample was maintained at a constant agitation for 2 h at 4 °C for the proteins' final extraction to the media. A centrifugation for 20 min at 15,000× g and 4 °C was performed, and the supernatant was placed on ice-water to make the aliquots (containing the soluble protein extract) in a fresh, chilled tube and to store the samples at −80 °C. The lupin drink was analyzed as it was in the bottle.

2.10. Gel Electrophoresis and Western Blots

The analyses of the protein extracts were achieved by mixing each sample individually with 6× protein sample buffer and heated to 95 °C for 5 min. The proteins were separated on commercial 4–20% gradient TGX gels (Bio-Rad, Hercules CA, USA). The molecular weight markers used for stained gels were Precision Plus Protein™ Dual Colour Standards (BioRad). The separated protein bands were visualized in a Gel Doc™ EZ Imager (BioRad). The proteins were electrophoretically transferred from gel to PVDF membranes. Prior to the transference, the membranes were blocked for 2 h at room temperature (RT) with 5% non-fat dry milk in PBST (phosphate-buffered saline, 0.05% Tween-20) followed by the incubation with the first antibody, goat IgG anti-beta conglutin (dilution 1:2000), overnight at 4 °C in continuous agitation. After washing 5 times with PBST, the membrane was incubated with secondary antibody goat anti-IgG rabbit conjugated with horseradish peroxidase (dilution 1:10,000) in 2% non-fat dry milk in PBST for 2 h at RT. The membrane was washed 5 times with PBST and the chemiluminescence signal was developed by membrane incubation with ECL Plus chemiluminescence substrate following the manufacturer's instructions (BioRad). The light signal was detected by exposure of the membrane to C-Digit Blot Scanner (LI-COR).

2.11. ELISA Test for the Detection and Quantification of β-Conglutin Allergen Proteins in Lupin-Containing Food

The protein standards used for Lup an 1 identification and quantification was performed using purified allergen β-conglutin proteins. Coating the wells was performed by using purified conglutin β1 (50 μg). The plate wells containing purified proteins were used as blanks (controls without β-conglutin protein) and were simultaneously incubated (triplicate samples) overnight at 4 °C. The wells on the plates were washed 5 times with 200 μL of PBS for each well. In order to avoid the unspecific protein-binding sites, the samples were blocked in the coated wells by adding 200 μL blocking buffer (5% non-fat dry milk/PBS) per well, and incubated for 2 h at RT. The wells of the plates were then washed five times with 200 μL PBS. The first anti-IgG β-conglutin antibody (dilution 1:1000) was incubated in each well for 2 h at RT. The solution containing the antibody was removed and each well was washed five times with 200 μL PBS. The solutions on each well containing washing buffer were eliminated by flicking the plate over. The remaining drops on each well were eliminated by patting the plate with a paper towel. The wells on the plates were then incubated with a goat anti-IgG rabbit HRP conjugated antibody. The incubations were developed for 1 h and 30 min at RT, and the washes were performed five times with 200 μL PBS on each well. The development of the signal was made using a compatible substrate, 100 μL of TMB which was added to each well and incubated for 5 min at RT.

The reaction was stopped using 1M Sulphuric Acid for 10 min. The signal was read in a microplate readed iMak (Bio-Rad) at 450 nm.

Protein detection and quantification following this ELISA method was performed in identical conditions as described above using protein extracts obtained from natural and processed food samples, and the extraction buffer alone for control samples.

3. Results and Discussion

3.1. Conglutin β1 Protein Structure Modelling and Antigenicity Assessment

The NLL β-conglutin isoform sequences were retrieved from Uniprot database and an alignment was made to analyze the sequence features (Supplementary Figure S1). The length of these sequences ranged from 580 to 637 amino acids, and the variability was calculated (Figure S1), finding sequences with 77.4 to 97.8% of identity. The N-terminal sequence corresponding with the first 200 amino acids comprises a mobile arm (Figure S2), showing the highest variability among the seven isoforms. A globular structure was built for the rest of the β-conglutin sequence integrated by 2 β-barrels of antiparallel β-sheets (Figure 1) with large conserved areas, where it was included in the sequence used to make the synthetic peptide (Figure S1) and the anti-β-conglutin antibody used in this ELISA method.

The crystallographic structures of particular seed storage proteins were used as templates to facilitate the correlative study between structure-functionality and antigenicity of these proteins [34]. To the best of the author's knowledge, epitopes characterization through the 2-D and 3D comparative analysis of different members of vicilin protein family have been only performed in a few organisms, such as peanut [35], or lupin [21,36]. Furthermore, the use of computational homology modelling has allowed this study to uncover particular 3D structure features for the antigenic epitope belonging to this family of proteins (Figure 1A). In order to uncover these epitopic regions, an accurate modelling of conglutin β1 protein sequence was achieved following a work-flow, where best structural templates were used to build each domain of this protein. After obtaining a first model, a refinement process, including energy minimization and structural discordances were corrected [37].

A structural assessment for conglutin β1 model accuracy was performed to improve the stereo-chemical and energy minimization following a comparative analysis with the proteins' templates (PDBs accession numbers 1uij, 2phl, 3s7e, and 2eaa, respectively).

The analysis of the templates showed z-values (normalized QMEAN4 scores) of −0.17, −0.56, −2.12 and 0.11, respectively, for the Q-mean parameter and −4.08 for the conglutin β1 built model, displaying comparative good z-scores. The overall quality of the structures was assessed using ProSA, showing a z-score of −5.31 for conglutin β1 model and −6.17, −6.2, −6.50, and −5.68 respectively for the individual crystallographic structural templates. Both, Q-mean and ProSA parameters showed comparable values between the conglutin β1 model and the PDB structure templates, which validated the accuracy and structure quality of the conglutin β1 protein model [38].

Therefore, a stereochemistry analysis was developed using Procheck based on the Ramachandran plot showing that the template models contained 92.2%, 92.9%, 94.8%, and 95.6%, respectively, of their residues located in favourable regions; 6.8, 6.6, 5.0, and 4.1, respectively, in allowed regions; 0.9, 0.6, 0.1 and 0.4, respectively, in generally allowed regions; and 0, 0, 0 and 0%, respectively, in disallowed regions. These values calculated for the conglutin β1 model showed 91.3%, 7.0%, 1.7%, 0%, respectively, finding even more residues located in favourable regions, less residues in allowed regions, and a similar number of residues in generally allowed and non-favourable regions. These parameters confirm the accuracy and reliability of the structural models built for the conglutin β1 protein sequence [39].

The antigenicity study of the β-conglutin proteins was made in order to identify one of the best common shared sequences on the protein with high antigenicity to be used for a synthetic peptide synthesis and further, for anti-β-conglutin antibody development. Bioinformatics tools were used based on biochemical principles as hydrophilicity/hydrophobicity, polarity and the volume of side changes, the antigenicity index, and B-cell epitopes identification. A semi-empirical method based on the physicochemical properties of amino acids and their frequencies of occurrence in experimentally known segmental epitopes was also used to identify the best antigenic sequence to develop the anti-β-conglutin antibody. Figure S1 highlights the antigenic peptide Nt-VDEGEGNYELVGIR-Ct chosen for experimental animal immunization. Figure S2 shows how this peptide has a combination of hydrophilic and hydrophobic amino acids (Kyte-Doolitle, Hope and Woods scales), which is one of the requirements to be an immunogenic peptide [40], constituting a high antigenicity sequence, where most of the amino acids are accessible to the solvent (Figure S2). In addition, this sequence integrates a structural loop (Figure S2), 2-D structure exerting good antigenic properties in a protein [41]. Furthermore, Figure 1B, Figures S1 and S2 show how the peptide chosen in this study corresponds to a sequential β-cell epitope with a high antigenicity index (Figure 1B,C) [40].

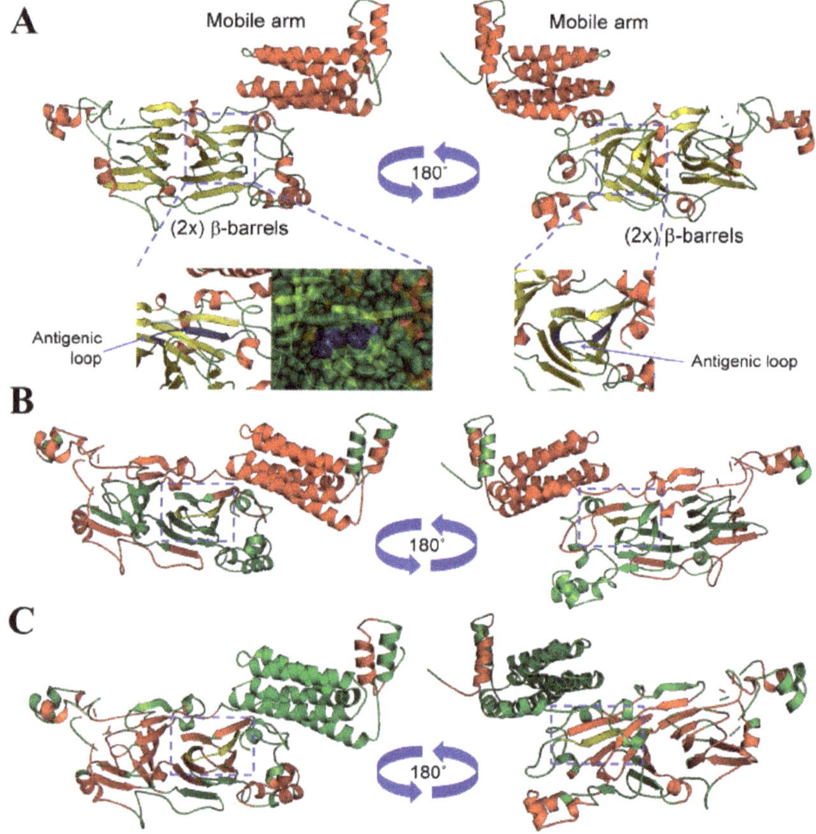

Figure 1. Structural and antigenic analysis of conglutin β1. (**A**) Three-dimensional structure of NLL conglutin β1 (Uniprot accession number F5B8V9) showing the mobile arm and globular domains. The structures were depicted as a cartoon diagram integrated by α-helices, β-sheets and coils (red, yellow and green colour, respectively). Two views (rotated 180 around the x-axis) are provided together with detailed views (carton diagram and surface) of the domain integrated by a coil-α-helix structure (blue colour), where the antigenic peptide chosen to develop the anti-β-conglutin antibody is located. (**B**) Three-dimensional structure depicted as a cartoon diagram of NLL conglutin β1 showing the antigenic β-cell sequential epitope regions (red colour), where the domain integrated by a coil-α-helix structure (yellow colour) are included. Two views (rotated 180 around the x-axis) are provided. (**C**) Three-dimensional structure depicted as a cartoon diagram of NLL conglutin β1 showing the Kolaskar and Tongaonkar antigenicity regions (red colour), where the domain integrated by a coil-α-helix structure (yellow colour) are also included. Two views (rotated 180 around the x-axis) are provided.

3.2. Purification and Reactivity of Conglutin β1 Protein

The purification of Lup an 1 was accomplished following the protocol set up by Jimenez-Lopez et al. [42]. Protein elution was accomplished with a linear gradient of imidazole (25–350 mM). The SDS–PAGE analysis of the eluted fractions showed a single protein band of ~70 kDa (Figure 2A), being the level of purity higher than 95%, with a typical yield of 27 mg/mL. An analysis by immunoblotting using the anti-β-conglutin protein antibody confirmed the identity of the purified β-conglutin protein in a unique reactive band (Figure 2B).

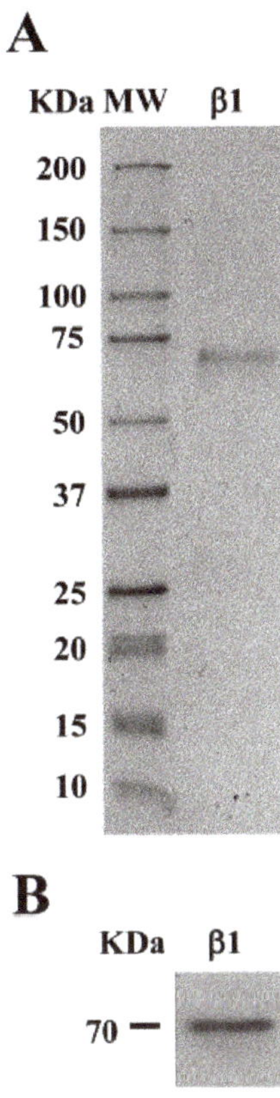

Figure 2. Purification of lupin main allergen β-conglutin (Lup an 1) protein. (**A**) The Coomassie-stained shows the purified β1-conglutin protein. (**B**) Immunoblotting shows the purified β1-conglutin protein identified by the anti-β-conglutin antibody. MW, molecular weight standard (kDa).

The protein extracts were analyzed, as well as the specificity of the anti-β-conglutin antibody reactivity by using the most commonly used species of the sweet lupin group (*L. albus*, *L. luteus* and *L. angustifolius*) to make foodstuffs. This study examined the differences in the protein profiles and composition coming from these lupin species, where the protein extracts were extracted from mature seeds, and showing proteins separated in SDS-PAGE and visualized with Coomassie Brilliant Blue staining (Figure 3, left panels).

Overall, in the three species lupin extracts, the proteins with molecular weights from ~12 to 75 kDa were present in different ratios depending on the lupin specie (Figure 3A–C). No significant amount of proteins was detected above 75 kDa, just a prominent band of approximately 100 kDa. Based on the abundance of the seed proteins, four main groups of polypeptides were able to be identified: group 1—

75 kDa and above; group 2—from 45 to 75 kDa; group 3—from 25 to 45 kDa; and group 4—from 10 to 25 kDa.

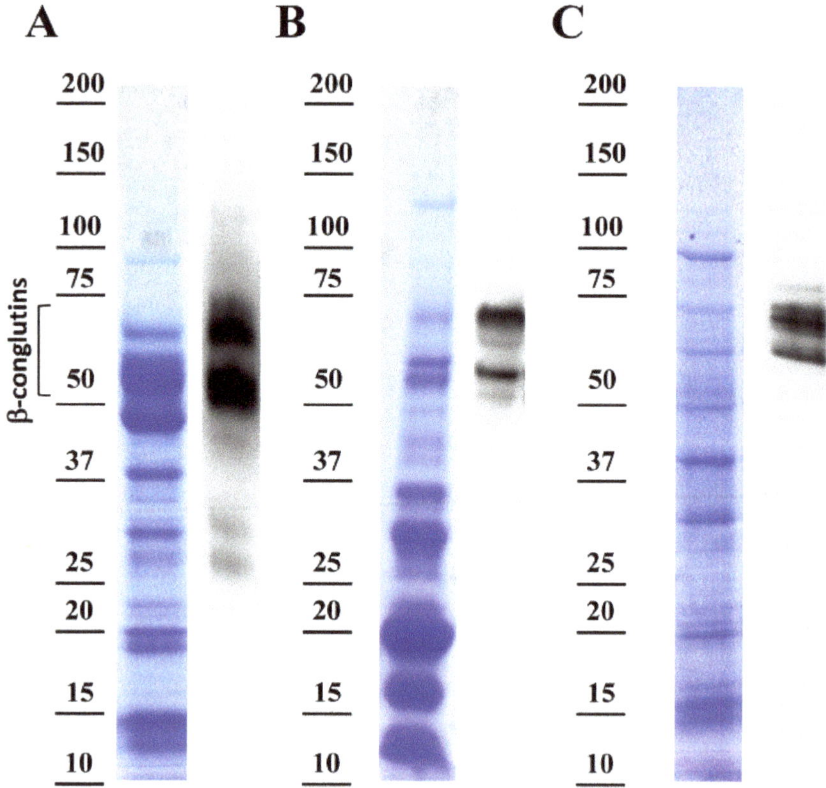

Figure 3. Specificity assessment of anti-β-conglutin antibody. Left panels show the Coomassie-stained SDS-PAGE protein bands from the total protein extracts, and the right panels show immunoblotting with the presence of β-conglutin proteins from (**A**) *Lupinus angustifolius*, (**B**) *Lupins albus*, and (**C**) *Lupinus lotus*.

Interestingly, *L. angustifolius* exhibited noticeable changes in the polypeptide composition between 50 and 75 kDa (Figure 2A). However, the three species displayed differential polypeptides content in this range of MW. These proteins localized in the expected size range (~70 kDa) related to β-conglutin polypeptides are present in the seeds from mature seeds onward [43] in *L. angustifolius*, *L. albus* and *L. luteus*, and particularly identified as these proteins obtained in the current study by the over-expression and purification (Lup an 1, F5B8V9, with theoretical MW ~70 kDa) (Figure S1). The major differences in polypeptides composition and abundance were observed in the range of MW concerning 25–45 kDa. On the other hand, protein profiles present in the range of 10–25 kDa were similar among the entire lupin species analyzed.

The right panels in Figure 3A–C showed polypeptide bands corresponding to β-conglutin by immunoblot assays using the anti-β-conglutin antibody, which is able to recognize the seven β-conglutin isoforms from *L. angustifolius* and the recently identified sequence of *L. albus* β-conglutin (Uniprot accession number Q6EBC1). This antibody denoted a high specificity since it showed specific binding to Lup an 1 while no signal from reactive bands was obtained in the control assays using pre-immune serum. In the present study, and due to the comparable proteolytic processing of β-conglutin proteins in mature seeds from the three different lupin species [44], the presence of comparable reactive bands were found in these three species using the anti-β-conglutin antibody, in the range of 50 to 75 kDa (Figure 3). This corresponds to the mature forms of β-conglutin and shows the high specificity of this antibody to be used in the ELISA method, and other potential related applications. Indeed, the ELISA method developed in this study positively detected and identified the lupin main β-conglutin allergen on each of the different lupin species. It shows the suitability of this anti-β-conglutin antibody to be used in the ELISA method for the identification of Lup an 1 in foodstuffs, independently of the proteins composition in these three species contained in foodstuffs. The identification and quantification of Lup an 1 using the antibody varied depending on the specie analyzed, where most abundantly detected (574.7918 ± 19.887 ng) in *L. angustifolius* seed protein extract, followed by *L. luteus* (457.8474 ± 15.8272 ng), and *L. albus* (299.3749 ± 19.7757 ng) (Table 2). These results are comparatively in agreement with β-conglutin protein levels found in these species [7].

Table 2. Detection and quantification of the NLL main seed allergen β-conglutin (Lup an 1) in natural and processed food samples. Each number corresponds to the following food stuff: 1: toasted bread; 2: biscuit; 3: lupin flour; 4: peanut butter; 5: carob spread; 6: white lupin seed; 7: yellow lupin seed; 8: blue lupin seed; 9: sauce; 10: fresh cheese; 11: wheat bread, 12: fermented food; 13: meat; 14: pickled lupine; 15: lupin drink; 16: Soy (Smoked TOFU); 17: Lentils; 18: chickpea; 19: Faba bean.

Natural/Food Samples	Absorbance (450 nm) [1]	Allergen β-Conglutin Proteins Quantity (ng) (Diluted Samples)	Allergen β-Conglutin Proteins Final Quantity (ng)	Allergen β-Conglutin Proteins Quantification (ppm)
1	0.0390 0.0380 0.0400	2.0313 1.9792 2.0833	8.1250 ± 0.1701	0.0406 ± 0.0009
2	0 0 0	0 0 0	0	0
3	1.8430 1.8430 1.9570	95.9896 95.9896 101.9271	391.8753 ± 11.1958	1.9594 ± 0.0560
4	0 0 0	0 0 0	0	0
5	0.3890 0.3860 0.3900	20.2604 20.1042 20.3125	80.9028 ± 0.3541	0.4045 ± 0.0018
6	1.3430 1.5670 1.4010	69.9479 81.6145 72.9688	299.3749 ± 19.7757	1.4969 ± 0.0989
7	2.2890 2.1030 2.2010	119.2188 109.5313 114.6354	457.8474 ± 15.8272	2.2892 ± 0.0791
8	2.8940 2.6910 2.6920	150.7292 140.1563 140.2083	574.7918 ± 19.8876	2.8740 ± 0.0994
9	0.0190 0.0200 0.0170	0.9896 1.0417 0.8854	3.8890 ± 0.2599	0.0194 ± 0.0013
10	1.0340 1.0380 1.0400	53.8542 54.0625 54.1667	216.1112 ± 0.5197	1.0806 ± 0.0026
11	0 0 0	0 0 0	0	0
12	0.7840 0.7060 0.7130	40.8334 36.7708 37.1354	152.9862 ± 7.3353	0.7649 ± 0.0367

Table 2. *Cont.*

Natural/Food Samples	Absorbance (450 nm) [1]	Allergen β-Conglutin Proteins Quantity (ng) (Diluted Samples)	Allergen β-Conglutin Proteins Final Quantity (ng)	Allergen β-Conglutin Proteins Quantification (ppm)
13	0.6430 0.6000 0.6170	33.4896 31.2500 32.1354	129.1667 ± 3.6839	0.6458 ± 0.0184
14	1.4020 1.5140 1.3860	73.2105 77.5442 70.8534	293.5833 ±11.0853	1.4252 ± 0.0554
15	0.2840 0.2930 0.2850	14.7917 15.2604 14.8438	59.8660 ± 0.8389	0.2993 ± 0.0042
16	0 0 0	0 0 0	0	0
17	0 0 0	0 0 0	0	0
18	0 0 0	0 0 0	0	0
19	0 0 0	0 0 0	0	0

[1] Triplicated experiments.

3.3. *β-Conglutin Proteins Main Allergen Detection Capability in Lupin—Derived Products*

The analytical methods capable to detect, identify and quantify the main lupin allergen protein in food with high specificity and sensitivity are of great importance to help the food industry in allergen management, and to guarantee the life quality of sensitized/allergic individuals. Although ELISA is a frequently used technique for the detection and identification of food allergen proteins, there is no currently available ELISA test to detect the seed allergen proteins in all three most common lupin species (*L. angustifolius*, *L. albus*, and *L. luteus*) used for foodstuff production. For this reason, an ELISA newly developed and assessed method capable of detecting and detecting the presence of the lupin main allergen β-conglutin protein (Lup an 1) in natural and foodstuffs samples containing these three lupin species which constitutes a molecular tool of great interest. In the present study, a broad range of food has been analyzed. Labelling information of lupin flour analyzed in the current study had non-defined specie composition. In addition to the three lupin species most commonly used in food and feed industry, this study analyzed non lupin based products such as peanut butter; legumes as chickpea, lentils, or faba beans; wheat toasted bread (negative control); other product that may content lupin traces as gluten free chocolate biscuits; and cooked food (roasted, boiling, heating in oven products) which may be an indicative of how this method can be implemented in processed (cooked) food; as well as fermented or pickle foodstuff; lupin milk and cheese; meat; along with positive and negative control foodstuffs. The data from the quantification of Lup an 1 has been summarized in Table 1. These samples were analyzed in order to check the suitability of the designed antibody (anti-β-conglutin) to quantify the main allergen β-conglutin from protein extracts obtained following a specific protocol for vicilin (β-conglutin) proteins extraction, being these proteins specifically soluble in solutions containing NaCl [33].

Some methods have been previously reported for the detection of lupin proteins in foodstuffs by enzyme-linked immunosorbent assay (ELISA) [4,45–47]. However, they exhibited main disadvantages in comparison to the current study:

(1) The protein extraction protocol in the current study is highly specific for the extraction of the vicilin family of proteins (β-conglutins), based on the presence of NaCl (0.25M) in the extraction buffer [33]. Previous studies made protein extractions using general protein extraction buffers [4,45,47], or using alternative methods from commercial kits [48] currently no longer available (Abnova, http://www.abnova.com/products/products_detail.asp?catalog_id=KA3310), displaying very limited or no information about: (i) The antibody design; (ii) antibody production and use in the ELISA detection method; (iii) the limited information about protocol for total proteins extraction, which may not be specific for β-conglutin extraction; (iv) no information about lupin species used [46,48] to obtain this protein extract. The last two are the main factors with high impact in the protein extract characteristics, such as a low amount or not of Lup an 1 content. These disadvantages may result in the increase of the number of false positives as a result of the detection of non-allergen proteins from a low specific antibody, or using non-appropriate protein extracts.

(2) A second advantage of the current method compared with previous ones and commercial kits is the design and the production of the antibody (anti-IgG β-conglutin proteins). In the current study, the experimental animal was immunized with a synthetic peptide commonly shared by the seven NLL β-conglutin protein isoforms. This synthetic peptide constitutes a highly antigenic epitope in these proteins probed in the current study, while also exhibits a high specificity to detect the lupin main allergen Lup an 1 in the most frequently used lupin species. On the contrary, previous methods [4,45–47] have used the whole crude protein extract from lupin flour to immunize the experimental animal and obtain the antibody. The method to produce this antibody makes this antibody non-specific, and detecting a wide range of proteins, including many non-allergenic proteins, may lead to false positive detections. The detection of false positives might be also enhanced due to the implementation of a buffer inappropriate for vicilin protein extraction from lupin-derived foodstuffs.

(3) The current method exhibited another advantage which was the type of standards (Figure S3) that was made for a specific quantification of Lup an 1. Previous methodological developments of standards for the quantification were based on lupin flour total protein extract [4,45–47]. This may induce variable immunization for a complex mix of proteins (allergenic and non-allergenic proteins) from the crude extract leading to an excess or lack of reactivity in the standard samples because a variable representation or content of Lup an 1 (over or under representation, depending on the extraction method).

Therefore, lupin allergy reactions could be developed with high severity from primary sensitization to lupin proteins or due to the cross-reactivity with proteins from other legumes. The abundance of these proteins in many (natural or processed) foodstuffs has led to the European Union (EU Regulation No. 1169/2011) [49] to include lupin in a list of allergens with mandatory identification in all lupin containing foodstuffs. Indeed, an EU Labelling Directive involves the mandatory declaration of manufacturers regarding the presence of 14 allergenic products on pre-packaged foods [50]. Since lupin is included among these, analytical tools should be developed for the detection of lupin allergen traces in complex and processed foodstuffs. Thus, a high accurate method to detect and identify the lupin main allergen in food would be very valuable, with a particular importance due to the increasing prevalence of lupin allergy [16] among the population. The current study used highly pure (>95%) recombinant purified Lup an 1 (Figure 2), with the combination of an antibody that was developed using a specific synthetic peptide (main antigenic epitopes of the Lup an 1), while making the identification and quantification of the Lup an 1 highly specific and accurate.

In order to evaluate the application of the developed method to actual foodstuffs, several commercial samples labelled as "it may contain traces", "with or without lupine" were tested for the presence of Lup an 1. The summarized ELISA results, together with the corresponding label information of samples, are described in Tables 1 and 2.

The Lup an 1 lower content value in the food samples analyzed was 0.0406 ± 0.0009 ppm for toasted bread (Table 2). This result is in the comparable range of an ELISA method developed to detect soy protein content in foodstuff [51]. In this regard, the lowest eliciting dose for allergic reactions to lupin, responsible for inducing mild symptoms in peanut-sensitized patients, was 0.5 mg of lupin flour [17]. More recently, the VITAL program of the Allergen Bureau of Australia and New Zealand (ABA) established 4 mg of protein as the reference allergenic dose for lupin [52]. Taking into consideration that conglutins are the most abundant protein in lupin, being 40% of the protein seeds content [7], our ELISA method is by far capable of detecting these allergenic reference doses of lupin proteins in foodstuffs as demonstrated in this study.

Overall, NLL exhibited the highest Lup an 1 value (574.7918 ± 19.8876 ng) compared to all the samples investigated. The analysis of the sample number 2 (biscuits), and these samples declaring the information "may contain traces of lupine", showed no Lup an 1 content (Table 2), which might be due to non-contamination with lupin flour or other lupin derived component. Despite this fact, the manufacturer has implemented the common practice of the precautionary labelling. The same result was obtained from samples as number 11 (Tables 1 and 2) labelled as "non lupin content".

The estimated content of Lup an 1 in commercial lupin flour displayed an intermediate value among the three lupin seed species analyzed (Tables 1 and 2), which is in agreement with the labelling composition as sweet lupin attributed to three main domesticated lupin species (*L. albus*, *L. luteus* and *L. angustifolius*). The samples number 5, 9, 10, and 12 to 16 containing "lupine flour" or "lupine protein" among their ingredients, Lup an 1 was detected, identified and quantified in a range of allergen in accordance with their labelling (Tables 1 and 2) and also, to the variable quantity of seed flour from these three different sweet lupin species.

3.4. *Specificity of the Anti-β-Conglutin Antibody Tested in Processed Foodstuffs and Potential Cross-Reactive*

Lupin allergen proteins have been identified to be stable towards thermal treatment [5,47] in studies concerning the impact of the type of food processing on the allergenicity. Indeed, comparable results have been obtained in the current study for soy and chickpea-containing products, as well as for food processed under fermentation, soaking, extrusion, cooking, boiling and microwave heating conditions. These cooking processes do not affect the allergenic potential of protein extracts, while the autoclaving process drastically reduces the binding capacity of antibodies tested with serum from atopic patients [53–55]. This may be due to the conformational epitopes disappearing in these proteins under these processed conditions.

The main factors with a significant influence in the allergen analysis with ELISA methods are food matrices, the level of food processing and the ELISA test kit selection [56]. The current study analyzed the different foodstuffs (Tables 1 and 2), which were differentially processed. The ELISA method was able to detect, specifically identify and quantify Lup an 1 allergen: toasted [bread (8.1250 ± 0.1701 ng)], fermented [Lupinen—Tempeh (152.9862 ± 7.3353 ng)], Pickle [Pickled lupin (293.5833 ± 11.0853 ng)], Pasteurized [Lupinen drink (59.8660 ± 0.8389 ng)], and cooked [BOLOGNESE SAUCE (3.8890 ± 0.2599 ng)] foodstuff. Furthermore, a previous study indicated that boiling affected the allergenicity of legumes, particularly soy (Alvarez-Alvarez et al., 2005). Despite this fact, this study was able to detect and quantify Lup an 1 in boiled lupin based foodstuffs like Lupinen-tempeh (152.9862 ± 7.3353 ng) and LupinenBurger-Mediterranean (129.1667 ± 3.6839 ng).

Allergy reactions to lupin seed proteins are often triggered in atopic patients sensitized to proteins from other legumes, such as pea, lentil, soya, chickpea and peanut [5,21,57]. These allergy reactions arise against seed storage proteins from the Leguminosae family included in foodstuffs, since these proteins share similar epitopic regions in these different legume seed proteins. In this regard, atopic patients sensitized to one legume allergen protein might develop cross-reactions to proteins from another legume [58,59]. The most frequently described and important cross-reactivity from a clinical point of view is between lupin and peanuts [5]. Additional cross-reactivity between lupin and other legumes (lentil, pea) has been also reported, but in low percentages in children [5,57]. Furthermore, at a molecular level, lupin IgE cross-reactivity has been reported for peanut, soya, lentil, chickpea and bean [5]: Lupin sensitized patients were reported to develop cross-reactivity between 59–72% [13,20,60] and 52–55% [20,60] for soya bean and pea, respectively. Allergy reactions to peanuts are currently the most frequently identified legume allergy, followed by soya bean [57].

In this regard, a high specific and reliable method to detect the presence of other legume proteins is of crucial importance in order to avoid the development of cross-reactivity allergy reactions. The method developed in the current study was able to detect the lupin main allergen Lup an 1 with high specificity compared to other previous methods. However, the method may show false negative results as a consequence of the low antibody specificity and/or the absence of detection of allergen proteins that were not extracted in enough quantity by these alternative previously developed methods. On the other hand, false positive results as consequence of the use of a non-high specific antibody developed against a whole protein extract may not have clinical relevance. This may further lead to cross-reactivity between the antibody and the target protein from a related species with similar detection results (lack of specificity), and in the end, this scenario has notable consequences for the quality of life of people.

In the current study, lupin cross-reactivity with other legumes could be clinically relevant. For that reason, the authors developed an accurate, reliable, and highly specific method to detect, identify and quantify the main lupin allergen Lup an 1 that may be responsible of cross-reactivity with other legume proteins [11,21,22,36]. The high specificity of our anti-β-conglutin antibody is capable of avoiding false positives as a result of the cross-reactivity (Table 2). An analysis of different food products containing multiple legume proteins, such as peanut butter, chickpea, lentil, faba bean, even cereals resulted in the negative detection of homologous proteins to Lup an 1. In comparison, other ELISA methods previously developed showed low specificity of their antibodies used in the ELISA methods, since cross-reactivity showed with other legumes: Either with pea, chickpea, peanut, lentil, and soy [45], with brown bean and fenugreek (Holden et al., 2007); black bean and soy [46]; or even with non-legume proteins such as sunflower seed, cashew, almond, and pumpkin seed [61]. Furthermore, Koeberl et al. [48] used three available commercial kits to analyze cross-reactivity between lupin and other legume proteins finding that all peanut samples tested showed cross-reactivity on ELISA test kit B and C [48]. In addition, cross-reactivity was also described for the entire lentil samples analyzed, thus promoting false positive results in all analyzed legume samples. These results highlight the importance of transparency in the information provided by developed kits, at least in the characteristics of ELISA antibodies and protocols for their production, and further for the proteins extraction protocols implemented in these kits. This could lead to more uncertainty about the link between the food allergen protein identification and quantification, with the development of clinical therapies.

4. Conclusions

A newly developed ELISA assay was assessed for its capability to detect, identify and quantify the lupin main allergen β-conglutin protein (Lup an 1) in natural products and processed foodstuffs (toasted, boiled, fermented, cooked, pickled, pasteurized lupin-based products). Cross-reactivity was tested using peanut and other legumes, obtaining negative results.

The standards were made using recombinant purified β-conglutin proteins, showing a highly specific method for particular allergen proteins compared to previous developments. This fact highlights the importance of particular molecular tools to develop a more reliable and highly specific analytical method for food allergen protein detection. This recombinant protein could constitute an available improvement for the quantification of food allergens, which should be approved as (certified) reference material for food allergens identification helping to implement the labelling mandatory EU regulations and as relevant proteins for clinical allergy research.

Finally, this study demonstrated that this ELISA method is based on accurate and reliable molecular tools, which can contribute to a more effective management of allergens by the food industry, the regulatory agencies and clinicians, thus helping to protect the health of sensitized/allergic consumers.

Supplementary Materials: The following are available online at http://www.mdpi.com/2304-8158/8/10/513/s1, Figure S1: Multiple alignment of the deduced amino acid sequences of β1 to β7., Figure S2: Antigenicity assessment of conglutin β1., Figure S3: Standard curve made of the conglutin β1 protein.

Author Contributions: Conceptualization: J.C.J.-L. and E.L.-C.; methodology: J.C.J.-L. and E.L.-C.; software: J.C.J.-L.; validation: J.C.J.-L. and E.L.-C.; data Analysis: J.C.J.-L. and E.L.-C. and J.D.A.; investigation: J.C.J.-L. and E.L.-C.; resources: J.C.J.-L. and J.D.A.; writing—original draft preparation: J.C.J.-L.; writing—review & editing: J.C.J.-L. and J.D.A.; funding acquisition: J.C.J.-L. and J.D.A.

Funding: This research was funded by the European Research Program MARIE CURIE (FP7-PEOPLE-2011-IOF) grant number PIOF-GA-2011-30155; the Spanish Ministry of Economy, Industry and Competitiveness grants numbers RYC-2014-16536 (Ramon y Cajal Research Program) and BFU2016-77243-P; the CSIC-Intramural grant number 201540E065; And The APC was funded by the Spanish Ministry of Economy, Industry and Competitiveness grants numbers RYC-2014-16536 (Ramon y Cajal Research Program).

Acknowledgments: J.C.J.-L. thanks the European Research Program MARIE CURIE (FP7-PEOPLE-2011-IOF) for through the grant ref. number PIOF-GA-2011-30155; to the Spanish Ministry of Economy, Industry and Competitiveness for the grant ref. number RYC-2014-16536 (Ramon y Cajal Research Program), and the grant ref. BFU2016-77243-P; and to CSIC-Intramural grant ref. 201540E065.

Conflicts of Interest: The authors declare no conflicts of interest.

References

1. Kurlovich, B.S.; Stankevich, A.K.; Stepanova, S.I. The review of the genus *Lupinus* L. In *Lupins (Geography, Classification, Genetic Resources and Breeding)*; Kurlovich, B.S., Ed.; OY International North Express: St. Petersburg, Russia; Pellosniemi, Finland, 2002; Chapter 2; pp. 11–38.
2. Uauy, R.; Gattas, V.; Yáñez, E. Sweet lupins in human nutrition. *World Rev. Nutr. Diet.* **1995**, *77*, 75–88.
3. Frick, K.M.; Kamphuis, L.G.; Siddique, K.H.M.; Singh, K.B.; Foley, R.C. Quinolizidine alkaloid biosynthesis in lupins and prospects for grain quality improvement. *Front. Plant Sci.* **2017**, *8*, 87. [CrossRef]
4. Ecker, C.; Cichna-Markl, M. Development and validation of a sandwich ELISA for the determination of potentially allergenic lupine in food. *Food. Chem.* **2012**, *130*, 759–766. [CrossRef]
5. Jappe, U.; Vieths, S. Lupine, a source of new as well as hidden food allergens. *Mol. Nutr. Food Res.* **2010**, *54*, 113–126. [CrossRef]
6. Delgado-Andrade, C.; Olias, R.; Jimenez-Lopez, J.C.; Clemente, A. Nutritional and beneficial effects of grain legumes on human health. *ARBOR Ciencia Pensamiento y Cultura* **2016**, *192*, 779.
7. Lima-Cabello, E.; Alche, V.; Foley, R.C.; Andrikopoulos, S.; Morahan, G.; Singh, K.B.; Alche, J.D.; Jimenez-Lopez, J.C. Narrow-leafed lupin (*Lupinus angustifolius* L.) β-conglutin proteins modulate the insulin signaling pathway as potential type 2 diabetes treatment and inflammatory-related disease amelioration. *Mol. Nutr. Food Res.* **2017**, *61*. [CrossRef]
8. Lima-Cabello, E.; Morales-Santana, S.; Foley, R.C.; Melser, S.; Alche, V.; Siddique, K.H.M.; Singh, K.B.; Alche, J.D.; Jimenez-Lopez, J.C. Ex vivo and in vitro assessment of anti-inflammatory activity of seed β-conglutin proteins from *Lupinus angustifolius*. *J. Funct. Foods* **2018**, *40*, 510–519. [CrossRef]
9. Lima-Cabello, E.; Morales-Santana, S.; Leon, J.; Alche, V.; Clemente, A.; Alche, J.D.; Jimenez-Lopez, J.C. Narrow-leafed lupin (*Lupinus angustifolius* L.) seed beta-conglutins reverse back the induced insulin resistance in pancreatic cells. *Food Funct.* **2018**, *9*, 5176–5188. [CrossRef]
10. Lima-Cabello, E.; Robles-Bolivar, P.; Alche, J.D.; Jimenez-Lopez, J.C. Narrow-leafed lupin Beta-conglutin proteins epitopes identification and molecular features analysis involved in cross-allergenicity to peanut and other legumes. *Genom. Comput. Biol.* **2016**, *2*, e29. [CrossRef]
11. Foley, R.C.; Jimenez-Lopez, J.C.; Kamphuis, L.G.; Hane, J.K.; Melser, S.; Singh, K.B. Analysis of conglutin seed storage proteins across lupin species using transcriptomic, protein and comparative genomic approaches. *BMC Plant Biol.* **2015**, *15*, 106. [CrossRef]
12. Cabello-Hurtado, F.; Keller, J.; Ley, J.; Sanchez-Lucas, R.; Jorrín-Novo, J.V.; Aïnouche, A. Proteomics for exploiting diversity of lupin seed storage proteins and their use as nutraceuticals for health and welfare. *J. Proteom.* **2016**, *143*, 57–68. [CrossRef]
13. De Jong, N.W.; Van Maaren, M.S.; Vlieg-Boersta, B.J.; Dubois, A.E.J.; De Groot, H.; van Wijk, R.G. Sensitization to lupine flour: Is it clinically relevant? *Clin. Exp. Allery* **2010**, *40*, 1571–1577. [CrossRef]
14. Ronchi, A.; Duranti, M.; Scarafoni, A. A real-time PCR method for the detection and quantification of lupin flour in wheat flour-based matrices. *Food Chem.* **2009**, *115*, 1088–1093.

15. Sirtori, E.; Resta, D.; Brambilla, F.; Zacherl, C.; Arnoldi, A. The effects ofvarious processing conditions on a protein isolate from *Lupinus angustifolius*. *Food Chem.* **2010**, *120*, 496–504. [CrossRef]
16. Sanz, M.L.; De Las Marinas, M.D.; Fernández, J.; Gamboa, P.M. Lupin allergy: A hidden killer in the home. *Clin. Exp. Allery* **2010**, *40*, 1461–1466. [CrossRef]
17. Peeters, K.A.; Koppelman, S.J.; Penninks, A.H.; Lebens, A.; Bruijnzeel-Koomen, C.A.; Hefle, S.L.; Taylor, S.L.; van Hoffen, E.; Knulst, A.C. Clinical relevance of sensitization to lupine in peanut-sensitized adults. *Allergy* **2009**, *64*, 549–555. [CrossRef]
18. Campbell, C.P.; Yates, D.H. Lupin allergy: A hidden killer at home, a menace at work; occupational disease due to lupin allergy. *Clin. Exp. Allery* **2010**, *40*, 1467–1472. [CrossRef]
19. Kaw, C.H.; Hefle, S.L.; Taylor, S.L. Sandwich Enzyme-Linked Immunosorbent Assay (ELISA) for the Detection of Lupine Residues in Foods. *Food Sci.* **2008**, *73*, T135–T140. [CrossRef]
20. Hieta, N.; Hasan, T.; Makinen-Kiljunen, S.; Lammintausta, K. Lupin allergy and lupin sensitization among patients with suspected food allergy. *Ann. Allergy Asthma Immunol.* **2009**, *103*, 233–237. [CrossRef]
21. Jimenez-Lopez, J.C.; Foley, R.C.; Brear, E.; Clarke, V.C.; Lima-Cabello, E.; Florido, J.F.; Singh, K.B.; Alché, J.D.; Smith, P.M.C. Characterization of narrow-leaf lupin (*Lupinus angustifolius* L.) recombinant major allergen IgE-binding proteins and the natural β-conglutin counterparts in sweet lupin seed species. *Food Chem.* **2018**, *44*, 60–70. [CrossRef]
22. Lima-Cabello, E.; Berral-Hens, J.M.; Florido, J.F.; Alche, J.D.; Jimenez-Lopez, J.C. The function of conglutin proteins from the "sweet lupin" group in food allergy. In *Legumes for Global Food Security*, 1st ed.; Nova Science Publishers, Inc.: New York, NY, USA, 2017; Chapter 10; pp. 249–275.
23. Goggin, D.E.; Mir, G.; Smith, W.B.; Stuckey, M.; Smith, P.M.C. Proteomic analysis of lupin seed proteins to identify conglutin β as an allergen, Lup an 1. *J. Agric. Food Chem.* **2008**, *56*, 6370–6377. [CrossRef]
24. Jimenez-Lopez, J.C.; Gachomo, E.W.; Sharma, S.; Kotchoni, S.O. Genetic sequencing and next-generation sequence data analysis: A comprehensive compilation of bioinformatics tools and databases. *Am. J. Mol. Biol.* **2013**, *3*, 115–130. [CrossRef]
25. Kyte, J.; Doolittle, R.F. A simple method for displaying the hydropathic character of a protein. *J. Mol. Biol.* **1982**, *157*, 105–132. [CrossRef]
26. Emini, E.A.; Hughes, J.V.; Perlow, D.S.; Boger, J. Induction of hepatitis A virus-neutralizing antibody by a virus-specific synthetic peptide. *J. Virol.* **1985**, *55*, 836–839.
27. Karplus, P.A.; Schulz, G.E. Prediction of chain flexibility in proteins. *Naturwissenschaften* **1985**, *72*, 212–213. [CrossRef]
28. Hopp, T.P.; Woods, K.R. Prediction of protein antigenicdeterminants from amino acid sequences. *Proc. Natl. Acad. Sci. USA* **1981**, *78*, 3824–3828. [CrossRef]
29. Welling, G.W.; Weijer, W.J.; van der Zee, R.; Welling-Wester, S. Prediction of sequential antigenic regions in proteins. *FEBS Lett.* **1985**, *188*, 215–218. [CrossRef]
30. Parker, K.C.; Bednarek, M.A.; Coligan, J.E. Scheme for rankingpotential HLA-A2 binding peptides based on independent binding ofindividual peptide side-chains. *J. Immunol.* **1994**, *152*, 163–175.
31. Kolaskar, A.S.; Tongaonkar, P.C. A semi-empirical method forprediction of antigenic determinants on protein antigens. *FEBS Lett.* **1990**, *276*, 172–174. [CrossRef]
32. Fox, B.G.; Blommel, P.G. Autoinduction of Protein Expression. *Curr. Protoc. Protein Sci.* **2009**, *56*, 5–23.
33. Nadal, P.; Canela, N.; Katakis, I.; O'Sullivan, C.K. Extraction, Isolation, and Characterization of Globulin Proteins from *Lupinus albus*. *J. Agric. Food Chem.* **2011**, *59*, 2752–2758. [CrossRef]
34. Wang, Y.; Wu, W.; Negre, N.N.; White, K.P.; Li, C.; Shah, P.K. Determinants of antigenicity and specificity in immune response for protein sequences. *BMC Bioinform.* **2011**, *12*, 251. [CrossRef]
35. Palladino, C.; Breiteneder, H. Peanut allergens. *Mol. Immunol.* **2018**, *100*, 58–70. [CrossRef]
36. Jimenez-Lopez, J.C.; Lima-Cabello, E.; Melser, S.; Foley, R.C.; Singh, K.B.; Alché, J.D. Lupin allergy: Uncovering structural features and epitopes of β-conglutin proteins in *Lupinus angustifolius* L. with a focus on cross-allergenic reactivity to peanut and other legumes. *Bioinform. Biomed. Eng.* **2015**, *9043*, 96–107.
37. Jimenez-Lopez, J.C.; Robles-Bolivar, P.; Lopez-Valverde, F.J.; Lima-Cabello, E.; Kotchoni, S.O.; Alché, J.D. Ole e 13 is the unique food allergen in olive: Structure-functional, substrates docking, and molecular allergenicity comparative analysis. *J. Mol. Graph. Model.* **2016**, *66*, 26–40. [CrossRef]

38. Jimenez-Lopez, J.C.; Kotchoni, S.O.; Hernandez-Soriano, M.C.; Gachomo, E.W.; Alché, J.D. Structural functionality, catalytic mechanism modeling and molecular allergenicity of phenylcoumaran benzylic ether reductase, an olive pollen (Ole e 12) allergen. *J. Comput. Aided Mol. Des.* **2013**, *27*, 873–895. [CrossRef]
39. Jimenez-Lopez, J.C.; Kotchoni, S.O.; Rodríguez-García, M.I.; Alché, J.D. Structure and functional features of olive pollen pectin methylesterase using homology modeling and molecular docking methods. *J. Mol. Model.* **2012**, *18*, 4965–4984. [CrossRef]
40. Jimenez-Lopez, J.C.; Rodríguez-García, M.I.; Alché, J.D. Analysis of the effects of polymorphism on pollen profilin structural functionality and the generation of conformational, T- and B-cell epitopes. *PLoS ONE* **2013**, *17*, e76066. [CrossRef]
41. Jimenez-Lopez, J.C.; Gachomo, E.W.; Ariyo, O.A.; Baba-Moussa, L.; Kotchoni, S.O. Specific conformational epitope features of pathogenesis-related proteins mediating cross-reactivity between pollen and food allergens. *Mol. Biol. Rep.* **2012**, *39*, 123–130. [CrossRef]
42. Jimenez-Lopez, J.C.; Melser, S.; DeBoer, K.; Thatcher, L.F.; Kamphuis, L.G.; Foley, R.C.; Singh, K.B. Narrow-Leafed Lupin (*Lupinus angustifolius*) β1- and β6-Conglutin Proteins Exhibit Antifungal Activity, Protecting Plants against Necrotrophic Pathogen Induced Damage from *Sclerotinia sclerotiorum* and *Phytophthora nicotianae*. *Front. Plant Sci.* **2016**, *9*, 1856. [CrossRef]
43. Magni, C.; Scarafoni, A.; Herndl, A.; Sessa, F.; Prinsi, B.; Espen, L.; Duranti, M. Combined 2D electrophoretic approaches for the study of white lupin mature seed storage proteome. *Phytochemistry* **2007**, *68*, 997–1007. [CrossRef] [PubMed]
44. Duranti, M.; Sessa, F.; Carpen, A. Identification, purification and properties of the precursor of conglutin β, the 7S storage globulin of *Lupinus albus* L. seeds. *J. Exp. Bot.* **1992**, *43*, 1373–1378. [CrossRef]
45. Holden, L.; Faeste, C.K.; Egaas, E. Quantitative sandwich ELISA for the determination of lupine (*Lupinus* spp.) in foods. *J. Agric. Food Chem.* **2005**, *53*, 5866–5871. [CrossRef]
46. Taylor, S.L.; Nordlee, J.A.; Niemann, L.M.; Lambrecht, D.M. Allergen immunoassays–considerations for use of naturally incurred standards. *Anal. Bioanal. Chem.* **2009**, *395*, 83–92.
47. Ecker, C.; Ertla, A.; Pulverera, W.; PalSzekely, A.N.; Petraschc, A.; Linsberger-Martin, G.; Cichna-Markl, M. Validation and comparison of a sandwich ELISA, two competitive ELISAs and a real-time PCR method for the detection of lupine in food. *Food Chem.* **2013**, *141*, 407–418. [CrossRef]
48. Koeberl, M.; Sharp, M.F.; Tian, R.; Buddhadasa, S.; Clarke, D.; Roberts, J. Lupine allergen detecting capability and cross-reactivity of related legumes by ELISA. *Food Chem.* **2018**, *256*, 105–112. [CrossRef]
49. European Parliament Regulation (EU). *No 1169/2011 of the European Parliament and of the Council of 25 October 2011 on the Provision of Food Information to Consumers, Amending Regulations (EC) No 1924/2006 and (EC) No 1925/2006 of the European Parliament and of the Council, and Repealing Commission Directive 87/250/EEC, Council Directive 90/496/EEC, Commission Directive 1999/10/EC, Directive 2000/13/EC of the European Parliament and of the Council, Commission Directives 2002/67/EC and 2008/5/EC and Commission Regulation (EC) No 608/2004 Text with EEA Relevance L304/18 C.F.R.*; EU: Brussel, Belgium, 2011.
50. EFSA. Directive 2007/68/EC Amending Annex IIIa of Directive 2000/13/EC of the European Parliament and of the Council (2007). *Off. J. Eur. Union* **2007**, *310*, 11–14.
51. Pedersen, M.H.; Holzhauser, T.; Bisson, C.; Conti, A.; Jensen, L.B.; Skov, P.S.; Bindslev-Jensen, C.; Brinch, D.S.; Poulsen, L.K. Soybean allergen detection methods—A comparison study. *Mol. Nutr. Food Res.* **2008**, *52*, 1486–1496. [CrossRef]
52. Taylor, S.L.; Baumert, J.L.; Houben, G.F.; Crevel, R.W.R.; Brooke-Taylor, S.; Allen, K.J.; The Allergen Bureau of Australia & New Zealand; Houben, G. Establishment of reference doses for residues of allergenic foods: Report of the VITAL expert panel. *Food Chem. Toxicol.* **2014**, *63*, 9–17.
53. Alvarez-Alvarez, J.; Guillamon, E.; Crespo, J.F.; Cuadrado, C.; Burbano, C.; Rodriguez, J.; Muzquiz, M. Effects of extrusion, boiling, autoclaving, and microwave heating on lupine allergenicity. *J. Agric. Food Chem.* **2005**, *53*, 1294–1298. [CrossRef]
54. Gupta, R.K.; Gupta, K.; Sharma, A.; Das, M.; Ansari, I.A.; Dwivedi, P.D. Health Risks and Benefits of Chickpea (Cicerarietinum) Consumption. *J. Agric. Food Chem.* **2017**, *65*, 6–22. [CrossRef]
55. Verma, A.K.; Kumar, S.; Das, M.; Dwivedi, P.D. A Comprehensive Review of Legume Allergy. *Clin. Rev. Allergy Immunol.* **2013**, *45*, 30–46. [CrossRef] [PubMed]

56. Török, K.; Hajas, L.; Horváth, V.; Schall, E.; Bugyi, Z.; Kemény, S.; Tömösközi, S. Identification of the factors affecting the analytical results offood allergen ELISA methods. *Eur. Food Res. Technol.* **2015**, *241*, 127–136. [CrossRef]
57. Peeters, K.A.; Nordlee, J.A.; Penninks, A.H.; Chen, L.; Goodman, R.E.; Bruijnzeel-Koomen, C.A.; Knulst, A.C. Lupine allergy: Not simply cross-reactivity with peanut or soy. *J. Allery Clin. Immunol.* **2007**, *120*, 647–653. [CrossRef] [PubMed]
58. Berkner, H.; Neudecker, P.; Mittag, D.; Ballmer-Weber, B.K.; Schweimer, K.; Vieths, S.; Rosch, P. Cross-reactivity of pollen and food allergens: Soybean Gly m 4 is a member of the Bet v 1 superfamily and closely resembles yellow lupine proteins. *Biosci. Rep.* **2009**, *29*, 183–192. [CrossRef]
59. Shaw, J.; Roberts, G.; Grimshaw, K.; White, S.; Hourihane, J. Lupin allergy in peanut-allergic children and teenagers. *Allergy* **2008**, *63*, 370–373. [CrossRef]
60. Hoffmann, B.; Münch, S.; Schwägele, F.; Neusüß, C.; Jira, W. A sensitive HPLC-MS/MS screening method for the simultaneous detection of lupine, pea, and soy proteins in meat products. *Food Control* **2017**, *71*, 200–209. [CrossRef]
61. Holden, L.; Moen, L.H.; Sletten, G.B.; Dooper, M.M. Novel polyclonal-monoclonal-based ELISA utilized to examine lupine (*Lupinus* species) content in food products. *J. Agric. Food Chem.* **2007**, *55*, 2536–2542. [CrossRef]

Article

Comparison of Various Soybean Allergen Levels in Genetically and Non-Genetically Modified Soybeans

Ayato Matsuo [1], Kaho Matsushita [1], Ayano Fukuzumi [1], Naoki Tokumasu [1], Erika Yano [1], Nobuhiro Zaima [1,2] and Tatsuya Moriyama [1,2,*]

1 Department of Applied Biological Chemistry, Graduate School of Agriculture, Kindai University, Nara 631-8505, Japan; ayato.m1223@gmail.com (A.M.); 1511430087w@nara.kindai.ac.jp (K.M.); ayano.fukuzumi.0123@gmail.com (A.F.); naoki.tokumasu.1205@gmail.com (N.T.); pyonchangoods@yahoo.co.jp (E.Y.); zaima@nara.kindai.ac.jp (N.Z.)

2 Agricultural Technology and Innovation Research Institute, Kindai University, Kindai University, Nara 631-8505, Japan

* Correspondence: tmoriyama@nara.kindai.ac.jp; Tel.: +81-742-43-8070

Received: 27 February 2020; Accepted: 16 April 2020; Published: 21 April 2020

Abstract: Several analyses of allergen levels have been reported as part of the safety assessment of genetically modified (GM) soybean; however, few comprehensive analyses have included new allergens. Thus, in this study the levels of eight major soybean allergens, including Gly m 7 (a newly reported soybean allergen), were semi-quantitatively detected in six GM soybeans and six non-GM soybeans using antigen-immobilized ELISA and immunoblotting. We also analyzed the IgE-reactivity to these soybeans through immunoblotting, using sera from three soybean-allergic patients. The results showed that there were no significant differences in the levels of the major soybean allergens in the GM and non-GM soybeans. Moreover, there were no significant differences in the serum IgE-reactive protein profiles of the patients, as analyzed using immunoblotting. These results indicate that, in general, CP4-EPSPS-transfected GM soybeans are not more allergenic than non-GM soybeans.

Keywords: soybean; allergens; allergenicity; genetically modified; Gly m 7

1. Introduction

Food resource problems associated with climate change, environmental destruction, and population growth are of increasing concern. As a means to overcome these concerns, scientists have developed "genetic modification technology", which alters the properties of agricultural products. Using this technology, genetically modified (GM) crops have been developed that are not only more resistant to herbicides but also contain beneficial traits such as drought tolerance, delayed ripening, bacterial disease resistance, high oleic acid levels, and pest resistance, to prepare for an increase in global demand [1–4]. The total area of cultivation of GM crops has increased worldwide, to 189,800,000 ha in 2017, with four major GM crops: soybeans (50%), corn (31%), cotton (13%), and rapeseed (5%) [5]. Soybeans are not only used as a raw material for soybean oil but are also widely used as foods such as tofu, fermented soybeans (natto), miso, soy sauce, and soy milk, and as additives in various processed foods in the form of soy protein isolate (SPI).

However, ingestion of soybeans can cause allergic reactions, and various soybean allergens have been identified to date [6]. Soybean allergies can be divided into class 1 food allergies and class 2 food allergies based on differences in sensitization routes [7,8]; 7S globulin (Gly m 5) [9,10], 11S globulin (Gly m 6) [11], Gly m 7 [12], Gly m Bd 30K [13,14], Kunitz-type trypsin inhibitor [15], oleosin [16], etc. have been identified as class 1 food allergens causing class 1 allergy. It has been reported that these allergens mainly cause systemic symptoms such as urticaria, diarrhea, vomiting, atopy, and anaphylaxis. Gly m 3 (profilin) [17,18] and Gly m 4 (starvation-associated message 22: SAM22) [18–21] from soybeans have been reported as class 2 allergens causing class 2 allergy (i.e., pollen–food allergy syndrome [PFAS]). Both Gly m 3 and Gly m 4 are homologues of Bet v 2 and Bet v 1, which are birch pollen allergens that mainly cause oral allergy syndromes (OASs) [18], although severe cases of anaphylaxis with facial swelling, airway narrowing, and breathing difficulties have also been reported [8].

Analyses of variation in the relative levels of known endogenous allergens is required to verify whether genetic transformation or transgenes adversely affect human health and whether the level of endogenous allergens is altered by genetic modification. Therefore, studies using sera from soybean allergy patients with IgE antibodies have been conducted using GM soybeans and non-GM soybeans. Lua et al. conducted an IgE-immunoblot and IgE-ELISA using GM soybeans and a closely related variety of non-GM soybeans; they found that GM soybeans had similar allergenicity to non-GM soybeans and identified no changes in the immunoblot results attributable to genetic modification [2]. Kim et al. reported that IgE-inhibition ELISA using patient serum showed equivalent inhibition in both non-GM soybean and GM soybean extracts, with IgE-immunoblots detecting the most 33 kDa bands in 50% (7/14) of the sera tested and in lanes applied with GM soybean extract and non-GM soybean extract, which were identical to P34 proteins (Gly m Bd 30K). They also argued that the allergenic risk of GM and non-GM soybeans is the same as the allergenic risk of wild-type soybeans because no specific IgE antibodies were detected against the recombinant protein that was genetically integrated into the soybeans, EPSPS (5-*enol*pyruvylshikimate-3-phosphate synthase) [22], which confers resistance to the herbicide glyphosate [23,24]. Tsai et al. also reported no significant differences in Gly m 4 levels in non-GM soybean cultivars and GM soybeans (transfected with EPSPS genes and CaMV 35S promoters) [25]. In addition, there are several reports describing the allergen levels of GM crops [26–31]. Thus, while some studies have explored the allergenicity of GM soybeans, there have been no studies focusing on a wide variety of soybean allergen components, and none that explore differential allergenicity of Gly m 7, a recently discovered soybean allergen. Therefore, in this study, to update the allergenicity assessment of GM soybeans, we analyzed the variability of various soybean class 1 food allergens and soybean class 2 (pollinosis-related) food allergen levels in GM soybeans and non-GM soybeans in vitro and compared the patterns of IgE-binding proteins using sera from soybean allergenic patients.

2. Materials and Methods

2.1. Materials

Horseradish peroxidase (HRP)-labeled anti-rabbit and anti-mouse IgGs were obtained from Thermo Fisher Scientific (Waltham, MA, USA), and HRP-labeled anti-guinea pig IgG was obtained from Jackson ImmunoResearch (West Grove, PA, USA). HRP-labeled anti-human IgE was obtained from Kirkegaard and Perry Laboratories, Inc. (Gaithersberg, MD, USA). ECL^{TM} Western blotting reagent and Hyperfilm^{TM}-MP X-ray films were obtained from GE Healthcare (Piscataway, NJ, USA). PVDF membrane (Immobilon^{TM}-P) was obtained from Millipore (Billerica, MA, USA).

2.2. Soybean Sample Extraction

GM soybeans and non-GM soybeans (controls) were obtained from an anonymous seed company. Each sample (approximately 2.5 g) was mixed with distilled water (25 mL), soaked at room temperature (25 °C) for 4 h, and crushed for 30 s in a mixer. Thereafter, the mixture was squeezed with quadruple gauze to obtain a protein extract. The extract was diluted 20- and 800-fold with distilled water for detection of the allergen levels.

2.3. Immunochromatography

To confirm genetic modification of the GM soybean samples, the transgene CP4-EPSPS (EPSPS derived from Agrobacterium CP4 strain) was detected using the Reveal for CP4 Strip Test Kit (Neogen) according to the instruction manual.

2.4. Electrophoresis and Immunoblotting

The extracted soybean proteins were subjected to SDS–PAGE. Proteins in the 12.5% gel were stained with Coomassie brilliant blue (CBB) (CBB R-350, GE Healthcare) to visualize the total protein patterns. The immunoblotting analysis was conducted by transferring the SDS–PAGE gel to an Immobilon-P^{TM} PVDF membrane (Millipore) using a semi-dry blotting method [32]. The membrane was incubated in 10 mM PBS (pH = 7.5) containing 0.1% Tween-20 (PBST) and 5% skim milk for blocking. The membrane was then incubated for 1 h at room temperature (25 °C) in a blocking buffer containing allergen-specific antibodies. After the membranes were washed four times with PBST for 10 min, the bound primary antibodies were detected by using HRP-conjugated goat anti-rabbit, anti-mouse, or anti-guinea pig IgG and an ECL^{TM} Western blotting kit (GE Healthcare). The resultant chemiluminescent signals were detected on X-ray film (Hyperfilm^{TM} MP, GE Healthcare). Immunoblotting experiments were performed three times, and band densities were determined using Alpha Ease^{TM} software (Alpha Innotech, San Leandro, CA, USA). The immunoblot results were expressed relative to the value of the control No. 1 sample (C1).

2.5. Antibodies Against the Major Soybean Allergens

Mouse monoclonal antibody against Gly m Bd 30 K [6] was kindly provided by Dr. Tadashi Ogawa (Professor Emeritus at Kyoto University). The rabbit polyclonal antibodies against Gly m 5 (7S globulin; β-conglycinin; α′, α, and β subunits) were obtained as previously described [33]. The mouse polyclonal antibodies against Gly m 6 (11S globulin; glycinin) were obtained by immunizing mice with purified 11S globulin in our laboratory. The rabbit polyclonal antibodies against Gly m 4 were obtained by immunizing rabbit with recombinant Gly m 4 in our laboratory. The rabbit polyclonal antibodies against oleosin were also obtained by immunizing rabbit with purified oleosin in our laboratory. The guinea pig polyclonal antibodies against Gly m 3 were also obtained by immunizing guinea pigs with recombinant Gly m 3 in our laboratory [34]. The rabbit antibody against soybean trypsin inhibitor was obtained from Rockland (Gilbertsville, PA, USA). Gly m 7 is a seed biotinylated protein with a single binding site for biotin. Two methods were used to detect Gly m 7: rabbit-derived peptide antibodies (unpublished data) that can detect the peptide moiety of Gly m 7, and HRP-labeled streptavidin that binds specifically to biotin. The specific reactivities of these antibodies have been confirmed in our previous studies. The soybean allergens detected in this study are listed in Table 1.

Table 1. Various soybean allergens detected in this study.

Allergen Name	Molecular Mass (kDa)	Features	References
	Soybean food allergens (class 1 food allergens)		
Gly m 5 (7S globulin)	72, 68, 50	Major storage protein, glycoprotein	[9,10]
Gly m 6 (11S globulin)	34, 20	Major storage protein	[11]
Gly m 7	76	Seed-specific biotinylated protein (SBP)	[12]
Gly m Bd 30K (P34)	34 (30)	Homologous to papain	[13,14]
Kunitz-type trypsin inhibitor	18	trypsin inhibitor	[15]
Oleosin	22–24	Oil body associated protein	[16]
	Pollen-related soybean allergens (class 2 food allergens)		
Gly m 3	14	Profilin, Bet v 2 homolog	[17,18]
Gly m 4	17	PR-10 family, Bet v 1 homolog	[18–21]

2.6. ELISA Using Allergen-Specific Antibodies

ELISA was used to evaluate the allergen levels of non-GM (control) and GM soybeans. ELISA plates were coated with sequentially diluted (100- to 1,000,000-fold diluted) soybean extracts. After sample coating, plates were blocked with Blocking one (nacarai tesque, Kyoto, Japan, dilution 1:5) for 1 h at room temperature (25 °C) and then washed with PBST three times. Next, diluted allergen-specific antibodies were added to the wells, and samples were incubated for 1 h at 37 °C. Plates were then washed with PBST five times and HRP-labeled secondary antibodies were added to the wells. Plates were then incubated for 1 h at 37 °C and then washed with PBST five times. The bound HRP-labeled secondary antibodies were visualized by reaction with 100 µL of tetramethylbenzidine (TMB) peroxidase substrate (KPL, Gaithersburg, MD, USA) for 5–15 min. The reaction was stopped by adding 100 µL of 1 M phosphoric acid to provide a stable endpoint color. The absorption was measured at 450 nm using an ARVOsx-1 1420 multilabel counter (PerkinElmer Life Sciences, Boston, MA, USA). Measurements were performed three times, and the mean absorbance values were calculated.

2.7. Detection of IgE-Binding Proteins using Patient Sera

A total of 3 commercially available soybean allergy patient sera were purchased from Kokusai Bio Co., ltd (Tokyo, Japan). The patients were all soybean class 1 food allergy patients from the United States. Immunoblotting and ELISA were performed using patient sera. HRP-labeled anti-human IgE antibody was used as a secondary antibody. The detected IgE-binding protein bands were exposed to X-ray films and captured by a scanner; band densities were determined using Alpha Ease^{TM} software (Alpha Innotech, San Leandro, CA, USA).

2.8. Statistical Analysis

Results are expressed as the mean ± standard deviation (SD). Data was analyzed by Student's t-test with Excel Statistics software (SSRI Co., Tokyo, Japan). P value < 0.05 was considered statistically significant.

3. Results

3.1. Confirmation of Genetically Modified and Non-Genetically Modified Soybeans

Proteins were extracted from 12 kinds of soybeans including six kinds of GM soybeans and six kinds of non-GM soybeans. Detailed information about each GM and non-GM soybean cultivar was not available for blind testing. However, these soybeans were all popular cultivars in the world. Immunochromatographic analysis of soybean extracts detected CP4-EPSPS in all six kinds of GM soybeans, confirming that the samples were GM soybeans, and did not detect CP4-EPSPS in all six kinds of non-GM soybeans (Figure 1a). SDS–PAGE of protein extracts from GM soybeans and non-GM soybeans as visualized by CBB showed no obvious differences (Figure 1b). Immunoblotting of the soybean protein extracts using antibodies against CP4-EPSPS, the recombinant gene product, detected CP4-EPSPS in the six GM soybean species but not in the six non-GM soybean species (Figure 1c).

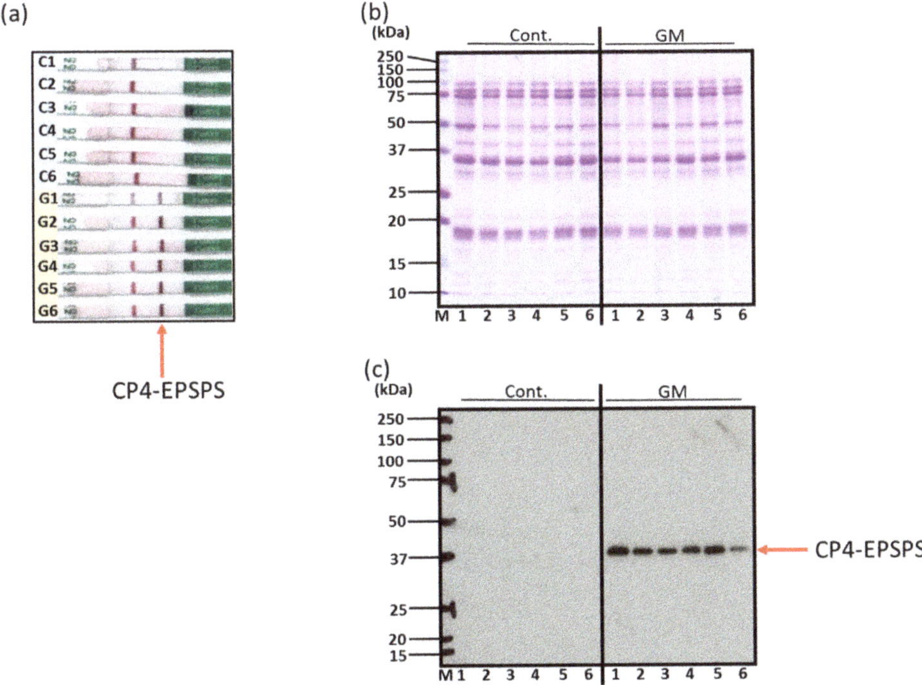

Figure 1. Characterization and confirmation of genetically modified (GM) and non-genetically modified (non-GM) soybeans. Indicated soybeans were extracted and subjected to immunochromatography (**a**), SDS–PAGE followed by CBB (CBB R-350, GE Healthcare) staining (**b**), and immunoblotting for detection of recombinant protein CP4-EPSPS (**c**). (C1–C6), non-GM soybeans; (G1–G6), GM soybeans.

3.2. Comparative Levels of Pollinosis-Related Soybean Allergens (Gly m 4 and Gly m 3) in GM-Soybean and Non-GM Soybean Extracts

Pollinosis-related soybean allergens Gly m 3 and Gly m 4 were detected by ELISA and immunoblotting in soybean extracts, as shown in Figures 2 and 3. Allergen content varied by individual sample; however, no significant difference in allergen content was found between the non-GM soybean and GM soybean groups for Gly m 3 or Gly m 4 by ELISA (Figure 2a,b, Figure 3a,b). Immunoblotting analysis revealed unique bands at approximately 13 kDa and 17 kDa identifying Gly m 3 (Figure 2c,e) and Gly m 4 (Figure 3c,e) in soybean extracts, respectively. No significant difference in allergen content was found between the non-GM soybean and GM soybean groups for Gly m 3 or Gly m 4 by immunoblotting (Figures 2d and 3d).

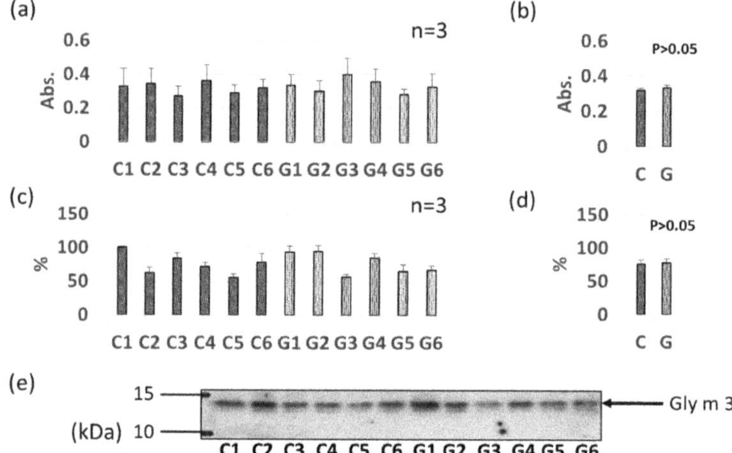

Figure 2. Comparison of Gly m 3 levels in GM-and non-GM soybeans by ELISA (**a**,**b**) and immunoblotting (**c**–**e**) using guinea pig-derived polyclonal antibodies. Soybean protein extracts were evaluated by ELISA (**a**,**b**) and immunoblotting (**c**–**e**) for detection of Gly m 3 levels. The ELISA data are presented in absorbance values (Abs). The individual data from six GM-and non-GM soybeans (**a**,**c**) are presented as the mean ± SD of three independent replicates. The collated data (**b**,**d**) are presented as the mean ± SD of all individual data points from the control (C1–C6, six non-GM soybeans) or experimental (G1–G6, six GM soybeans) groups relative to the value of control number 1 (C1). A representative immunoblot is also provided (**e**).

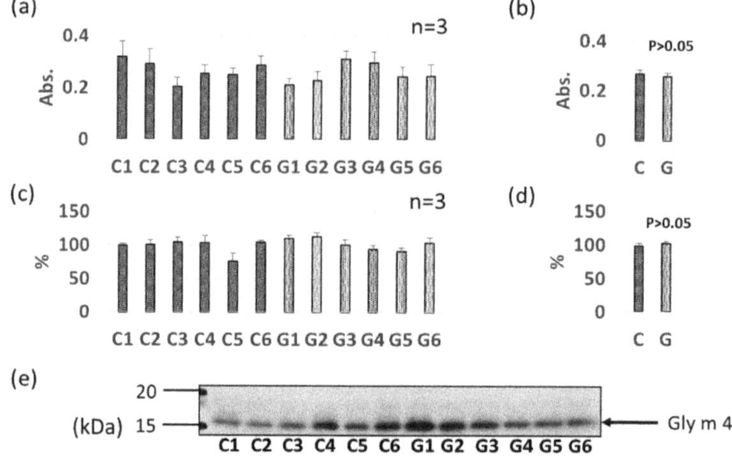

Figure 3. Comparison of Gly m 4 levels in GM-and non-GM soybeans by ELISA (**a**,**b**) and immunoblotting (**c**–**e**) using rabbit-derived polyclonal antibodies. Soybean protein extracts were evaluated by ELISA (**a**,**b**) and immunoblotting (**c**–**e**) for detection of Gly m 4 levels. The ELISA data are presented in absorbance values (Abs). The individual data from six GM-and non-GM soybeans (**a**,**c**) are presented as the mean ± SD of three independent replicates. The collated data (**b**,**d**) are presented as the mean ± SD of all individual data points from the control (C1–C6, six non-GM soybeans) or experimental (G1–G6, six GM soybeans) groups relative to the value of control number 1 (C1). A representative immunoblot is also provided (**e**).

3.3. Comparative Levels of Other Soybean Allergens in GM-Soybean and non-GM Soybean

As with Gly m 3 and Gly m 4, other soybean allergen levels in control (non-GM soybeans) and GM soybeans were detected and compared relatively based on ELISA and immunoblotting (Figures 4–9). There were no significant differences in levels of oleosin between the control (non-GM soybeans) and GM soybean groups (Figure 4) as determined by ELISA and immunoblotting. Trypsin inhibitor (Figure 5a–e) and Gly m Bd 30 K (Figure 6a–e) levels were also similar between the groups, with no significant differences between the non-GM soybean and GM soybean groups. The major storage protein, Gly m 5 (7S globulins), is composed of three subunits: α, α′, and β subunit. ELISA results showed no significant differences in the banding intensity of Gly m 5 between the control and experimental group (Figure 7a,b) and no significant differences in the banding intensities of each of the three subunits (α + α′, and β subunit) by immunoblotting (Figure 7c–f). Additionally, the major storage protein Gly m 6 (11S globulin) is composed of an acidic subunit (AS) and basic subunit (BS). There were no significant differences in the levels of Gly m 6 between the control non-GM soybean and the GM soybean groups as determined by ELISA and immunoblotting (Figure 8). Levels of Gly m 7, a recently discovered soybean allergen, were also examined; there were no significant differences in Gly m 7 levels in the non-GM soybean and the GM soybean groups as determined by ELISA (Figure 9a,b) or by immunoblotting using two different anti-Gly m 7 antibodies (Figure 9c–f).

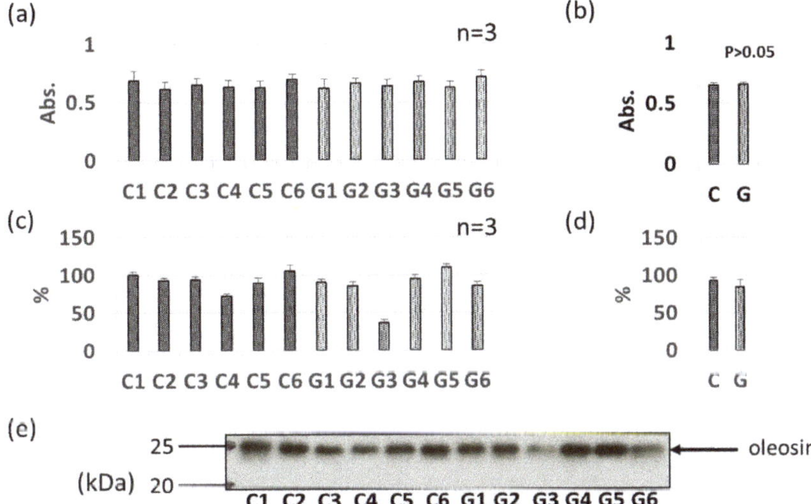

Figure 4. Comparison of oleosin levels in GM-and non-GM soybeans by ELISA (**a**,**b**) and immunoblotting (**c**–**e**) using rabbit-derived polyclonal antibodies. Soybean protein extracts were evaluated by ELISA (**a**,**b**) and immunoblotting (**c**–**e**) for detection of oleosin levels. The ELISA data are presented in absorbance values (Abs). The individual data from six GM-and non-GM soybeans (**a**,**c**) are presented as the mean ± SD of three independent replicates. The collated data (**b**,**d**) are presented as the mean ± SD of all individual data points from the control (C1–C6, six non-GM soybeans) or experimental (G1–G6, six GM soybeans) groups relative to the value of control number 1 (C1). A representative immunoblot is also provided (**e**).

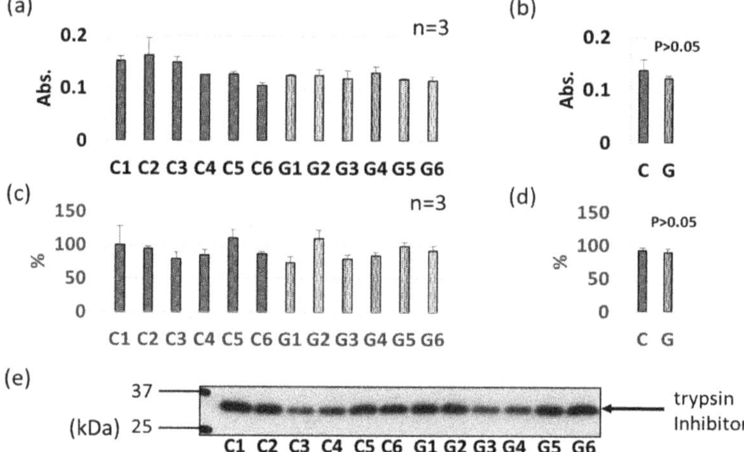

Figure 5. Comparison of trypsin inhibitor levels in GM-and non-GM soybeans by ELISA (**a**,**b**) and immunoblotting (**c**–**e**) using a rabbit-derived antibody. Soybean protein extracts were evaluated by ELISA (**a**,**b**) and immunoblotting (**c**–**e**) for detection of trypsin inhibitor levels. The ELISA data are presented in absorbance values (Abs). The individual data from six GM-and non-GM soybeans (**a**,**c**) are presented as the mean ± SD of three independent replicates. The collated data (**b**,**d**) are presented as the mean ± SD of all individual data points from the control (C1–C6, six non-GM soybeans) or experimental (G1–G6, six GM soybeans) groups relative to the value of control number 1 (C1). A representative immunoblot is also provided (**e**).

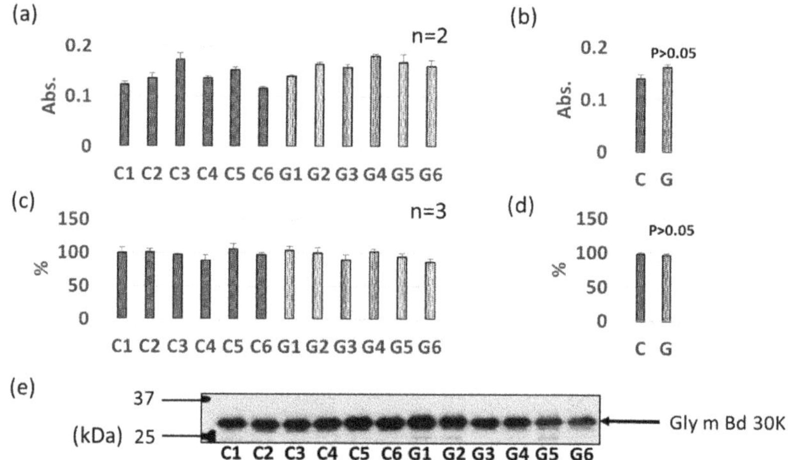

Figure 6. Comparison of Gly m Bd 30K levels in GM-and non-GM soybeans by ELISA (**a**,**b**) and immunoblotting (**c**–**e**) using a mouse-derived monoclonal antibody. Soybean protein extracts were evaluated by ELISA (**a**,**b**) and immunoblotting (**c**–**e**) for detection of Gly m Bd 30K levels. The ELISA data are presented in absorbance values (Abs). The individual data from six GM-and non-GM soybeans (**a**,**c**) are presented as the mean ± SD of three independent replicates. The collated data (**b**,**d**) are presented as the mean ± SD of all individual data points from the control (C1–C6, six non-GM soybeans) or experimental (G1–G6, six GM soybeans) groups relative to the value of control number 1 (C1). A representative immunoblot is also provided (**e**).

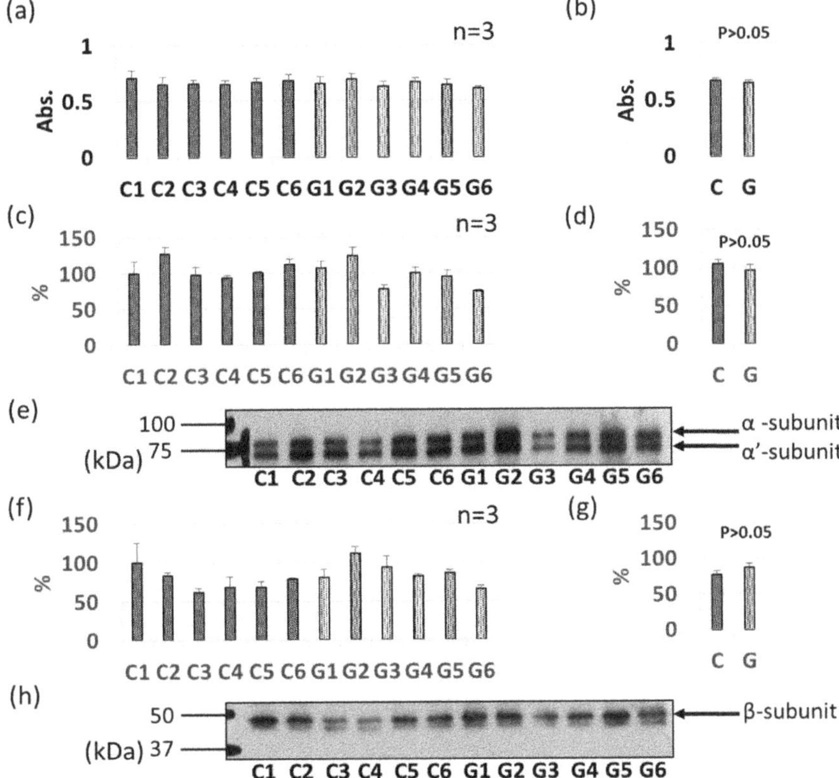

Figure 7. Comparison of Gly m 5 levels in GM-and non-GM soybeans by ELISA (**a**,**b**) and immunoblotting (**c**–**h**) using rabbit-derived polyclonal antibodies. Soybean protein extracts were evaluated by ELISA (**a**,**b**) and immunoblotting (**c**–**h**) for detection of Gly m 5 levels. The ELISA data are presented in absorbance values (Abs). The α- and α′-subunits of Gly m 5 were detected (**c**–**e**) separately from the β-subunit of Gly m 5 (**f**–**h**). The individual data from six GM-and non-GM soybeans (**a**–**f**) are presented as the mean ± SD of three independent replicates. The collated data (**b**,**d**,**g**) are presented as the mean ± SD of all individual data points from the control (C1–C6, six non-GM soybeans) or experimental (G1–G6, six GM soybeans) groups relative to the value of control number 1 (C1). Representative immunoblots are also provided (**e**,**h**).

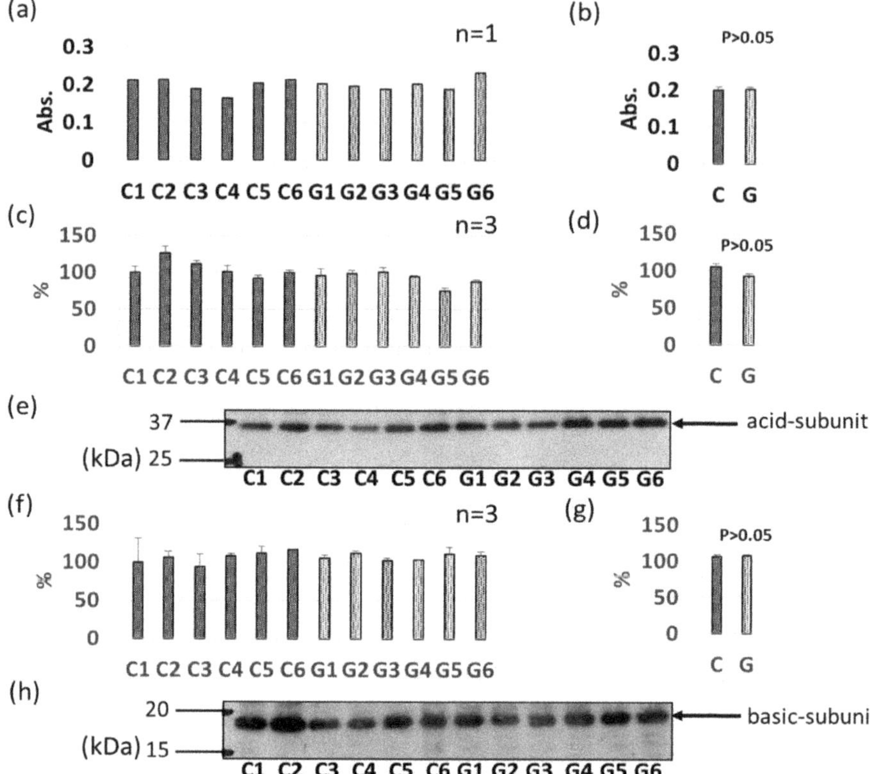

Figure 8. Comparison of Gly m 6 levels in GM-and non-GM soybeans by ELISA (**a**,**b**) and immunoblotting (**c**–**h**) using mouse-derived polyclonal antibodies. Soybean protein extracts were evaluated by ELISA (**a**,**b**) and immunoblotting (**c**–**h**) for detection of Gly m 6 levels. The ELISA data are presented in absorbance values (Abs). The acidic subunit of Gly m 6 was detected (**c**–**e**) separately from the basic subunit of Gly m 6 (**f**–**h**). The individual data from six GM-and non-GM soybeans (**c**,**f**) are presented as the mean ± SD of three independent replicates. The collated data (**b**,**d**,**g**) are presented as the mean ± SD of all individual data points from the control (C1–C6, six non-GM soybeans) or experimental (G1–G6, six GM soybeans) groups relative to the value of control number 1 (C1). Representative immunoblots are also provided (**e**,**h**).

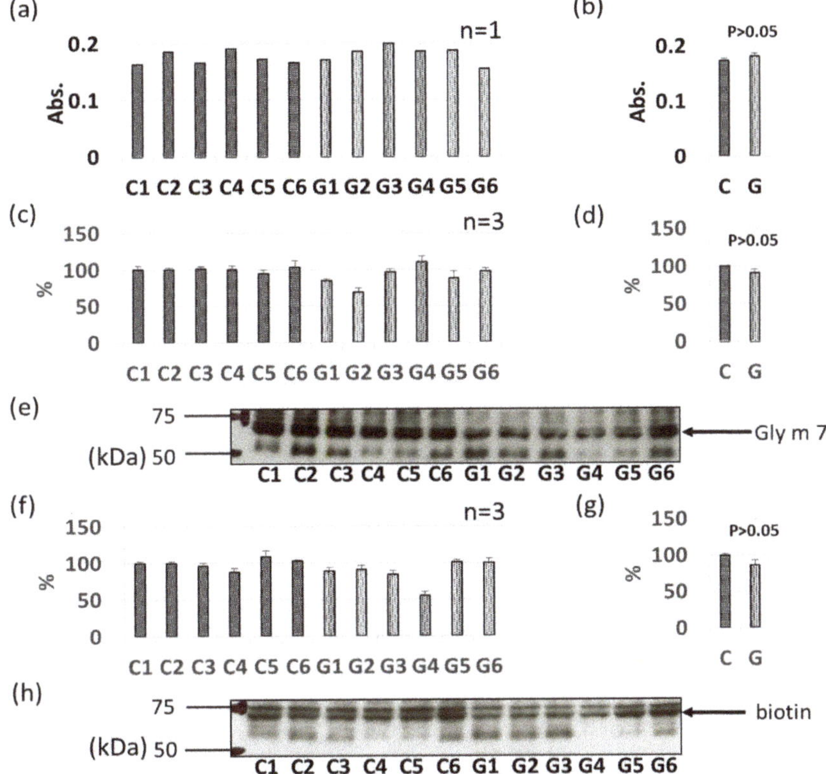

Figure 9. Comparison of Gly m 7 levels in GM-and non-GM soybeans by ELISA (**a**,**b**) and immunoblotting (**c**–**h**). Soybean protein extracts were evaluated by ELISA (**a**,**b**) and immunoblotting (**c**–**h**) for detection of Gly m 7 levels. The ELISA data are presented in absorbance values (Abs). Immunoblotting for Gly m 7 levels was performed using a rabbit-derived peptide-antibody (**c**–**e**) and streptavidin–HRP for the biotin moiety of Gly m 7 (**f**–**h**). The individual data from six GM-and non-GM soybeans (**c**,**f**) are presented as the mean ± SD of three independent replicates. The collated data (**b**,**d**,**g**) are presented as the mean ± SD of all individual data points from the control (C1–C6, six non-GM soybeans) or experimental (G1–G6, six GM soybeans) groups relative to the value of control number 1 (C1). Representative detections are also provided (**e**,**h**). ELISA and immunoblotting were performed using animal-derived antibodies or streptavidin–HRP as described in Section 2.5.

3.4. IgE-ELISA and IgE-Immunoblotting using Patient Serum

The allergenicity of non-GM soybeans and GM soybeans was then compared by IgE-ELISA and immunoblotting using the sera of three commercial soybean-allergic patients (Figures 10–13). Soybean strains exhibited varying levels of allergenicity to serum IgE as determined by IgE-ELISA, but there was no significant difference between the allergenicity of the non-GM soybean and the GM soybean groups in the sera of all three patients (Figure 10). Furthermore, IgE-binding patterns were evaluated by IgE-immunoblotting and analyzed both visually and by densitometric analysis. IgE-immunoblotting revealed qualitative differences in IgE-binding patterns for different soybean strains in the sera of different patients. For example, in patient serum 1, the peak densitometric intensities of IgE-bound proteins were found at approximately 72 kDa, 37 kDa, and 18 kDa (Figure 11). In patient serum 2, the peak densitometric intensities of IgE-bound proteins were found at approximately 50 kDa and 18 kDa (Figure 12). In patient serum 3, the peak densitometric intensities of IgE-bound proteins were found at approximately 75–50 kDa and 30 kDa (Figure 13). The peak densitometric intensities of IgE-bound proteins should supposedly correlated with the molecular weight of major soybean allergens, such as Gly m 5 and Gly m 6 (full list in Table 1). There were no qualitative differences in the IgE-immunoblotting results of the non-GM soybean and the GM soybean groups for all three patients' sera; specifically, IgE-binding bands were not increased or decreased in GM soybeans (Figures 11–13). These results indicate that the patient-serum IgE does not specifically bind to the transgene product (CP4-EPSPS).

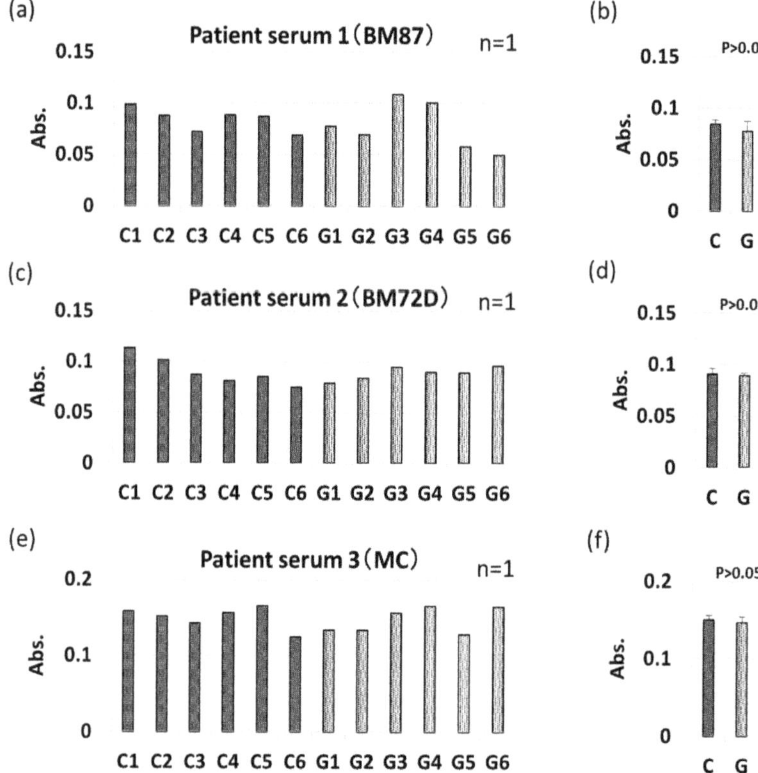

Figure 10. IgE-ELISA of GM-and non-GM soybeans using patient sera. IgE-ELISA performed using sera from three soybean-allergenic patients is shown as follows: patient serum 1 (**a**,**b**), patient serum 2 (**c**,**d**), patient serum 3 (**e**,**f**). The collated data (**b**,**d**,**f**) are presented as the mean ± SD of all individual data points from the control (C1–C6, non-GM soybeans) or experimental (G1–G6, GM soybeans) groups for each serum sample.

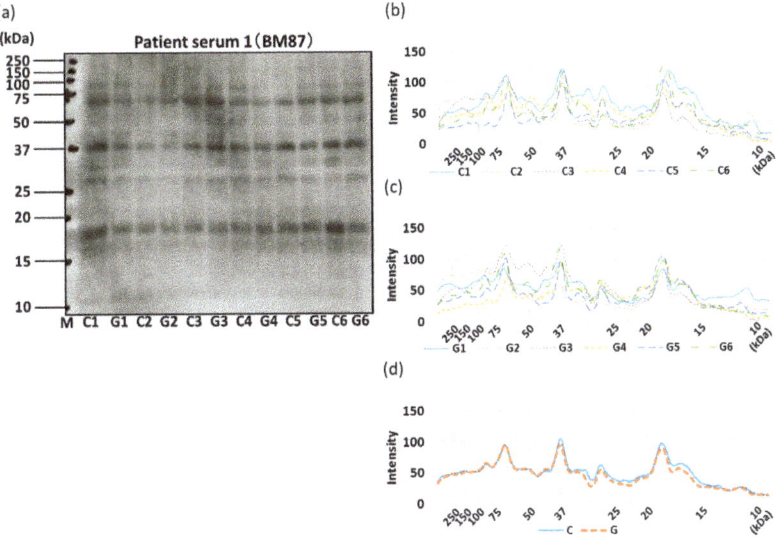

Figure 11. IgE-immunoblotting of GM-and non-GM soybeans using patient serum 1. A representative IgE-immunoblot of GM soybeans (G1–G6) and non-GM soybeans (C1–C6) is shown (**a**). Immunoblots were analyzed on each soybean sample once by densitometry and separated into non-GM soybean (**b**) and GM soybean groups (**c**). The average IgE-immunoblot profiles of non-GM soybeans and GM soybeans were calculated (**d**).

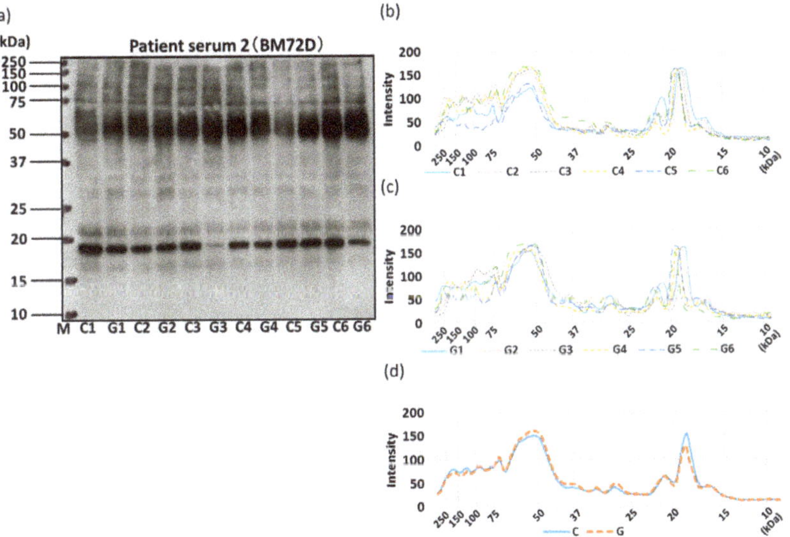

Figure 12. IgE-immunoblotting of GM-and non-GM soybeans using patient serum 2. A representative IgE-immunoblot of GM soybeans (G1–G6) and non-GM soybeans (C1–C6) is shown (**a**). Immunoblots were analyzed on each soybean sample once by densitometry and separated into non-GM soybean (**b**) and GM soybean groups (**c**). The average IgE-immunoblot profiles of non-GM soybeans and GM soybeans were calculated (**d**).

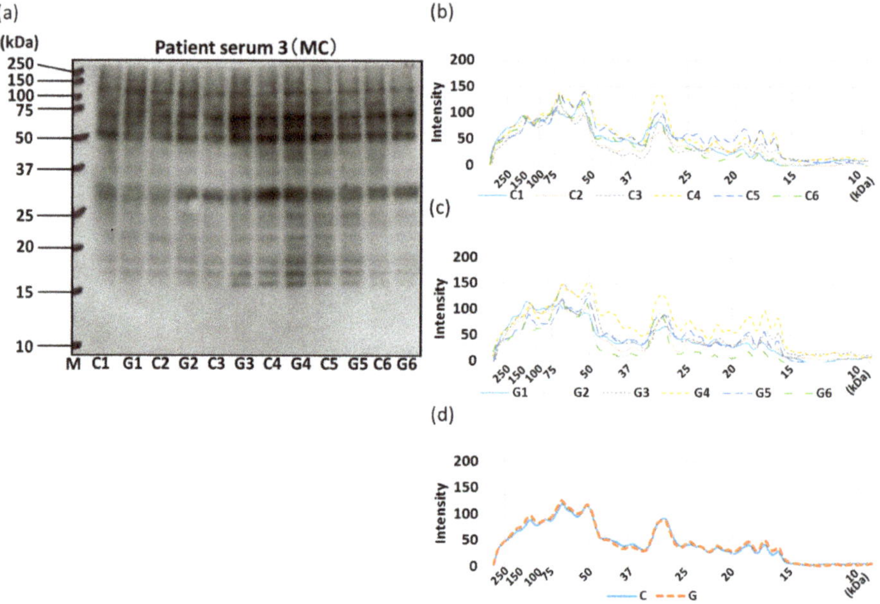

Figure 13. IgE-immunoblotting of GM-and non-GM soybeans using patient serum 3. A representative IgE-immunoblot of GM soybeans (G1–G6) and non-GM soybeans (C1–C6) is shown (**a**). Immunoblots were analyzed on each soybean sample once by densitometry and separated into non-GM soybean (**b**) and GM soybean groups (**c**). The average IgE-immunoblot profiles of non-GM soybeans and GM soybeans were calculated (**d**).

4. Discussion

In immunochromatography, two lines specific to genetic recombination were detected from six GM soybeans. Twelve soybean strains (six non-GM and six GM strains) were evaluated to determine whether genetic modification affected expression of previously identified allergens or IgE allergenicity. Immunoblotting using antibodies to detect the recombinant gene product CP4-EPSPS revealed expression of EPSPS in the six GM soybean strains but not in the non-GM strains. These results confirmed that all six GM soybean species used in this study had been genetically modified and demonstrated that all six non-GM soybean species had not been genetically modified to express EPSPS (Figure 1a–c). SDS–PAGE and CBB staining showed no visible differences in the protein expression profiles of GM soybean and non-GM soybean groups as a whole; and it was speculated that no new protein bands detectable at the CBB staining levels were found to be generated, increased, decreased, or eliminated by the introduction of CP4-EPSPS.

One of the major storage proteins, Gly m 5 (7S globulin: β-conglycinin), consists of three subunits (α subunit, approximately 68 kDa; α′-subunit, approximately 72 kDa; β-subunit, approximately 50 kDa); the α subunit was first identified as an allergen, and subsequent studies using IgE antibodies from the sera of soybean-allergic patients revealed that the α′and β subunits were also allergens [35]. Structural homology between these three subunits is relatively high. Gly m 5, which is found in tofu, a processed soybean food, is stable against pepsin-digestion, and has been reported to be responsible for food-dependent exercise-induced anaphylaxis (FDEIA) [36]. Gly m 6 (11S globulin) is also known to be a major soybean allergen [11]. Both Gly m 5 and Gly m 6 are seed storage proteins that account for about 70% of all seed proteins [37]. In this study, we found that the allergen levels of these two major seed storage proteins do not differ significantly between GM soybeans and non-GM soybeans.

The newly discovered soybean allergen Gly m 7 is a unique seed-specific biotinylated protein (SBP) that belongs to the late embryogenesis (LEA) protein family. It was discovered by Riascos et al. in a study evaluating the allergenicity of boiled lentils. The authors generated full-length cDNA clones encoding SBPs identified in lentils from developing soybean seeds and successfully expressed the protein as His-tagged recombinant proteins (rSBP) in *Escherichia coli*. They succeeded in purification of naturally-derived soybean SBP (nSBP-soy, later named Gly m 7) and confirmed IgE-positive and basophil-stimulating effects between soybean and peanut-allergic sera, suggesting that Gly m 7 may cause IgE-mediated allergic reactions [12]. In the present study, we detected and compared the levels of this novel allergen by two methods (peptide-antibody and biotin-detection) and found no significant differences between its expression levels in the GM soybean and non-GM soybean groups in either case.

Gly m Bd 30K is the predominant allergen found in soybeans. It is a 32 kDa protein also known as the vacuolar protein p34 in soybeans. It has been identified as an oil-body associated component of soybean seeds [13]. A Kunitz-type trypsin inhibitor was identified as a soybean allergen in 1980 [15] and was reported to be an occupational inhalant allergen [38]. There were no significant differences in protein levels between the GM soybean and non-GM soybean groups for any of these classical allergens.

The soybean allergen Gly m 4, which belongs to the pathogenesis-related protein10 (PR-10) family, is a homolog of the pollen-antigen Bet v 1 of birch. Gly m 4 has been widely reported to cross-react with food PR-10 proteins. Berkner et al. reported immunoblot inhibition assays using recombinant (r)Gly m 4 indicating that rBet v 1 was most inhibited by IgE-binding to rGly m 4 (100%), followed by rGly m 4, apple (rMal d 1), and cherry (rPru av 1) [21]. Gly m 3 (profilin) also cross-reacts with Bet v 2, another birch pollen-antigen, and is an actin-binding protein present in all organisms (including plants and animals), with more than 70% homology between Gly m 3 and Bet v 2. Rihs et al. reported that there were common IgE-binding epitopes in rGly m 3 and rBet v 2 as determined by EAST (enzyme allergosorbent test) inhibition assays using sera from non-soybean-allergic patients with cypress pollinosis; furthermore, preincubation of sera with rGly m 3 completely inhibited IgE binding to rBet v 2 [17]. Quantification of Gly m 3 in some soy products by indirect ELISA was also reported [39].

The immunoblotting and ELISA assays in this study indicated that the levels of these two pollinosis-related soybean allergens (Gly m 3 and Gly m 4) were not increased or decreased by genetic modification (Figures 2 and 3). Interestingly, the levels of these two allergens have been reported to be significantly increased by worm wounding [28]. In particular, since Gly m 4 is a pathogenesis-related protein, it is known that its expression is induced by stresses such as disease. Therefore, it is suggested that the level of these allergens is greatly affected by the cultivation environment. Our results suggest that the level of PR protein is unlikely to be increased by the genetic modification process.

Next, IgE-binding was evaluated using the serum of three soybean-allergic patients in order to evaluate the allergenic capacity of GM and non-GM soybeans from a clinical perspective. The IgE-ELISA results showed no significant differences between the GM soybean and the non-GM soybean groups when testing the sera of all three patients (Figure 10). IgE-immunoblotting revealed qualitative differences in IgE-binding patterns for different soybean strains in the sera of different patients. These IgE-binding proteins were supposed to be soybean-major allergens such as Gly m 5, Gly m 6 (Figure 11a–c, Figure 12a–c, Figure 13a–c). There were no qualitative differences in the IgE-immunoblotting results of the non-GM soybean and the GM soybean groups for all three patients' sera, indicating that the allergen-candidate molecules did not differ between GM soybeans and non-GM soybeans (Figure 11d, Figure 12d, Figure 13d). These results also indicate that patient-serum IgE does not specifically bind to the transgene product (CP4-EPSPS).

Taken together, it was concluded that the CP4-EPSPS transfected GM soybeans used in this study had similar allergen abundance levels and allergen reactivity to the non-GM soybeans. These results are similar to other GM soybean allergenicity studies conducted thus far. In this study, we found that GM technology did not increase or decrease the level of endogenous soybean allergen proteins, nor did it induce the appearance of new soybean allergens. However, further investigation of the allergenicity of GM soybeans will be necessary for more rigorous evaluation. Research on more soybean varieties, changes in allergenicity due to changes in the cultivation environment, and sensitization potencies, as well as the allergen levels of GM- and non-GM soybeans should be considered.

Author Contributions: T.M. designed the study. A.M., K.M., and A.F. performed the experiments. A.M. and T.M. wrote the paper. N.T. and E.Y. supported the experiments. T.M. and N.Z. reviewed and edited the manuscript. All authors read and approved the manuscript.

Funding: This work was supported by the Japan Society for the Promotion of Science KAKENHI (Grant-in-Aid for Scientific Research(C)) (grant numbers JP16K07756 and 19K05919 to T.M.). This study was also supported in part by a grant from NPO, "Science of Food Safety and Security (SFSS)" and by a grant from Agricultural Technology and Innovation Research Institute (ATIRI), Kindai University.

Acknowledgments: We would like to thank Editage (www.editage.jp) for English language editing.

Conflicts of Interest: The authors declare no conflict of interest.

References

1. Chrispeels, M.J. Global production and consumption of genetically engineered crops. *J. Huazhong Agric. Univ.* **2014**, *4*, 120–132.
2. Lua, M.; Jina, Y.; Weber, B.B.; Goodman, R.E. A comparative study of human IgE binding to proteins of a genetically modified (GM) soybean and six non-GM soybeans grown in multiple locations. *Food Chem. Toxicol.* **2018**, *112*, 216–223. [CrossRef] [PubMed]
3. Ichim, M.C. The Romanian experience and perspective on the commercial cultivation of genetically modified crops in Europe. *Transgenic Res.* **2019**, *28*, 1–7. [CrossRef] [PubMed]
4. Shukla, M.; Al-Busaidi, K.T.; Trivedi, M.; Tiwari, R.K. Status of research, regulations and challenges for genetically modified crops in India. *GM Crops Food* **2018**, *9*, 173–188. [CrossRef] [PubMed]
5. ISAAA. *Global Status of Commercialized Biotech/GM Crops: Biotech Crop Adoption Surges as Economic Benefits Accumulate in 22 Years*; ISAAA: Ithaca, NY, USA, 2017; pp. 100–104, ISBN 978-1-892456-67-2.
6. Ogawa, T.; Samoto, M.; Takahashi, K. Soybean allergens and hypoallergenic soybean products. *J. Nutr. Sci. Vitaminol.* **2000**, *46*, 271–279. [CrossRef]
7. Yagami, T. Allergies to Cross-Reactive Plant Proteins. Latex-fruit syndrome is comparable with pollen-food allergy syndrome. *Int. Arch. Allergy Immunol.* **2002**, *128*, 271–279. [CrossRef] [PubMed]
8. Amlot, P.L.; Kemeny, D.M.; Zachary, C.; Parkes, P.; Lessof, M.H. Oral allergy syndrome (OAS): Symptoms of IgE-mediated hypersensitivity to foods. *Clin. Allergy* **1987**, *17*, 33–42. [CrossRef]
9. Ogawa, T.; Bando, N.; Tsuji, H.; Nishikawa, K.; Kitamura, K. α-Subunit of β-Conglycinin, an allergenic protein recognized by IgE Antibodies of soybean-sensitive patients with atopic dermatitis. *Biosci. Biotechnol. Biochem.* **1995**, *59*, 831–833. [CrossRef]
10. Maruyama, N.; Sato, S.; Cabanos, C.; Tanaka, A.; Ito, K.; Ebisawa, M. Gly m 5/Gly m 8 fusion component as a potential novel candidate molecule for diagnosing soya bean allergy in Japanese children. *Clin. Exp. Allergy* **2018**, *48*, 1726–1734. [CrossRef]
11. Holzhauser, T.; Wackermann, O.; Ballmer-Weber, B.K.; Bindslev-Jensen, C.; Scibilia, J.; Perono-Garoffo, L.; Utsumi, S.; Poulsen, L.K.; Vieths, S. Soybean (Glycine max) allergy in Europe: Gly m 5 (β-conglycinin) and Gly m 6 (glycinin) are potential diagnostic markers for severe allergic reactions to soy. *J. Allergy Clin. Immunol.* **2009**, *123*, 452–458. [CrossRef]
12. Riascos, J.J.; Weissinger, S.M.; Weissinger, A.K.; Kulis, M.; Burks, A.W.; Pons, L. The Seed Biotinylated Protein of Soybean (Glycine max): A BoilingResistant New Allergen (Gly m 7) with the Capacity to Induce IgE Mediated Allergic Responses. *J. Agric. Food Chem.* **2016**, *64*, 3890–3900. [CrossRef] [PubMed]
13. Ogawa, T.; Tsuji, H.; Bando, N.; Kitamura, K.; Zhu, Y.L.; Hirano, H.; Nishikawa, K. Identification of the soybean allergenic protein, Gly m Bd 30K, with the soybean seed 34-kDa oil-body-associated protein. *Biosci. Biotechnol. Biochem.* **1993**, *57*, 1030–1033. [CrossRef] [PubMed]
14. Ogawa, T.; Bando, N.; Tsuji, H.; Okajima, H.; Nishikawa, K.; Sasaoka, K. Investigation of the IgE-binding protein in soybeans by immunoblotting with the sera of the soybean-sensitive patient with atopic dermatitis. *J. Nutr. Sci. Vitaminol.* **1991**, *37*, 555–565. [CrossRef] [PubMed]
15. Moroz, L.A.; Yang, W.H. Kunitz Soybean Trypsin Inhibitor—A Specific Allergen in Food Anaphylaxis. *N. Engl. J. Med.* **1980**, *302*, 1126–1128. [CrossRef]

16. Cao, Y.; Zhao, L.; Ying, Y.; Kong, X.; Hua, Y.; Chen, Y. The characterization of soybean oil body integral oleosin isoforms and the effects of alkaline pH on them. *Food Chem.* **2015**, *177*, 288–294. [CrossRef]
17. Rihs, H.P.; Chen, Z.; Ruëff, F.; Petersen, A.; Rozynek, P.; Heimann, H.; Baur, X. IgE binding of the recombinant allergen soybean profilin (rGly m 3) is mediated by conformational epitopes. *J. Allergy Clin. Immunol.* **1999**, *104*, 1293–1301. [CrossRef]
18. Yagami, A.; Inaba, Y.; Kuno, Y.; Suzuki, K.; Tanaka, A.; Sjolander, S.; Saito, H.; Matsunaga, K. Two cases of pollen-food allergy syndrome to soy milk diagnosed by skin prick test, specific serum immunoglobulin E and microarray analysis. *J. Dermatol.* **2009**, *36*, 50–55. [CrossRef]
19. Kleine-Tabbe, J.; Vogel, L.; Crowell, D.N.; Haustein, U.F.; Vieths, S. Severe oral allergy syndrome and anaphylactic reactions caused by a Bet v 1- related PR-10 protein in soybean, SAM22. *J. Allergy Clin. Immunol.* **2002**, *110*, 797–804. [CrossRef]
20. Mittag, D.; Vieths, S.; Vogel, L.; Becker, W.M.; Rihs, H.P.; Helbling, A.; Wüthrich, B.; Ballmer-Weber, B.K. Soybean allergy in patients allergic to birch pollen: Clinical investigation and molecular characterization of allergens. *J. Allergy Clin. Immunol.* **2004**, *113*, 148–154. [CrossRef] [PubMed]
21. Berkner, H.; Neudecker, P.; Mittag, D.; Ballmer-Weber, B.K.; Schweimer, K.; Vieths, S.; Rösch, P. Cross-reactivity of pollen and food allergens: Soybean Gly m 4 is a member of the Bet v 1 superfamily and closely resembles yellow lupine proteins. *Biosci. Rep.* **2009**, *29*, 183–192. [CrossRef] [PubMed]
22. Kim, S.H.; Kim, H.M.; Ye, Y.M.; Kim, S.H.; Nahm, D.H.; Park, H.S.; Ryu, S.R.; Lee, B.O. Evaluating the Allergic Risk of Genetically Modified Soybean. *Yonsei Med. J.* **2006**, *47*, 505–512. [CrossRef]
23. Wu, H.; Wang, X.; Zhou, X.; Zhang, Y.; Huang, M.; He, J.; Shen, W. Targeting the middle region of CP4-EPSPS protein for its traceability in highly processed soy-related products. *J. Food Sci. Technol.* **2017**, *54*, 3142–3151. [CrossRef] [PubMed]
24. Funke, T.; Healy-Fried, M.L.; Han, H.; Alberg, D.G.; Bartlett, P.A.; Schönbrunn, E. Differential Inhibition of Class I and Class II 5-Enolpyruvylshikimate-3-phosphate Synthases by Tetrahedral Reaction Intermediate Analogues. *Biochemistry* **2007**, *46*, 13344–13351. [CrossRef] [PubMed]
25. Tsai, J.J.; Chang, C.Y.; Liao, E.C. Comparison of Allergenicity at Gly m 4 and Gly m Bd 30K of Soybean after Genetic Modification. *J. Agric. Food Chem.* **2017**, *65*, 1255–1262. [CrossRef]
26. Selb, R.; Wal, J.M.; Moreno, F.J.; Lovik, M.; Mills, C.; Hoffmann-Sommergruber, K.; Fernandez, A. Assessment of endogenous allergenicity of genetically modified plants exemplified by soybean—Where do we stand? *Food Chem. Toxicol.* **2017**, *101*, 139–148. [CrossRef]
27. Goodman, R.E.; Panda, R.; Ariyarathna, H. Evaluation of Endogenous Allergens for the Safety Evaluation of Genetically Engineered Food Crops: Review of Potential Risks, Test Methods, Examples and Relevance. *J. Agric. Food Chem.* **2013**, *61*, 8317–8332. [CrossRef]
28. Graf, L.; Hayder, H.; Mueller, U. Endogenous allergens in the regulatory assessment of genetically engineered crops. *Food Chem. Toxicol.* **2014**, *73*, 17–20. [CrossRef]
29. Fernandez, A.; Mills, E.N.; Lovik, M.; Spoek, A.; Germini, A.; Mikalsen, A.; Wal, J.M. Endogenous allergens and compositional analysis in the allergenicity assessment of genetically modified plants. *Food Chem. Toxicol.* **2013**, *62*, 1–6. [CrossRef]
30. Panda, R.; Ariyarathna, H.; Amnuaycheewa, P.; Tetteh, A.; Pramod, S.N.; Taylor, S.L.; Ballmer-Weber, B.K.; Goodman, R.E. Challenges in testing genetically modified crops for potential increases in endogenous allergen expression for safety. *Allergy* **2013**, *68*, 142–151. [CrossRef]
31. EFSA Panel on Genetically Modified Organisms (GMO). Guidance on allergenicity assessment of genetically modified plants. *EFSA J.* **2017**, *15*, e04862. [CrossRef]
32. Kyhse-Andersen, J. Electroblotting of multiple gels: A simple apparatus without buffer tank for rapid transfer of proteins from polycrylamide to nitrocellulose. *J. Biochem. Biophys. Methods* **1984**, *10*, 203–209. [CrossRef]
33. Hei, W.; Li, Z.; Ma, X.; He, P. Determination of beta-conglycinin in soybean and soybean products using a sandwich enzyme-linked immunosorbent assay. *Anal. Chim. Acta* **2012**, *734*, 62–68. [CrossRef] [PubMed]
34. Hanafusa, K.; Murakami, H.; Ueda, T.; Yano, E.; Zaima, N.; Moriyama, T. Worm wounding increases levels of pollen-related food allergens in soybean (Glycine max). *Biosci. Biotechnol. Biochem.* **2018**, *82*, 1207–1215. [CrossRef] [PubMed]
35. Krishnan, H.B.; Kim, W.S.; Jang, S.; Kerley, M.S. All three subunits of soybean beta-conglycinin are potential food allergens. *J. Agric. Food Chem.* **2009**, *57*, 938–943. [CrossRef] [PubMed]

36. Adachi, A.; Horikawa, T.; Shimizu, H.; Sarayama, Y.; Ogawa, T.; Sjolander, S.; Tanaka, A.; Moriyama, T. Soybean beta-conglycinin as the main allergen in a patient with food-dependent exercise-induced anaphylaxis by tofu: Food processing alters pepsin resistance. *J. Clin. Exp. Allergy* **2009**, *39*, 167–173. [CrossRef]
37. Natarajan, S.; Khan, F.; Song, Q.; Lakshman, S.; Cregan, P.; Scott, R.; Shipe, E.; Garrett, W. Characterization of Soybean Storage and Allergen Proteins Affected by Environmental and Genetic Factors. *J. Agric. Food Chem.* **2016**, *64*, 1433–1445. [CrossRef]
38. Quirce, S.; Fernández-Nieto, M.; Polo, F.; Sastre, J. Soybean trypsin inhibitor is an occupational inhalant allergen. *J. Allergy Clin. Immunol.* **2002**, *109*, 178. [CrossRef]
39. Amnuaycheewa, P.; Gonzalez de Mejia, E. Purification, characterisation, and quantification of the soy allergen profilin (Gly m 3) in soy products. *Food Chem.* **2010**, *119*, 1671–1680. [CrossRef]

Review

Seed Protein of Lentils: Current Status, Progress, and Food Applications

Hamid Khazaei [1,*], Maya Subedi [1,†], Mike Nickerson [2], Cristina Martínez-Villaluenga [3], Juana Frias [3] and Albert Vandenberg [1]

1 Department of Plant Sciences, University of Saskatchewan, Saskatoon, SK S7N 5A8, Canada
2 Food and Bioproduct Sciences, University of Saskatchewan, Saskatoon, SK S7N 5A8, Canada
3 Institute of Food Science, Technology and Nutrition (ICTAN), Spanish National Research Council (CSIC), Jose Antonio Novais 10, 28040 Madrid, Spain
* Correspondence: hamid.khazaei@usask.ca or hamid.khazaei@gmail.com; Tel.: +1-306-966-5859
† Current address: Agriculture and Agri-Food Canada, Lethbridge, AB T1J 4B1, Canada.

Received: 20 July 2019; Accepted: 30 August 2019; Published: 4 September 2019

Abstract: Grain legumes are widely recognized as staple sources of dietary protein worldwide. Lentil seeds are an excellent source of plant-based proteins and represent a viable alternative to animal and soybean proteins for food processing formulations. Lentil proteins provide not only dietary amino acids but are also a source of bioactive peptides that provide health benefits. This review focuses on the current knowledge of seed protein, extraction and isolation methods, bioactive peptides, and food applications of lentil protein. Lentil is the most rapidly expanding crop for direct human consumption, and has potential for greater impact as a protein source for food processing applications. Improvements in lentil protein quality, amino acid composition, and processing fractions will enhance the nutritional quality of this rapidly expanding crop globally.

Keywords: *Lens*; protein; amino acid; legume; functionality; bioactive peptides

1. Introduction

The importance of food legumes in sustainable agriculture and food security is increasing worldwide. Cultivation of food legumes delivers generous environmental and economic benefits based on their ability to fix nitrogen to replace synthetic fertilizers and thereby reduce greenhouse gas emissions. In contrast with animal-based protein, which has a huge environmental cost, seeds of grain legumes are widely consumed as staple sources of dietary proteins. They are the main source of dietary protein for over one billion people and provide opportunities for greater use in new plant-based protein foods and animal feeds. Protein confers food value, handling properties, and gustatory qualities [1].

Cultivated lentil (*Lens culinaris* Medik.), one of the ancient crops, is a quick-cooking and nutritious staple legume grown in more than 70 countries and consumed globally in whole, dehulled, and split form [2]. Its seeds are lens-shaped and have a wide range of seed coat colours (green, tan, brown, gray, white, and black) and patterns (marbled, dotted, spotted, complex, and unpatterned [3]). The cotyledons can be yellow, red, or green. Red cotyledon lentils are a source of staple protein and nutritious food in many parts of the Indian subcontinent and eastern Mediterranean regions, where they are consumed mainly in dehulled form as split cotyledons [4]. The major commercial market classes of lentil are red (based on cotyledon colour of dehulled seeds) and green (based on seed coat colour). Large green lentils (with yellow cotyledon) are primarily marketed in Europe, and parts of the Middle East and South America, mostly consumed as whole seeds [5].

Lentil seeds are typically rich in protein, dietary fiber, complex carbohydrates, and essential micronutrients such as iron, zinc, and vitamin B complex [6]. Its seeds also have high antioxidant activity compared to other grain legume species mainly due to specific phenolic compounds [7]. Tahir et al. [8] reported only small differences between green and red lentils for protein concentration.

Lentil is in high demand globally and has, by far, the highest growth rate in consumption in comparison to other major pulses (Table 1). Since the beginning of the green revolution (1960–1965), the annual increase of production of faba bean and pea have not kept pace with the growth of the human population (2.4% annual average growth). Chickpea and common bean have increased at the same rate as population growth, while lentil production has grown at more than 10% per year, second only to soybean. The main reason for the high growth rate is presumed to be the fast cooking time of lentils relative to the other main pulses, in spite of the fact that lentil is often more expensive than most of the other pulses.

Table 1. Global trends of production of selected annual pulse crops in comparison to increases in soybean production and the global human population since 1960.

Crop	Global Production				Global Growth Trends	
	1960–1965		2012–2017			
	Annual Mean (Mt)	% of Total	Annual Mean (Mt)	% of Total	50 Years % Growth	Annual % Growth
Lentil	1.0	2.8	5.9	9.2	515	10.3
Faba bean	5.1	14.7	4.5	7.0	−12	−0.2
Pea	10.5	30.0	13.1	20.5	25	0.5
Chickpea	6.3	17.9	12.7	19.8	104	2.1
Common bean	12.1	34.5	27.9	43.5	132	2.6
Total pulses [1,2]	34.9	100	64.2	90	84	1.7
Soybean (total)	28.6		319.0		1115	22.3
Soybean eaten directly [3]	10.0		67.0		670	13.4
People (1967) [4]	3465 million		7600 million		119	2.4

[1] Total includes only the pulses listed in table. [2] Source: FAOSTAT (Food and Agriculture Organization of the United Nations) [9]. [3] Estimated direct human consumption. [4] Source: UN.

Nearly one third of the world's population, particularly children in low-income countries, is protein deficient [10,11]. The high protein content of lentil, its fast-cooking time, and increased production are contributing to lentil gaining in importance as a staple food for combating human protein malnutrition globally. There is a great need to accelerate genetic improvement for high-yielding lentil cultivars with high quantity and quality of protein. This review focuses on the current knowledge of lentil protein quantity and quality, and the gap between current knowledge, and what knowledge is needed to alleviate nutritional and environmental concerns.

2. Protein Content

Like seeds of most cultivated legumes, lentils are a rich source of high-quality protein. Lentil seeds contain on average about 26% crude protein (see Table 2). Several studies have reported genetic variation for protein content in lentil seeds (Table 2). A Russian bulletin published in 1930 reported that the protein content among lentil varieties varied from 27.5%–31.7% [12]. In the 1970s, an evaluation of an extensive global collection of 1688 accessions of lentil for protein content, reported a greater range of protein content from of 23.4%–36.4% [13]. A subsequent evaluation reported an even wider range when a larger germplasm set (1816 accessions) was investigated [14]. Kumar et al. [15] reported a lower average protein content for lentil with broader variability among lentil species.

Table 2. Genetic variation and methods used measuring seed the crude protein content in cultivated lentil.

Protein Content (Range, % DM [1])	Number of Accessions	Environment	Method	Reference
27.5–31.7	-	Russia	A micro Kjeldahl Method	Barulina [12]
23.4–36.4	1688	Egypt	A micro Kjeldahl Method	Hawtin et al. [13]
25.5–28.9	24	Lebanon and Syria	NIR [2] using a Neotec model FQA51 A analyzer	Erskine et al. [16]
19.6–29.8 and 18.6–30.2	829 and 987	Tel Hadya (Syria)	NIR	Hamdi et al. [14]
23.9–26.3	58	Australia	A Kjeldahl N × 6.25 on an oven-dry basis method	Stoddard et al. [17]
23.0–32.0	-	-	-	Hedley [18]
24.3–30.2	4	-	NIR using a NIR Systems 6500 analyzer calibrated against the Dumas method	Wang and Daun [19]
23.8–29.3	22	Saskatoon, Canada	A Dumus Combustion method to determine Nitrogen percentage using a method in 46-30.01	Tahir et al. [8]
22.7–31.88	46	Turkey	A Kjeldahl method AOAC, Official Method of Analysis	Karaköy et al. [20]
21.8–27.1	14	Italy	A Kjeldahl method	Zaccardelli et al. [21]
25.3–29.3	35	Saudi Arabia	A Kjeldahl method of Association of Official Analytical Chemists (AOAC)	Alghamdi et al. [22]
24.6–30.0	23	Multiple	Multiple methods	Heuzé et al. [23]
10.5–27.1	45	India	A Kjeldahl method	Kumar et al. [15]

[1] DM, dry matter. [2] NIR, near-infrared reflectance spectroscopy.

3. Lentil Seed Storage Proteins

Seed storage proteins of lentil are located in the cotyledons, representing up to 80% of total proteins. Seed storage proteins primarily provide nitrogen, carbon, and sulphur during seed germination and seedling growth and development [24]. They are also involved in plant defence mechanisms, e.g., for bruchids in legumes [25], and in antimicrobial activity (reviewed in Cândido et al. [26]). Seed storage proteins are classifiable based on their solubility in different solvents. The first report of protein components in lentil was that of Osborne and Campbell [27] who isolated globulins from lentil seeds. Later, Danielson [28] grouped the globulins into two classes, 7S (vicilin and convicilin-type) and 11S (legumin-like) based on their sedimentation coefficients. The predominant lentil storage proteins, similar to other legume species, are salt-soluble globulins and water-soluble albumins (Table 3). Both the globulins and albumins of lentils are heterogeneous [29–31]. Legumes in general contain relatively large concentrations of globulins [32]. The 7S/11S ratio is an important characteristic for describing seed nutritional quality [33] and is reported to be very high in lentil, close to three [31]. The ratio of three in lentil is twelve-fold higher than in seeds of pea and *Medicago truncatula*. These results suggest that lentil may meet certain criteria for specific end uses based on protein quality characteristics.

Table 3. Summary of reported analyses of cultivated lentil seed storage protein fractions (%) based on their solubility in different solvents.

Salt Soluble	Water Soluble	Acid Soluble	Ethanol Soluble	Reference
Globulins	Albumins	Glutelins	Prolamins	
44%	26%	20%	2%	Saint-Clair [34]
47%	4%	15%	3%	Bhatty et al. [35] [1]
54%	20%	-	-	Bhatty [36]
42%	11%	47% [2]	-	Neves and Lourenco [32]
27%	61%	3%	2%	Sulieman et al. [37]
70%	16%	11%	3%	Boye et al. [38]

[1] About a quarter of the meal proteins were not solubilized by the solvents. [2] Prolamins, glutelins, non-nitrogen protein, and residual nitrogen. The quantitative data presented in this table are based on peptide bands and their molecular weight on SDS-PAGE (sodium dodecyl sulfate–polyacrylamide gel electrophoresis). The albumin, glutelin, and prolamin fractions contained 13, 4, and 10 polypeptides, with molecular weight (MW) of about 20, 17–46, and 16–64 kDa, respectively [39]. Globulins contain legumin- and vicilin-like proteins. The native globulin, with a molecular weight of 375 kDa, has twelve polypeptides and MW ranging from 14–61 kDa [32].

Table 3 shows the reported high variation for each specific lentil seed storage fraction, particularly for albumins. The variation may reflect genotypic variation in storage protein profiles, or the extraction conditions employed in each individual study. For instance, Bhatty et al. [35] reported that, in addition to the albumin proteins, direct extraction of lentil meal with water solubilizes some non-protein nitrogen and may solubilize some salt-soluble proteins. Exhaustive extraction of the meal with a salt solution followed by precipitation of the salt-soluble proteins by dialysis provides a better estimate of this protein fraction.

Metabolic proteins (enzymes and structural proteins) are another major type of protein found in lentil seeds. Several non-storage proteins such as enzymes involved in DNA replication, proteins involved in various physiological processes and house-keeping proteins have been identified [31]. Sulieman et al. [37] showed that lentil protein fractions are altered quantitatively and qualitatively due to cooking, and the effect was most pronounced in prolamin fractions.

4. Amino Acids (AA)

Cultivated lentil proteins, like those of other grain legumes, are rich in endogenous amino acids (arginine, aspartic and glutamic acids, and leucine-more than half of total AA), low in some essential amino acids (EAA) like threonine, methionine, phenylalanine, tryptophan, histidine, valine, isoleucine, and leucine-excluding lysine), and poor in sulphur-containing amino acids (methionine and cysteine, see Table 4). For the WHO/FAO/United Nations University [40]. AA requirement patterns also showed low levels of both sulphur-containing amino acids and tryptophan in lentil seeds.

Table 4. Amino acid composition ± standard deviation (g/16 g N) in seeds of six major grain legumes.

Amino Acid	Lentil [1]	Faba Bean [2]	Pea [3]	Soybean [4]	Chickpea [5]	Common Bean [6]	SEM [7]
Alanine	4.2 ± 0.4	4.1 ± 0.2	4.3 ± 0.2	4.2 ± 0.3	4.1 ± 0.5	3.8 ± 0.3	0.07
Arginine	7.8 ± 1.0	10.2 ± 1.1	8.2 ± 0.7	7.4 ± 0.6	9.0 ± 1.2	6.5 ± 0.7	0.51
Aspartic acid	10.7 ± 1.1	11.0 ± 1.6	11.3 ± 0.5	11.3 ± 0.7	11.6 ± 0.7	10.6 ± 1.3	0.15
Cysteine	1.1 ± 0.3	1.4 ± 0.3	1.3 ± 0.3	1.3 ± 0.4	1.3 ± 0.1	1.1 ± 0.2	0.06
Glutamic acid	16.1 ± 2.6	16.7 ± 2.2	16.4 ± 0.7	17.8 ± 1.2	16.8 ± 2.2	15.6 ± 2.1	0.31
Glycine	4.1 ± 0.7	4.3 ± 0.2	4.3 ± 0.2	4.1 ± 0.4	3.6 ± 0.6	4.2 ± 0.5	0.10
Histidine	2.4 ± 0.5	2.6 ± 0.2	2.3 ± 0.3	2.6 ± 0.1	2.6 ± 0.5	2.7 ± 0.2	0.05
Isoleucine	4.1 ± 0.5	4.0 ± 0.4	4.1 ± 0.5	4.4 ± 0.6	3.8 ± 0.4	4.2 ± 0.3	0.09
Leucine	7.2 ± 0.4	7.7 ± 0.6	7.3 ± 0.8	7.5 ± 0.4	7.0 ± 0.4	7.5 ± 0.7	0.11
Lysine	6.7 ± 0.6	6.4 ± 0.1	7.6 ± 1.2	6.4 ± 0.6	6.5 ± 0.8	6.3 ± 0.5	0.19

Table 4. *Cont.*

Amino Acid	Lentil [1]	Faba Bean [2]	Pea [3]	Soybean [4]	Chickpea [5]	Common Bean [6]	SEM [7]
Methionine	0.9 ± 0.2	0.7 ± 0.1	1.0 ± 0.1	1.3 ± 0.3	1.4 ± 0.3	1.0 ± 0.4	0.11
Phenylalanine	5.0 ± 0.6	4.2 ± 0.2	4.8 ± 0.5	4.8 ± 0.3	5.5 ± 0.5	4.4 ± 0.7	0.18
Proline	3.8 ± 0.4	4.1 ± 0.5	4.4 ± 0.9	5.1 ± 0.3	4.4 ± 0.4	3.8 ± 0.4	0.19
Serine	4.7 ± 0.7	4.6 ± 0.4	4.9 ± 0.5	5.1 ± 0.5	4.8 ± 0.9	5.2 ± 0.7	0.08
Threonine	3.7 ± 0.4	3.5 ± 0.2	3.8 ± 0.3	3.9 ± 0.4	3.7 ± 0.6	4.0 ± 0.2	0.07
Tryptophan	0.8 ± 0.1	0.9 ± 0.1	1.2 ± 0.6	1.4 ± 0.3	1.0 ± 0.1	1.1 ± 0.4	0.09
Tyrosine	2.5 ± 0.7	3.1 ± 0.3	3.3 ± 0.5	3.4 ± 0.6	2.9 ± 0.5	3.7 ± 0.6	0.18
Valine	4.7 ± 0.4	4.4 ± 0.4	4.5 ± 0.5	4.7 ± 0.5	4.0 ± 0.4	4.9 ± 0.5	0.13

[1] Mean of data extracted from Kahn and Baker [41]; Chatterjee and Abrol [42]; Bhatty et al. [35]; Sosulski [43]; Bhatty and Christison [44]; Shekib et al. [45]; Pirman et al. [46]; Porres et al. [47]; Zia-Ul-Haq et al. [48], and Grela et al. [7]. [2] Mean of data extracted from Kaldy and Kasting [49]; Bhatty and Christison [44]; Lisiewska et al. [50]; Schumacher et al. [51], and Grela et al. [7]. [3] Mean of data extracted from Bhatty and Christison [44]; Leterme et al. [52]; Pownall et al. [53]; Schumacher et al. [51], and Grela et al. [7]. [4] Mean of data extracted from Kuiken et al. [54]; Tkachuk and Irvine [55]; Cho and Bayley [56]; Cavins et al. [57]; Wang and Cavins [58]; Zarkadas et al. [59], and Sotak-Peper et al. [60]. [5] Mean of data extracted from Wang and Daun [61]; Alajaji and El-Adawy [62]; Wang et al. [63]; El-Beltagi et al. [64], and Grela et al. [7]. [6] Mean of data extracted from Wu et al. [65]; Słupski [66] and, Grela et al. [7]. The mean of AA for all species includes data from https://www.feedtables.com. [7] Standard error of means.

Lentil has relatively a similar AA profile in comparison to other grain legume species (Table 4). Principal component analysis (PCA) of the AA from the studied legume species (Figure 1) revealed lentil and pea have very similar AA composition compared to other studied species and were characterized by high lysine. Principal component 1 alone explained over 99% of total variation. Faba bean was characterized by high arginine content. The PCA analysis also revealed that AA were grouped based on their AA amount. The average AA composition of *M. truncatula* was found to be very close to pea in various growing conditions [67]. Soybean and common bean had also similar AA profile (Figure 1).

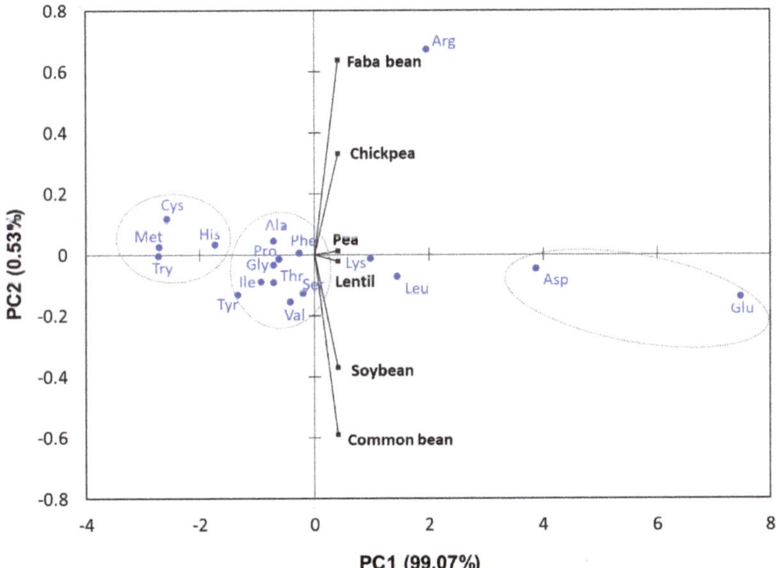

Figure 1. Principal component analysis (PCA) of amino acid data (meta-analysis) from Table 2. The biplot shows amino acid data and legume crops as vectors. Vectors that are close together are correlated in terms of the observed amino acid pool for each crop. PCA analysis was employed to illustrate relationships between amino acids and legume species using the R statistical package (R Development Core Team, 2018, www.Rproject.org).

Lentil protein is low in methionine (0.9%), especially when compared with animal-based proteins (>2.2%, [68]). Methionine is typically lower in plant-based proteins compared with animal-based proteins [69], and in general, plant proteins are only low in a few EAA. Combining plant proteins that are lower in lysine and higher in methionine (e.g., wheat, rice and hemp) with plant proteins that are higher in lysine and lower in methionine (grain legumes, including lentil) may balance the anabolic properties of plant-based protein intake [68].

The main non-protein AAs in lentil seeds are trigonelline [70], erythro-γ-hydroxyarginine 2(S), 4(R)-4-hydroxyarginine [71], γ-hydroxyarginine, γ-hydroxyornithine, and homoarginine [72].

5. Wild Lentil Taxa Protein Properties

The genus *Lens* has seven closely related taxa, namely *L. culinaris* (cultivated lentil), *L. orientalis*, *L. tomentosus* (primary gene pool); *L. odemensis*, *L. lamottei*, (secondary gene pool); *L. ervoides* (tertiary gene pool); and *L. nigricans* (quaternary gene pools) [73]. Bhatty [74] reported similar protein content for three wild lentil species from different gene pools (*L. orientalis*, *L. ervoides*, and *L. nigricans*) in comparison to cultivated lentil (range of 24.2%–26.2%). However, a much wider range (18.1%–32.7%) for protein content was reported recently for *L. orientalis*, *L. tomentosus*, *L. odemensis*, *L. ervoides*, and *L. nigricans*. The highest protein content was found in the *L. ervoides* accession, ILWL 47, with 32.7% protein content [15]. *L. ervoides* accessions also showed the highest variation compared to other studied species.

With respect to the storage protein fractions, all the wild species except *L. nigricans* had greater amounts of albumin and globulin than cultivated lentil. *L. nigricans* also had higher non-protein nitrogen than *L. orientalis* and *L. ervoides* [74]. In contrast, Rozan et al. [70] findings indicated that *L. ervoides* contained less non-protein AA than other *Lens* species.

The protein AA composition profiles of wild lentil species *L. orientalis*, *L. ervoides*, and *L. nigricans* were identical and similar to cultivated lentil [74]. A more recent study revealed that *L. orientalis* seeds had relatively higher AAsthan the other *Lens* species, including cultivated lentil [70].

6. Anti-Nutritional Factors Affecting Lentil Protein

Nutritional value of seed protein is determined primarily by the amount of anti-nutritional factors (ANFs) and the AA digestibility [75]. Removal of ANFs is necessary to improve nutritional quality. In general, whole lentils are low in ANFs. Lentil has relatively high tannin content compared to other grain legumes [7]. Tannins are primarily located in the seed coat and can be removed by processing (e.g., dehulling). Tannins can reduce protein digestibility by reacting with lysine and methionine and making them available in a smaller amount during digestion [76]. Zero tannin lentils are now available [77]. The zero tannin trait in lentil is controlled by a single recessive gene (*tan*) that results in a phenotype characterized by green stems, white flowers, and thin and transparent seed coats, a consequence of a major reduction in most of the seed coat polyphenols [78].

Trypsin inhibitors are low molecular weight proteins found in a wide range of plants including legumes [79] that irreversibly inhibit physiological trypsin enzyme. They induce hypersecretion of pancreatic enzymes (trypsin), thereby stimulating pancreatic hypertrophy, which leads to reduced digestion and absorption of amino acids and, hence, their bioavailability [79–81]. The bioavailability changes result in lower retention of nitrogen and sulphur, and impaired growth [82]. Trypsin inhibitor content of lentil is significantly lower than other grain legumes excluding pea (e.g., [83–85]). The trypsin inhibitors can be markedly reduced by soaking, cooking, and germination of lentil seeds [85].

7. Environmental Effects on Seed Proteins

The seed filling stage involves mobilization and transport processes required for importing various seed constituents. Environmental stresses (such as drought) can impair seed filling due to the disruption of metabolic pools downward of sucrose in starch synthesis [86], resulting in an increase of seed protein. Studies in many crops report increased content of seed protein in response to drought stress, for example, in wheat [87,88], chickpea [89], and soybean [90,91]. In contrast, some other studies reported reduction of protein content in response to other environmental stresses [92]. These differences may be attributable to the intensity and duration of stresses imposed on plants. Additionally, at the beginning of seed filling stage under unfavorable conditions, proteins related to protection against stress are probably synthesized (increasing protein content), whereas a reduction in protein content is due to their hydrolysis and degradation.

In lentil, heat stress is reported to reduce protein content (26%–41% [93]). Heat stress also inhibited the accumulation of globulins, albumins, glutelins, and prolamins. Excluding proline, glycine, alanine, isoleucine, leucine, and lysine, which increased under heat stress, the rest of the amino acids significantly decreased [93]. The decrease in lentil storage proteins and most of the AA composition profile resulting from high temperature stresses may be explained by the inactivity of biosynthetic enzymes [94], and by changes in nitrogen content [95]. The increase in some AA (e.g., proline and glycine) under stress conditions may be the effects of osmoregulation mechanisms [96]. Further investigation is required to determine the impacts of environmental stresses on protein content, protein fractions and AA composition, especially in legume crops such as lentil.

8. Yield and Protein Relationships and Stability

For lentil breeders, a major challenge is the simultaneous increase of both yield and protein content while maintaining progress in the development of resistance to biotic and abiotic stresses. Hamdi et al. [14], using two large sets of ICARDA lentil germplasm (829 + 987 accessions), found negative correlation between seed protein content and seed yield. They also reported high heritability (0.84) for protein content. Erskine et al. [16] showed the same trend in lentil for a smaller germplasm set. Later, Stoddard et al. [17] showed lack of correlation between protein and yield. More recently, Lizarazo et al. [97] reported negative relationships between protein concentration and seed yield in 14 lentil cultivars grown in a boreal growing environment. This may suggest that independent selection of both characters during breeding is challenging, i.e., the rate of gain in one trait being reduced by that in the other.

Barulina [12] indicated that the protein content varied little across locations among lentil accessions. Similar results are reported for lentil from different authors [17,35,97] and for other legume crops (e.g., [67,98]). These observations suggest there is low G × E interaction for protein content and AA composition. This supports the hypothesis that the nitrogen fixing ability of legumes makes their protein concentration relatively stable across environments [17]. It is known that the protein content of the seed is highly affected by soil nitrogen level, so in legumes the *Rhizobium* bacteria may greatly enhance the percentage seed protein. For example, Ivanov [99] reported large differences between seed protein content of non-inoculated and inoculated chickpea plants, 12.6% and 31.2%, respectively.

9. Agronomic Protein Yield of Lentil

Protein yield is calculated as protein fraction × grain yield. In lentil, a value of 0.33 t ha^{-1} protein was reported by Erskine et al. [16]. Khatun et al. [100] reported 0.2 t ha^{-1} for lentil protein yield grown in Bangladesh. Lizarazo et al. [97] reported protein yield of 0.4 t ha^{-1} for lentil grown in northern Europe, which was less than that for faba and (1.6 t ha^{-1}) and narrow-leafed lupin (1.1 t ha^{-1}). In pea, protein yield has been reported at 0.7 t ha^{-1} [101] and 0.9 t ha^{-1} [102].

The potential protein yield of lentil in specific environments has most likely not been fully explored due to limitations imposed by the narrow genetic base of most lentil breeding programs, which are not able to fully exploit the genetic potential due to adaptation bottlenecks. With the recent increasing emphasis on genomics within breeding programs, it may be possible to more fully explore the genetic potential for improvements in protein quantity and quality. The AGILE (Application of Genomic Innovation in the Lentil Economy) project (https://knowpulse.usask.ca/study/2675314), which evaluated phenotypic influences of temperature and photoperiod of 324 diverse lentil genotypes in replicated trials in the three main global agro-ecological regions of lentil production, may provide deeper understanding of aspects of protein quality and protein yield potential of lentil. Genomic information of the 324 sequenced lentil genotypes grown across the three major agro-ecological zones for lentil production may further provide the understanding of the underlying genetics of protein quality and protein yield in lentil.

10. Seed Crude Protein Determination in Lentil

The two most prevalent protein determination methods, Kjeldahl and Dumas combustion, are commonly used for grain and seed protein analysis in crops including lentil [12,17,19]. The methods rely on the release of nitrogen from the amine groups found in the peptide bonds of the polypeptide chains of protein. The traditional Kjeldahl method is based on oxidation to release nitrogen, while the Dumas combustion method breaks down the bonds in the peptide chains, permitting the release of nitrogen through complete combustion of the sample [103]. The released nitrogen content is multiplied by a factor to measure protein content [104]. The factor varies between crop species depending on nitrogen content of protein between 13% and 19%. For pulses, the average nitrogen (N) content of protein was found to be about 16%, which led to use the calculation N × 6.25 to convert nitrogen content into protein content [105].

In a series of experiments, most researchers employed a Kjeldahl method to determine crude protein in lentil. For example, Barulina [12] and Hawtin et al. [13] determined seed protein content in lentil accessions using a macro–Kjeldahl (Table 2). Recently, different versions of a modified Kjeldahl method were adopted to measure protein content in lentil seeds [15,17]. Although the Kjeldahl method was more precise and frequently used to analyze protein, this method is under threat by the challenge of safer, clearer, and faster instruments employed in the Dumas Combustion method [103] which resulted in faster, safer, and more reliable data for protein content in seeds compared to the Kjeldahl method. The Dumas combustion method was adopted to determine lentil seed protein content by Tahir et al. [8] when they compared lentil protein with other pulses in Canada using AACC (American Association of Cereal Chemists) method 46–30 to determine percent crude protein (CP; N × 6.25) through the use of a LECO CNS-2000 Nitrogen Analyzer (LECO Corporation, St. Joseph, MI, USA, Model No. 602-00-500). A rapid test method using near–infrared (NIR) spectroscopy as a complement to current protein determination using either the Kjeldahl or Dumas combustion method was also successfully applied to estimate protein content in lentil seeds using a different model of analyzer [14,16,19]. Protein measured through NIR was validated by calibrated value against measurements obtained through either Kjeldahl or Dumas methods using representative samples. The NIR method was found to be a rapid, low cost, and green complementary technique as it does not use chemicals and reagents [106]. NIR is a useful high throughput method for estimating protein content for lentil breeding programs if calibrated curves are used to validate the method. The value of protein concentration using three different methods in lentil research are illustrated in Table 2.

11. Protein Isolation Methods and Extraction

Lentils are traditionally consumed as whole seed, dehulled split seeds, or as footballs (cotyledons remain attached) in salads and soups or stews commonly known as 'dal' [4,107]. Diverse and novel applications are needed to identify ways to increase the use of lentils in the food industry. Nutritional components in lentil seeds such as dietary fiber, starch and protein concentrates or isolates can be extracted and separated [38]. These can be used as ingredients in the preparation of diverse value-added food products.

Isolation or separation of seed proteins from pulses is possible using wet or dry processes [108–110]. Dry processes, such as pin milling and then air classification, are designed to differentiate fractions of starch and protein based on size and density. Air classification separates milled lentils into a light to fine fraction (the protein concentrate) and a heavy or coarse fraction (the starch concentrate) [111,112]. Protein concentrates produced by air classification through dry processes generally contain 38%–68% protein [113,114]. In the past, air classification processes were well adapted to extraction of isolates of lentils and peas because of the large diameter and fairly uniform distribution of starch granules [112]. The dry method is a relatively easy and simple process, however, efficacy of separation is not high enough to yield high protein concentration. Currently, the wet method for extraction is more widely adopted for legume protein extraction [108]. The extraction of pulse proteins through wet methods may be relatively easy and reliable, as they are highly soluble under alkaline and acidic conditions. In wet methods, protein is extracted by solubilization in an alkaline solution by dispersing pulse flour in water at pH 8–10, followed by stirring of the dispersion. Then, the insoluble material is removed by centrifugation and proteins are recovered by adjusting the, supernatant pH to a value around 4.5, where proteins are precipitated isoelectrically. The precipitation is usually carried out at the isoelectric point of the protein at which its solubility is the lowest, which for lentil protein is around pH 4.5. The final concentrate or isolated protein is then dried using spray-drum or freeze-drying methods [115]. Lentil isolates prepared with an alkaline process yield overall 80% of protein [108,116,117]. Many researchers have used various wet fractionation methods to isolate lentil protein [108,109,116–118] under single or multiple isoelectric pH conditions using diluted sodium hydroxide [36] (Table 5). For example, Boye et al. [38] reported they extracted lentil protein isolates from red and yellow cotyledon lentils using isoelectric precipitation at pH 9 and 25 °C using a 1:10 solid to solvent ratio, resulting in protein concentrates between 78.2% and 88.6%. Similarly, Joshi et al. [119] extracted lentil protein isolate by alkaline extraction at pH 8 using 1:10 solid to solvent ratio at room temperature when they studied physicochemical characteristics of the isolated protein that obtained from three drying methods (freeze, spray and vacuum drying). In another study, Kaur et al. [120] revealed that yield of lentil protein isolate ranged from 81.7%–83.5% for Indian cultivars when they performed the protein isolation using isoelectric precipitation pH 4.5. In contrast, Alsohaimy et al. [117] found the highest protein recovery (93% and 100%) from lentil isolate at isoelectric pH of 12 with ammonium sulphate and alcohol precipitation solvent, with a 5:100 solid to solvent ratio. They compared seven different pH values ranging from 6 to 12 with three different protein recovery methods—isoelectric precipitation, ammonium sulphate preparation, and alcohol precipitation. Similarly, Lee et al. [115] reported pH 9 at 30 °C as the optimum extraction condition for green lentil that yielded 56.6% protein, and pH 8.5 at 35 °C for red lentil that yielded 59.3% protein when they compared five pH levels (distilled water, pH 8, 8.5, 9, and 9.5) and four temperatures (22, 30, 35, and 40 °C). Johnston et al. [121] used a modified isoelectric precipitation procedure by adjusting initial pH to 9 initially and then collecting lentil protein isolate at pH 4.6 with 1:10 solid to solvent ratio. Cultivar, particle size of the flour, type of solubilizing agent, temperature, and pH of extraction medium influenced the protein yield and quality [110].

Table 5. Conditions for wet fractionation methods for extraction of lentil protein in six recent studies.

Conditions for Lentil Protein Extraction	% Protein Yield	% Protein in Final Extracts	Reference
pH: 6, 7, 8, 9, 10, 11, and 12 Temperature: room Solid to solvent ratio: 5:100	80.0	21.5	Alsohaimy et al. [117]
pH: 8, 8.5, 9, 9.5 Temperature: 22 °C, 30 °C, 35 °C, and 40 °C Solid to solvent ratio: 1:10	56.6–59.3	-	Lee et al. [115]
pH: 9 Temperature: 25 °C Solid to solvent ratio: 1:10	50.3–69.1	-	Boye et al. [38]
pH: 8 Temperature: room Solid to solvent ratio: 1:10	-	-	Joshi et al. [119]
pH: 4.6 Temperature: room Solid to solvent ratio: 1:10	82.0	14.5	Johnston et al. [121]
pH: 8, 9 and 10 Temperature: room Solid to solvent ratio: 1:10	70.3–85.7	12.3–16.5	Jarpa-Parra et al. [122]

In addition to air classification, a new emerging dry fractionation method, triboelectrostatic separation has emerged recently as a novel solvent-free approach to separate protein isolates in the food industry [123–125]. This method relies on differences in dielectric properties of flour particles instead of their size and density. The basic principle of this technique is that proteins can be electrostatically charged more than carbohydrates, because of the ionizable N-terminus and C-terminus groups in their amino acid residues [126]. Thus, an electric field can separate protein and carbohydrate rich fractions depending upon their types and magnitudes of charge. Like air classification, the main advantages of this method are that it does not use chemical reagents that render the concentrates unsafe for consumption, or that induce changes in functional character. This method is more energy efficient and effective than air classification because it effectively separates particles that are similar in size and density, but different in charge. The major limitation of this approach is that different components may exhibit similar charges under certain conditions, which reduces concentrate purity, and gravitational force may cause airborne particles [125].

Triboelectrostatic separation of legume flours pneumatically conveys the milled particles through tubes or fluidized/vibrating beds, thus imparting a positive or negative charge to the surface of the constituent protein and carbohydrate particles depending on their tribo-charging behavior and the contact medium. Upon contact charging, the oppositely charged particles are separated in a strong electric field [126]. The amount of charge gained on a particulate from triboelectric charging depends on factors such as surface conditions, area of contact, speed of rubbing, the materials involved, and humidity [123]. Electrostatic separation methods have been widely used in the mining and pharmaceutical industries, and the effectiveness of this approach is under investigation in the legume industry. For example, the development of optimization of triboelectric bio-separation using a single-stage separation of navy bean flour was successful [126]. Later, single- and multi-stage tribo-electrostatic bioseparation processes for dry fractionation of protein concentrate found that the two-stage approach resulted in a protein-rich fraction yield of 38% accounting for 60% of the total protein which was a significantly higher than that of the optimized single-stage triboelectrostatic separation [127]. Similarly, Jafari et al. [124] and Tabtabaei et al. [128] examined the physiochemical and functional properties of navy bean protein concentrated using triboelectrostatic separation. They found that electrostatically separated increased from 25.4% to 43.0% total protein yield of original navy bean flour, the protein fractions protein fractions exhibited superior solubility, superior emulsion stability, foam expansion and foam volume stability compared to the wet-fractionated navy bean protein isolate. This method has been extensively explored for fractionation of protein isolated from navy bean and other legumes, but there is limited information in the literature regarding the fractionation of protein from lentil flour using triboelectrostatic processes. However, the variation in turbocharging behavior of proteins and carbohydrates of flour extends its scope to protein isolation of lentil flours.

12. Bioactive Peptides

Food proteins not only provide dietary amino acids but also supply health benefits because of the presence of bioactive peptides, short fragments of 2–20 amino acid residues that are encrypted and inactive within the sequence of the precursor protein. Bioactive peptides play an important role in human health, being released during digestion or food processing (enzymatic hydrolysis, cooking, germination, fermentation, and ripening of foods), then absorbed in the intestine and transported to target tissues where they exert specific physiological effects [129].

Proteins from pulses are considered a good source of bioactive peptides. As an example, lentil convicilin was investigated as a source of bioactive peptides using the predictive tools in the BIOPEP database (http://www.uwm.edu.pl/biochemia, see Supplementary Table S1). This storage protein contains a total of 126 peptides encrypted in the amino acid sequence. The following array of biological effects was reported: (i) inhibition of angiotensin I converting enzyme (ACE), dipeptidyl aminopeptidase III and IV, calmodulin-dependent nucleotide phosphodiesterase (CaNPDE), and renin; (ii) stimulation of the release of vasoactive substances and the uptake of glucose; (iii) antioxidant activity, and (iv) regulation of secretion of gastric mucosa. This fact encourages more research studies exploring the biotechnological production of bioactive peptides from lentil proteins for functional food or nutraceutical applications.

A few studies are reported with the aim of ensuring efficient peptide release from lentil proteins. Critical processing parameters, such as pH, temperature and time require optimization, and the enzymes or microorganisms used for peptide release require evaluation for efficacy, reproducibility and stability. Enzymatic hydrolysis of lentil proteins has been performed using a wide number of proteolytic enzymes including Savinase, Alcalase, Protamex, Neutrase, Flavourzyme, bromelain, and papain [130,131]. Savinase® 16L was reportedly the most effective enzyme to produce bioactive peptides from cultivated lentil concentrates [130].

Bioactive peptides derived from lentil proteins reportedly exhibit antihypertensive, antioxidant, and antifungal activities. Table 6 summarizes the in vitro effect of protein hydrolysates and bioactive peptides produced during gastrointestinal digestion, enzymatic hydrolysis, germination and fermentation of different lentil-based raw materials. Most studies performed to date have shown that enzymatic hydrolysis of lentil proteins by food grade commercial proteases (savinase, papain, alcalase, flavourzyme, and bromelain), digestive enzymes (pepsin, trypsin, α-chymotrypsin, pancreatin) or germination of lentil seeds (30–40 °C for 5 days) produce peptides with the ability to inhibit angiotensin I converting enzyme (ACE, EC. 3.4.15.1) (see Table 6). ACE is a carboxypeptidase involved in the cleavage of angiotensin I into angiotensin II, a vasoactive peptide that binds with receptors on the vascular wall to cause vasoconstriction, therefore, inhibition of ACE may reduce systolic and diastolic blood pressure [132].

Cultivated lentil proteins treated with Savinase® produce multifunctional peptides with dual antioxidant and ACE inhibitory activities [133]. Three peptides were identified to have the highest potencies for inhibiting ACE and delay oxidation of proteins in the presence of oxygen radicals in vitro (LLSGTQNQPSFLSGF, NSLTLPILRYL, TLEPNSVFLPVLLH). The gastrointestinal digestion of these peptides greatly improved their dual biological activity, indicating that smaller peptide fragments with higher biological potency are produced at the gastrointestinal level. The antioxidant/antihypertensive activity of lentil peptides was linked to the primary structure of the C-terminal heptapeptide [133]. In particular, the ACE inhibition relies on the formation of hydrogen bonds between C-terminal residues of peptides and residues of the ACE catalytic site. The ability of these peptides to inhibit ACE is consistent with earlier studies showing that hydrophobic or aromatic residues or proline residue at the C-terminus positively contribute to the improvement of ACE inhibitory potency [134].

Lentil proteins are also sources of antifungal peptides with potential application as ingredients in the bakery industry. Recently, Rizzello et al. [135] produced a hydrolysate from a legume flour blend consisting of lentil, pea and faba bean by the combination of fermentation with *Lactobacillus plantarum* 1A7 and enzymatic hydrolysis with Veron. Among the antifungal compounds of the hydrolysate, four were identified as antifungal peptides derived from lentil lectin. These were purified to confirm their capacity to inhibit the development of *Penicillium roqueforti* conidia at a minimum inhibitory concentration of 7–9 mg/mL. Similarly, Wang and Ng [136] isolated a natural antifungal peptide from a red lentil protein extract by chromatographic fractionation. It was able to inhibit 50% of the mycelial growth of *Mycosphaerella arachidicola* at a concentration of 36 μM.

Although bioactive peptides have been identified and isolated from lentil for potential use in functional food and nutraceutical applications, none are currently available in the market for human use. The principal obstacle to the regulatory approval of health claims is the lack of in vivo studies supporting the health and safety claims of bioactive peptides [137]. To overcome these challenges, future research should focus more on generating data on the safety, efficacy, mechanisms of action, interactions of bioactive peptides with other drugs, absorption, distribution, metabolism, and excretion of bioactive peptides in clinical trials.

Table 6. In vitro biological activity of cultivated lentil protein hydrolysates and peptides.

Biological Activity	Raw Material	Processing Conditions	Peptide Sequence	Effect Observed	Reference
Antioxidant and antihypertensive	Protein concentrate	Enzymatic hydrolysis with Savinase 16 L (0.1 U/mg protein, pH 8, 40 °C, 2 h)	LLSGTQNQPSFLSGF [1]	ACE [2] inhibition: IC_{50} [3] = 120 µM ORAC [4]: 0.013 µmol Trolox eq./µmol	Garcia-Mora et al. [133]
			NSLTLPILRYL	ACE inhibition: IC_{50} = 77.14 µM ORAC: 1.432 µmol Trolox eq./µmol	
			TLEPNSVFLPVLLH	ACE inhibition: IC_{50} = 117.81 µM ORAC: 0.139 µmol Trolox eq./µmol	
Antihypertensive	Sprouts	Germination (30–40 °C for 5 days, 98% humidity)	Unknown	ACE inhibition: IC_{50} = 0.044 and 0.034 mg/mL	Mamilla and Mishra [138]
Antifungal	Flour	Fermentation with *Lactobacillus plantarum* (7 log cfu/g) and enzymatic hydrolysis with Veron PS (E/S [5] of 1/400, 30 °C, 24 h)	HIGIDVNSIK	Inhibition of germination of *Penicillium roqueforti* DPPMAF1 conidia: MIC [6] = 7–9 mg/mL	Rizzello et al. [135]
			NLIFQGDGYTTK		
			FSPDQQNLIFQGDGYTTK		
			HIGIDVNSIK		
Antihypertensive	Protein isolate	Enzymatic hydrolysis with pepsin (E/S of 1/100, pH 2, 37 °C for 18 h)	Unknown	ACE inhibition: IC_{50} = 606 µg/mL	Boschin et al. [139]
Antihypertensive	Protein isolate	Pepsin (250 U/mg, pH 2, 37 °C, 2 h) and pancreatin (0.7%, pH 7, 37 °C, 1 h)	KLRT	ACE inhibition: IC_{50} = 0.13–0.02 mg/mL for different peptide fractions	Jakubczyk and Baraniak [140]
			TLHGMV		
			VNRLM		
Antihypertensive	Red protein concentrates	Pepsin (E/S of 1/250, for 2 h, pH 2, 37 °C) + Trypsin and α-chymotrypsin (E/S of 1/250 for each enzyme, 2.5 h, pH 6.5, 37 °C)	Unknown	ACE inhibition: IC_{50} = 0.090 mg/mL	Barbana and Boye [131]
		Papain (E/S of 1/25, pH 6.5, 4 h, 40 °C)	Unknown	IC_{50} = 0.086 mg/mL	
		Alcalase (1/8 for E/S ratio, pH 7, 1 h, 50 °C) + Flavourzyme (E/S of 1/10, pH 8, 1.5 h at 50 °C)	Unknown	IC_{50} = 0.154 mg/mL	
		Bromelain (E/S of 1/4, pH 8, 8 h, 40 °C)	Unknown	IC_{50} = 0.190 mg/mL	
	Green protein concentrates	Pepsin (pepsin (E/S of 1/250, for 2 h, pH 2, 37 °C) + Trypsin and α-chymotrypsin (E/S of 1/250 for each enzyme, 2.5 h, pH 6.5, 37 °C)	Unknown	ACE inhibition: IC_{50} = 0.053 mg/mL	
		Papain (E/S of 1/25, pH 6.5, 4 h, 40 °C)	Unknown	IC_{50} = 0.080 mg/mL	
		Alcalase (1/8 for E/S ratio, pH 7, 1 h, 50 °C) + Flavourzyme (E/S of 1/10, pH 8, 1.5 h at 50 °C)	Unknown	IC_{50} = 0.152 mg/mL	
		Bromelain (E/S of 1/4, pH 8, 8 h, 40 °C)	Unknown	IC_{50} = 0.174 mg/mL	

Table 6. *Cont.*

Biological Activity	Raw Material	Processing Conditions	Peptide Sequence	Effect Observed	Reference
Antihypertensive	Red protein isolates	Pepsin (E/S of 4/10, pH 2, 37 °C, 2 h) and pancreatin (E/S of 0.5/10, pH 7, 37 °C, 2 h)	Unknown	ACE inhibition: IC_{50} = 0.008–0.33 mg/mL in gastric phase	Akıllıoğlu and Karakaya [141]
				IC_{50} = 0.26–0.89 mg/mL in intestinal phase	
Antifungal	Red lentil extract	Chromatographic fractionation	TETNSFSITKFSPDGNKLIFQGDGYTTKGK	Inhibition of mycelial growth in *Mycosphaerella arachidicola*: IC_{50} = 36 µM	Wang and Ng [136]

[1] Amino acids are coded according to their one letter abbreviation: A = alanine; C = cystine, D = aspartic acid, E = glutamic acid, F = phenylalanine, G = glycine, H = histidine, I = isoleucine, K = lysine, L = leucine, M = methionine, N = asparagine, P = proline, Q = glutamine, R = arginine, S = serine, T = threonine, V = valine, W = tryptophan, Y = tyrosine. [2] ACE, angiotensin I converting enzyme. [3] IC_{50}, inhibitory concentration that reduces 50% of the original enzymatic activity. [4] ORAC, oxygen radical absorbance capacity. [5] E/S, enzyme to substrate ratio. [6] MIC, minimal inhibitory concentration.

13. Food Applications of Lentil Proteins

Lentil protein ingredients represent a viable alternative to proteins from animal-derived sources and soybean, especially since the food industry aims to diversify their formulations because of cultural, religious or ethical dietary restrictions, growing populations, availability, and cost reduction. Although commonly sold as flours, lentil protein ingredients can also be further fractionated into higher protein enriched flours (<60% protein), concentrates (60%–85% protein) or isolates (>85% protein). Enriched flours and dry concentrates are usually produced by air classification methods [142], whereas concentrates/isolates are produced using wet extraction processes, such as by alkaline extraction followed by isoelectric precipitation [109]. These fractions are known to have good nutritional and functional value. In terms of functionality, the method and conditions used to produce the ingredients can have a big impact on their functionality within food applications.

13.1. Functionality

The majority of published studies have focused on the solubility and emulsifying properties of lentil proteins. Solubility of lentil proteins relates to the balance between protein-solvent and protein-protein interactions. Ladjal-Ettoumi et al. [143] reported a typical u-shaped pH-dependent solubility profile that was comparable to pea and chickpea proteins, where proteins assume a high surface charge away from their isoelectric point (net charge of 0 mV, pH 4.5) to promote more protein-solvent interactions (e.g., +30 mV at pH 2, and −40 mV at pH 8). Minimum solubility was found near the isoelectric point (~15%), and higher solubility at pH 2 and 8 (~65%). Can Karaca et al. [144] found lentil protein isolates at pH 7 to have high solubility (91%) when produced either from alkaline extraction-isoelectric precipitation or by salt extraction. Boye et al. [145] reported lentil protein concentrates produced by alkaline extraction-ultrafiltration showed greater solubility than those produced by alkaline extraction-isoelectric precipitation, with both showing the u-shaped pH-dependent profile.

Lentil protein isolates have demonstrated excellent emulsifying properties. During emulsion formation, proteins migrate to the oil-water interface and then rearrange to orient hydrophobic groups towards the oil phase and hydrophilic groups towards the water phase, lowering interfacial tension. Aggregation of absorbed proteins then creates a viscoelastic interfacial film to stabilize oil droplets from coalescence and gravitational separation. Ladjal-Ettoumi et al. [143] and Chang et al. [146] reported lentil protein-stabilized emulsions were most stable at pH levels away from their isoelectric point. Can Karaca et al. [144] indicated protein isolates prepared by alkaline extraction-isoelectric precipitation to have greater emulsion forming properties and stability than if salt extraction methods were used to prepare the isolates. Primozic et al. [147] examined the stabilizing effects of high-pressure homogenized lentil protein isolates relative to unmodified isolates. The authors found that homogenization acted to reduce the particle size, hydrophobicity, and interfacial storage moduli of the lentil proteins relative to unmodified proteins, but had no effect on their interfacial tension. Overall, modified lentil proteins showed better physical stability for prepared emulsions than unmodified proteins. Gumus et al. [148] showed that lentil protein-stabilized emulsions were also effective at inhibiting oxidative reactions within fish oil-in-water emulsions.

Water holding and fat/oil absorption capacities relates to the amount of water or oil that a gram of protein material can hold, and to the surface properties (hydrophilic vs. hydrophobic groups), protein/aggregate conformation and solvent conditions used. Aryee and Boye [149] reported both water holding and fat absorption capacities were improved with wet extraction (i.e., isolate), followed by cooked flour and then raw flour. Boye et al. [145] indicated that red lentil protein concentrates produced by alkaline extraction-ultrafiltration exhibited greater water holding and fat absorption capacities than concentrates produced by alkaline extraction-isoelectric precipitation. Water holding was similar to that of yellow pea, and greater than that of chickpea. Fat absorption capacity of red lentil protein concentrates (produced from alkaline extraction-ultrafiltration) was much greater than that of other pulses. In the case of foaming, proteins migrate to the air-water interface, re-align and aggregate similar as in emulsions, to form a viscoelastic lamella that entraps gas bubbles [39]. Toews and Wang [150] reported that lentil protein concentrate produced the most stable foams and had the highest foaming capacity in comparison to pea, navy bean and chickpea. Lately, investigation of the foaming properties of the pulse cooking water (known as aquafaba) is gaining some interest as an egg replacer [151]. This highlights an opportunity for food technologists to apply lentil proteins in food applications.

13.2. Challenges

Like other pulses, lentil protein ingredients also have unwanted flavour compounds that limit their widespread use. Numerous flavour reduction strategies are available to reduce these compounds in pulses. For instance, Chang et al. [152] used various organic solvent (acetone, ethanol, and isopropanol) treatments to reduce flavour compounds found in lentil protein isolates, however, this resulted in negative effects on protein functionality. Shariati-Levari et al. [153] used infrared heating to reduce flavour compounds in lentils, and Ma et al. [154] reported pre-cooking (roasting and cooking) significantly reduced flavour compounds in green lentils.

13.3. Applications

Lentil protein concentrates have been used to replace eggs in production of protein-enriched doughnuts [155], angel food cake, and muffins [156]. Lentil flour was used to make gluten-free crackers [157], lentil flour with transglutaminase has been used as a binding agent to make protein-enriched restructured beef steaks or beef patties [158,159], and lentil protein isolates have been used as an emulsifier to produce salad dressings [160]. Lentil protein isolates have also been applied as stabilizers for nano emulsion systems [147,161], as encapsulation agents for delivery of omega-3 rich oils [162,163], and in combination with zein, as anti-microbial films [164], and used to produce nanofibers [165]. Keeping in mind the many technological functions of lentil proteins, niches are emerging for their inclusion in functional foods, in nutraceuticals, or even in cosmetics.

14. Conclusions

Lentil is the most rapidly expanding pulse crop for direct human consumption, and has potential for greater impact as a desirable protein source for food applications. Improvements in lentil protein quality, amino acid composition, and processing fractions will enhance the nutritional quality of this rapidly expanding pulse crop. Genetic strategies focused on increasing the concentration of limiting amino acids are required in lentil. The potential of lentil wild species in breeding programs by introgression of favourable genes for protein improvement may have potential as a long-term breeding strategy.

Supplementary Materials: The following are available online at http://www.mdpi.com/2304-8158/8/9/391/s1, Table S1: List of peptides with bioactive potential encrypted in lentil (*Lens culinaris* Medik.) convicilin (*see* http://www.uwm.edu.pl/biochemia).

Author Contributions: Conceptualization, H.K. and A.V.; writing—original draft preparation, H.K.; writing—review and editing, H.K., M.S., M.N., C.M-V., J.F., and A.V.

Funding: This study received support from the Natural Sciences and Engineering Research Council of Canada (NSERC) Industrial Research Chair Program, from the Saskatchewan Pulse Growers (SPG), and from Institute of Food Science, Technology and Nutrition - Spanish National Research Council (ICTAN-CSIC, Madrid, Spain, funded by project AGL2013-43247-R from AEI/FEDER UE) and from the Strategic Research Chair Program of the Saskatchewan Ministry of Agriculture.

Conflicts of Interest: The authors declare no conflict of interest.

References

1. Krishnan, H.B.; Coe, E.B., Jr. Seed storage proteins. In *Encyclopedia of Genetics*; Science Direct: Amsterdam, The Netherlands, 2001; pp. 1782–1787.

2. Nosworthy, M.G.; Neufeld, J.; Frohlich, P.; Young, G.; Malcolmson, L.; House, J.D. Determination of the protein quality of cooked Canadian pulses. *Food Sci. Nutr.* **2017**, *5*, 896–903. [CrossRef] [PubMed]
3. Vandenberg, A.; Slinkard, A.E. Genetics of seed coat color and pattern in lentil. *J. Hered.* **1990**, *81*, 484–488. [CrossRef]
4. Vandenberg, A. Lentil: Improvement in developing countries. In *The Lentil: Botany, Production and Uses*; Erskine, W., Muehlbauer, F.J., Sarker, A., Sharma, B., Eds.; CAB International: Oxford, UK, 2009; pp. 391–424.
5. Muehlbauer, F.J. Lentil: improvement in developing countries. In *The Lentil: Botany, Production and Uses*; Erskine, W., Muehlbauer, F.J., Sarker, A., Sharma, B., Eds.; CAB International: Oxford, UK, 2009; pp. 137–154.
6. Khazaei, H.; Caron, C.T.; Podder, R.; Kundu, S.S.; Diapari, M.; Vandenberg, A.; Bett, K.E. Marker-trait association analysis of iron and zinc concentrations in lentils (*Lens culinaris* Medik.) seeds. *Plant Genome.* **2017**, *10*, 1–8. [CrossRef] [PubMed]
7. Grela, E.R.; Kiczorowska, B.; Samolińska, W.; Matras, J.; Kiczorowski, P.; Rybiński, W.; Hanczakowska, E. Chemical composition of leguminous: Part I—Content of basic nutrients, amino acids, phytochemical compounds and antioxidant activity. *Eur. Food Res. Technol.* **2017**, *243*, 1385–1395. [CrossRef]
8. Tahir, M.; Lindeboom, N.; Båga, M.; Vandenberg, A.; Chibbar, R.N. Composition and correlation between major seed constituents in selected lentil (*Lens culinaris* Medik). *Can. J. Plant Sci.* **2011**, *91*, 825–835. [CrossRef]
9. Food and Agriculture Organization of the United Nations. FAOSTAT. Available online: http://faostat.fao.org (accessed on 2 September 2019).
10. Semba, R.D. The rise and fall of protein malnutrition in global health. *Ann. Nutr. Metab.* **2016**, *69*, 79–88. [CrossRef]
11. Arsenault, J.E.; Brown, K.H. Effects of protein or amino-acid supplementation on the physical growth of young children in low-income countries. *Nutr. Rev.* **2017**, *75*, 699–717. [CrossRef]
12. Barulina, H. Lentils of the U.S.S.R. and of other countries: A botanic-agronomical monograph. In *Bulletin of Applied Botany, of Genetics and Plant-Breeding*; Leningrad Vsesoiuznaia akademiia sel'sko khoziaistvennyk im: Leningrad (St. Petersburg), Russia, 1930; Suppl. S40, pp. 1–319.
13. Hawtin, G.C.; Rachie, K.O.; Green, J.M. Breeding strategy for the nutritional improvement of pulses. In *Nutritional Standards and Methods of Evaluation for Food Legume Breeders*; Hulse, J.H., Rachie, K.O., Billingsley, L.W., Eds.; IDRC: Ottawa, ON, Canada, 1977; pp. 43–51.
14. Hamdi, A.; Erskine, W.; Gates, P. Relationships among economic characters in lentil. *Euphytica* **1991**, *57*, 109–116. [CrossRef]
15. Kumar, J.; Singh, J.; Kanaujia, R.; Gupta, S. Protein content in wild and cultivated taxa of lentil (*Lens culinaris* ssp. *culinaris Medikus). Indian J. Genet. Plant Breed.* **2016**, *76*, 631–634.
16. Erskine, W.; Williams, P.C.; Nakkoul, H. Genetic and environmental variation in the seed size, protein, yield, and cooking quality of lentils. *Field Crops Res.* **1985**, *12*, 153–161. [CrossRef]
17. Stoddard, F.L.; Marshall, D.R.; Ali, S.M. Variability in grain protein concentration of peas and lentils grown in Australia. *Aust. J. Agric. Res.* **1993**, *44*, 1415–1419. [CrossRef]
18. Hedley, C.L. *Carbohydrate in Grain Legume Seeds*; CAB International: Wallingford, UK, 2001; pp. 1–11.
19. Wang, N.; Daun, J.K. Effects of variety and crude protein content on nutrients and anti-nutrients in lentils (*Lens culinaris*). *Food Chem.* **2006**, *95*, 493–502. [CrossRef]
20. Karaköy, T.; Erdem, H.; Baloch, F.S.; Toklu, F.; Eker, S.; Kilian, B.; Özkan, H. Diversity of macro-and micronutrients in the seeds of lentil landraces. *Sci. World J.* **2012**, *2012*, 1–9. [CrossRef] [PubMed]
21. Zaccardelli, M.; Lupo, F.; Piergiovanni, A.R.; Laghetti, G.; Sonnante, G.; Daminati, M.G.; Sparvoli, F.; Lioi, L. Characterization of Italian lentil (*Lens culinaris* Medik.) germplasm by agronomic traits, biochemical and molecular markers. *Genet. Resour. Crop Evol.* **2012**, *59*, 727–738. [CrossRef]
22. Alghamdi, S.S.; Khan, A.M.; Ammar, M.H.; El-Harty, E.H.; Migdadi, H.M.; Abd El-Khalik, S.M.; Al-Shameri, A.M.; Javed, M.M.; Al-Faifi, S.A. Phenological, nutritional and molecular diversity assessment among 35 introduced lentil (*Lens culinaris* Medik.) genotypes grown in Saudi Arabia. *Int. J. Mol. Sci.* **2014**, *15*, 277–295. [CrossRef]
23. Heuzé, V.; Tran, G.; Sauvant, D.; Bastianelli, D.; Lebas, F. Lentil (*Lens culinaris*). Feedipedia, a programme by INRA, CIRAD, AFZ and FAO. 2015. Available online: https://www.feedipedia.org/node/284 (accessed on 2 September 2019).
24. Shewry, P.R. Seed storage proteins: structures and biosynthesis. *Plant Cell Online* **1995**, *7*, 945–956.

25. Sales, M.P.; Gerhardt, I.S.; Grossi-de-Sá, F.; Xavier-Filho, J. Do legume storage proteins play a role in Defending seeds against bruchids? *Plant Physiol.* **2000**, *124*, 512–522. [CrossRef]
26. Cândido Ede, S.; Pinto, M.F.; Pelegrini, P.B.; Lima, T.B.; Silva, O.N.; Pogue, R.; Grossi-de-Sá, M.F.; Franco, O.L. Plant storage proteins with antimicrobial activity: Novel insights into plant defense mechanisms. *FASEB J.* **2011**, *25*, 3290–3305. [CrossRef]
27. Osborne, T.B.; Campbell, G.F. Proteins of lentils. *J. Am. Chem. Soc.* **1898**, *20*, 362–375. [CrossRef]
28. Danielson, C.E. An electrophoretic investigation of vicilin and legumin from seeds of peas. *Acta Chem. Scand.* **1950**, *4*, 762–771. [CrossRef]
29. Siddiqi, S.H. Chromatography on DEAE-cellulose and electrophoresis in polyacrylamide gel of protein fractions from lentil seeds. *Pak. J. Sci. Ind. Res.* **1982**, *25*, 167–169.
30. Bhatty, R.S. Composition and quality of lentil (*Lens culinaris*): A review. *Can. Inst. Food Sci. Technol. J.* **1988**, *21*, 144–160. [CrossRef]
31. Scippa, G.S.; Rocco, M.; Ialicicco, M.; Trupiano, D.; Viscosi, V.; Di Michele, M.; Arena, S.; Chiatante, D.; Scaloni, A. The proteome of lentil (*Lens culinaris* Medik.) seeds: Discriminating between landraces. *Electrophoresis* **2010**, *31*, 497–506. [CrossRef] [PubMed]
32. Neves, V.A.; Lourenco, E.J. Isolation and in vitro hydrolysis of globulin G1 from lentils (*Lens culinaris* Medik). *J. Food Biochem.* **1995**, *19*, 109–120. [CrossRef]
33. Bourgeois, M.; Jacquin, F.; Savois, V.; Sommerer, N.; Labas, V.; Henry, C.; Burstin, J. Dissecting the proteome of pea mature seeds reveals the phenotypic plasticity of seed protein composition. *Proteomics* **2009**, *9*, 254–271. [CrossRef] [PubMed]
34. Saint-Clair, P.M. Responses of Lens Esculenta Moench to Controlled Environmental Factors. Ph.D. Thesis, Committee of Agriculture, Wangeningen University, Wangeningen, The Netherlans, 1972; pp. 72–77.
35. Bhatty, R.S.; Slinkard, A.E.; Sosulski, F.W. Chemical composition and protein characteristics of lentils. *Can. J. Plant Sci.* **1976**, *56*, 787–794. [CrossRef]
36. Bhatty, R.S. In vitro hydrolysis of pea, faba bean and lentil meals and isolated protein fractions by pepsin and trypsin. *Can. Inst. Food Sci. Technol. J.* **1988**, *21*, 66–71. [CrossRef]
37. Sulieman, M.A.; Amro, B.H.; Gamaa, A.O.; Mohamed, M.E.T.; Elhadi, A.I.E.K.; Abdullahi, H.E.T.; Elfadil, E.B. Changes in total protein digestibility, fractions content and structure during cooking of lentil cultivars. *Pak. J. Nutr.* **2008**, *7*, 801–805.
38. Boye, J.I.; Aksay, A.; Roufik, S.; Ribereau, S.; Mondor, M.; Farnworth, E.; Rajamohamed, S.H. Comparison of the functional properties of pea, chickpea and lentil protein concentrates processed using ultrafiltration and isoelectric precipitation techniques. *Food Res. Int.* **2010**, *43*, 537–546. [CrossRef]
39. Jarpa-Parra, M. Lentil protein: A review of functional properties and food application. An overview of lentil protein functionality. *Int. J. Food Sci. Technol.* **2018**, *53*, 892–903. [CrossRef]
40. World Health Organization/Food and Agriculture—Organization of the United Nations/United Nations University. *Protein and Amino Acid Requirements in Human Nutrition*; Report of a Joint WHO/FAO/UNU Expert Consultation; WHO Technical Report Series 935; World Health Organization: Geneva, Switzerland, 2007.
41. Kahn, N.A.; Baker, B.E. The amino-acid composition of some Pakistani pulses. *J. Sci. Food Agric.* **1957**, *8*, 301–305. [CrossRef]
42. Chatterjee, S.R.; Abrol, Y.P. Amino acid composition of new varieties of cereals and pulses and nutritional potential of cereal-pulse combinations. *J. Food Sci. Technol.* **1975**, *12*, 221–227.
43. Sosulski, F.W. Legume protein concentration by air classification. In *Developments in Food Proteins—2*; Hudson, B.J.F., Ed.; Applied Science Publishers: Essex, UK, 1983.
44. Bhatty, R.S.; Christison, G.I. Composition and nutritional quality of pea (*Pisum sativum* L.), faba bean (*Vicia faba* L. spp. minor) and lentil (*lens culinaris Medik*) meals, protein concentrates and isolates. *Qual. Plant Plant Foods Hum. Nutr.* **1984**, *34*, 41–51. [CrossRef]
45. Shekib, L.A.H.; Zoueil, M.E.; Youssef, M.M.; Mohamed, M.S. Amino acid composition and in vitro digestibility of lentil and rice proteins and their mixture (Koshary). *Food Chem.* **1986**, *20*, 61–67. [CrossRef]
46. Pirman, T.; Stibilj, V.; Stekar, J.M.A.; Combe, E. Amino acid composition of beans and lentil. *Zb. Bioteh. Fak. Univ. Ljubl. Kmet. Zooteh.* **2001**, *78*, 57–68.
47. Porres, J.M.; Urbano, G.; Fernández-Fígares, I.; Prieto, C.; Pérez, L.; Aguilera, J.F. Digestive utilisation of protein and amino acids from raw and heated lentils by growing rats. *J. Sci. Food Agric.* **2002**, *82*, 1740–1747. [CrossRef]

48. Zia-Ul-Haq, M.; Ahmad, S.; Shad, M.A.; Iqbal, S.; Qayum, M.; Ahmad, A.; Luthria, D.L.; Amarowicz, R. Compositional studies of lentil (*Lens culinaris* Medik.) cultivars commonly grown in Pakistan. *Pak. J. Bot.* **2011**, *43*, 1563–1567.
49. Kaldy, N.S.; Kasting, R. Amino acid composition and protein quality of eight faba bean cultivars. *Can. J. Plant Sci.* **1974**, *54*, 869–871. [CrossRef]
50. Lisiewska, Z.; Kmiecik, W.; Słupski, J. Content of amino acids in raw and frozen broad beans (*Vicia faba* var. major) seeds at milk maturity stage, depending on the processing method. *Food Chem.* **2007**, *105*, 1468–1473. [CrossRef]
51. Schumacher, H.; Paulsen, H.M.; Gau, A.E.; Link, W.; Jürgens, H.U.; Sass, O.; Dieterich, R. Seed protein amino acid composition of important local grain *Lupinus angustifolius* L., *Lupinus luteus* L., *Pisum sativum* L. and *Vicia faba* L. *Plant Breed.* **2011**, *130*, 156–164. [CrossRef]
52. Leterme, P.; Monmart, T.; Baudart, E. Amino acid composition of pea (*Pisum sativum*) proteins and protein profile of pea flour. *J. Sci. Food Agric.* **1990**, *53*, 107–110. [CrossRef]
53. Pownall, T.L.; Udenigwe, C.C.; Aluko, R.E. Amino acid composition and antioxidant properties of pea seed (*Pisum sativum* L.) enzymatic protein hydrolysate fractions. *J. Agric. Food Chem.* **2010**, *58*, 4712–4718. [CrossRef] [PubMed]
54. Kuiken, K.; Lyman, C.M.; Bradford, M.; Trant, M.; Dieterich, S. Essential amino acid composition of soy bean meals prepared from twenty strains of soy beans. *J. Biol. Chem.* **1949**, *177*, 29–36. [PubMed]
55. Tkachuk, H.; Irvine, G.N. Amino acid composition of cereals and oilseed meals. *Cereal Chem.* **1969**, *16*, 206–218.
56. Cho, C.Y.; Bayley, H.S. Evaluations of rapeseed and soybean meals as protein sources for swine: Apparent digestibilities of amino acids. *Can. J. Anim. Sci.* **1970**, *50*, 521–528. [CrossRef]
57. Cavins, J.F.; Kwolek, W.E.; Inglett, G.E.; Cowan, J.C. Amino acid analysis of soybean meal: Interlaboratory study. *J. Asso. Off. Anal. Chem.* **1972**, *55*, 686–691.
58. Wang, H.L.; Cavins, J.F. Yield and amino acid composition of fractions obtained during tofu production. *Cereal Chem.* **1989**, *66*, 359–361.
59. Zarkadas, C.G.; Gagnon, C.; Gleddie, S.; Khanizadeh, S.; Cober, E.R.; Guillemette, R.J.D. Assessment of the protein quality of fourteen soybean (*Glycine max* (L.) Merr.) cultivars using amino acid analysis and two-dimensional electrophoresis. *Food Res. Int.* **2007**, *40*, 129–146. [CrossRef]
60. Sotak-Peper, K.M.; González-Vega, J.C.; Stein, H.H. Amino acid digestibility in soybean meal sourced from different regions of the United States and fed to pigs. *J. Anim. Sci.* **2017**, *95*, 771–778. [CrossRef]
61. Wang, N.; Daun, J.K. *The Chemical Composition and Nutritive Value of Canadian Pulses*; Canadian Grain Commission Report; Canadian Commision: Winnipeg, MB, Canada, 2004; pp. 19–29.
62. Alajaji, S.A.; El-Adawy, T.A. Nutritional composition of chickpea (*Cicer arietinum* L.) as affected by microwave cooking and other traditional cooking methods. *J. Food Comp. Anal.* **2006**, *19*, 806–812. [CrossRef]
63. Wang, X.; Gao, W.; Zhang, J.; Zhang, H.; Li, J.; He, X.; Ma, H. Subunit, amino acid composition and in vitro digestibility of protein isolates from Chinese kabuli and desi chickpea (*Cicer arietinum* L.) cultivars. *Food Res. Int.* **2010**, *43*, 567–572. [CrossRef]
64. El-Beltagi, H.S.; El-Senousi, N.A.; Ali, Z.A.; Omran, A.A. The impact of using chickpea flour and dried carp fish powder on pizza quality. *PLoS ONE* **2017**, *12*, 1–15. [CrossRef] [PubMed]
65. Wu, W.; Williams, W.P.; Kunkel, M.E.; Acton, J.C.; Huang, Y.; Wardlaw, F.B.; Grimes, L.W. Amino acid availability and availability-corrected amino acid score of red kidney beans (*Phaseolus vulgaris* L.). *J. Agric. Food Chem.* **1996**, *44*, 1296–1301. [CrossRef]
66. Słupski, J. Effect of cooking and sterilisation on the composition of amino acids in immature seeds of flageolet bean (*Phaseolus vulgaris* L.) cultivars. *Food Chem.* **2010**, *121*, 1171–1176.
67. Djemel, N.; Guedon, D.; Lechevalier, A.; Salon, C.; Miquel, M.; Prosperi, J.M.; Rochat, C.; Boutin, J.P. Development and composition of the seeds of nine genotypes of the *Medicago truncatula* species complex. *Plant Physiol. Biochem.* **2005**, *43*, 557–566. [CrossRef] [PubMed]
68. Van Vliet, S.; Burd, N.A.; van Loon, L.J. The skeletal muscle anabolic response to plant-versus animal-based protein consumption. *J. Nutr.* **2015**, *145*, 1981–1991. [CrossRef]
69. Gorissen, S.H.M.; Crombag, J.J.R.; Senden, J.M.G.; Waterval, W.A.H.; Bierau, J.; Verdijk, L.B.; van Loon, L.J.C. Protein content and amino acid composition of commercially available plant-based protein isolates. *Amino Acids* **2018**, *50*, 1685–1695. [CrossRef]

70. Rozan, P.; Kuo, Y.H.; Lambein, F. Amino acids in seeds and seedlings of the genus *Lens*. *Phytochemistry* **2001**, *58*, 281–289. [CrossRef]
71. Bell, E.A.; Lyddiard, J.R.A.; Malik, M.A.; Motevalli, M.; Nunn, P.B.; O'Brien, P.; Perera, K.P.W.C. X-ray single-crystal structure of 2(S),4(R)-4-hydroxyarginine from *Lens culinaris* seeds. *Phytochemistry* **1999**, *50*, 1201–1204. [CrossRef]
72. Sulser, H.; Sager, F. Identification of uncommon amino acids in the lentil seed (*Lens culinaris* Med.). *Experientia* **1976**, *32*, 422–423. [CrossRef]
73. Wong, M.M.L.; Gujaria-Verma, N.; Ramsay, L.; Yuan, H.Y.; Caron, C.; Diapari, M.; Vandenberg, A.; Bett, K.E. Classification and characterization of species within the genus *Lens* using genotyping-by-sequencing (GBS). *PLoS ONE* **2015**, *10*, e0122025. [CrossRef] [PubMed]
74. Bhatty, R.S. Protein subunits and amino acid composition of wild lentil. *Phytochemistry* **1986**, *25*, 641–644. [CrossRef]
75. Gatel, F. Protein-quality of legume seeds for non-ruminant animals—A literature-review. *Anim. Feed Sci. Technol.* **1994**, *45*, 317–348. [CrossRef]
76. Davis, K.R. Effect of processing on composition and Tetrahymena relative nutritive value on green and yellow peas, lentils and white pea beans. *Cereal Chem.* **1981**, *58*, 454–460.
77. Matus, A.; Slinkard, A.E.; Vandenberg, A. The potential of zero tannin lentil. In *New Crops*; Janick, J., Simon, J.E., Eds.; Wiley: New York, NY, USA, 1993; pp. 279–282.
78. Mirali, M.; Purves, R.W.; Stonehouse, R.; Song, R.; Bett, K.; Vandenberg, A. Genetics and biochemistry of zero-tannin lentils. *PLoS ONE* **2016**, *11*, e0164624. [CrossRef] [PubMed]
79. Savage, G.P. Antinutritional factors in peas. In *Recent Advances of Research in Antinutritional Factors in Legume Seeds*; Huisman, J., van der Poel, T.F.B., Liener, I.E., Eds.; Centre for Agricultural Publishing and Documentation: Wageningen, The Netherlands, 1989; pp. 342–350.
80. Liener, I.E.; Kakade, M.L. Protease inhibitors. In *Toxic Constituents in Plant Feedstuff*; Liener, I.E., Ed.; Academic Press: New York, NY, USA, 1980; pp. 7–71.
81. Pisulewska, E.; Pisulewski, P.M. Trypsin inhibitor activity of legume seed (peas, chickling vetch, lentils, and soya beans) as affected by the technique of harvest. *Anim. Feed Sci. Technol.* **2000**, *86*, 261–265. [CrossRef]
82. Norton, G. Proteinase inhibitors. In *Toxic Substances in Crop Plants*; D'Mello, J.P.F., Duffus, C.M., Duffus, J.H., Eds.; The Royal Society of Chemistry: Cambridge, UK, 1991; pp. 68–106.
83. Bhatty, R.S. Trypsin inhibitor activity in faba bean (*Vicia faba* var. minor), changes during germination and distribution. *Can. J. Plant Sci.* **1977**, *57*, 979–982. [CrossRef]
84. Elkowicz, K.; Sosulski, F.W. Antinutritional factors in eleven legumes and their air-classified protein and starch fractions. *Food Sci.* **1982**, *47*, 1301–1304. [CrossRef]
85. Shi, L.; Mu, K.; Arntfield, S.D.; Nickerson, M.T. Changes in levels of enzyme inhibitors during soaking and cooking for pulses available in Canada. *J. Food Sci. Technol.* **2017**, *54*, 1014–1022. [CrossRef]
86. Zinselmeier, C.; Jeong, B.R.; Boyer, J.S. Starch and the control of kernel number in maize at low water potentials. *Plant Physiol.* **1999**, *121*, 25–35. [CrossRef]
87. Gooding, M.J.; Ellis, R.H.; Shewry, P.R.; Schofield, J.D. Effects of restricted water availability and increased temperature on the grain filling, drying and quality of winter wheat. *J. Cereal Sci.* **2003**, *37*, 295–309. [CrossRef]
88. Rakszegi, M.; Darkó, E.; Lovegrove, A.; Molnár, I.; Láng, L.; Bedő, Z.; Molnár-Láng, M.; Shewry, P. Drought stress affects the protein and dietary fiber content of wholemeal wheat flour in wheat/Aegilops addition lines. *PLoS ONE* **2019**, *14*, e0211892. [CrossRef] [PubMed]
89. Behboudian, M.H.; Ma, Q.; Turner, N.C.; Palta, J.A. Reactions of chickpea to water stress: Yield and seed composition. *J. Sci. Food Agric.* **2001**, *81*, 1288–1291. [CrossRef]
90. Dornbos, D.L.; Mullen, R.E. Soybean seed protein and oil contents and fatty acid composition adjustments by drought and temperature. *J. Am. Oil Chem. Soc.* **1992**, *69*, 228–231. [CrossRef]
91. Mertz-Henning, L.M.; Ferreira, L.C.; Henning, F.A.; Mandarino, J.M.G.; Santos, E.D.; Oliveira, M.C.N.D.; Nepomuceno, A.E.L.; Farias, J.R.B.; Neumaier, N. Effect of water deficit-induced at vegetative and reproductive stages on protein and oil content in soybean grains. *Agronomy* **2018**, *8*, 3. [CrossRef]

92. Sehgal, A.; Sita, K.; Siddique, K.H.M.; Kumar, R.; Bhogireddy, S.; Varshney, R.K.; HanumanthaRao, B.; Nair, R.M.; Prasad, P.V.V.; Nayyar, H. Drought or/and heat-stress effects on seed filling in food crops: Impacts on functional biochemistry, seed yields, and nutritional quality. *Front. Plant Sci.* **2018**, *9*, 1705. [CrossRef] [PubMed]
93. Sita, K.; Sehgal, A.; Bhandari, K.; Kumar, J.; Kumar, S.; Singh, S.; Siddique, K.H.M.; Nayyar, H. Impact of heat stress during seed filling on seed quality and seed yield in lentil (*Lens culinaris* Medikus) genotypes. *J. Sci. Food Agric.* **2018**, *98*, 5134–5141. [CrossRef]
94. Liu, T.; Staden, J. Partitioning of carbohydrates in salt-sensitive and salt-tolerant soybean cultures under salinity stress and its subsequent relief. *Plant Growth Regul.* **2001**, *33*, 13–17. [CrossRef]
95. Zhao, H.; Dai, T.; Jiang, D.; Cao, W. Effects of high temperature on key enzymes involved in starch and protein formation in grains of two wheat cultivars. *J. Agron. Crop Sci.* **2008**, *194*, 47–54. [CrossRef]
96. Mossé, J.; Huet, J.C.; Baudet, J. The amino acid composition of wheat grain as a function of nitrogen content. *J. Cereal Sci.* **1985**, *3*, 115–130. [CrossRef]
97. Lizarazo, C.I.; Lampi, A.M.; Liu, J.; Sontag-Strohm, T.; Piironen, V.; Stoddard, F.L. Nutritive quality and protein production from grain legumes in a boreal climate. *J. Sci. Food Agric.* **2015**, *95*, 2053–2064. [CrossRef] [PubMed]
98. Warsame, A.; O'Sullivan, D.M.; Tosi, P. Seed storage proteins of faba bean (*Vicia faba* L): Current status and prospects for genetic improvement. *J. Agric. Food Chem.* **2018**, *66*, 12617–12626. [CrossRef] [PubMed]
99. Ivanov, N.N. The causes of chemical variability in the seeds of chickpea in different geographical areas. In *Monograph of the Genius*; Quoted in Van der Maesen, 1972; Cicer, L.A., Ed.; Agriculture University Communication: Wageningen, The Netherlands, 1933.
100. Khatun, A.M.; Bhuiyan, A.H.; Dey, T.K. Nitrogen uptake and protein yield in lentil as influenced by seed collection from different parts of plants. *Bangladesh J. Agric. Res.* **2010**, *35*, 515–523. [CrossRef]
101. Ali-Khan, S.T. Seed yield, seed weight, percent protein and protein yield of field peas as affected by seeding dates. *Can. J. Plant Sci.* **1977**, *57*, 17–20. [CrossRef]
102. Karjalainen, R.; Hovinen, S. Variation in protein content of peas under Finnish conditions. *J. Sci. Agric. Soc. Finl.* **1981**, *53*, 228–238. [CrossRef]
103. Krotz, L.; Ciceri, E.; Giazzi, G. Protein Determination in Cereals and Seeds. Food Quality and Safety. 2008. Available online: http/Foodqualityandsafety.com (accessed on 2 September 2019).
104. Moore, J.C.; de Vries, W.; Lipp, M.; Grifiths, J.C.; Abernethy, D.R. Total protein methods and their potential utility to reduce the risk of food protein adulteration. *Compr. Rev. Food Sci. Food Saf.* **2010**, *9*, 330–351. [CrossRef]
105. FAO. Pulses. 2016. Available online: www.fao.com (accessed on 2 September 2019).
106. Mihaljev, A.Ž.; Jakšić, M.S.; Prica, N.B.; Ćupić, Ž.N.; Živkov-Baloš, M. Comparison of the Kjeldahl method, Dumas method and NIR method for total nitrogen determination in meat and meat products. *J. Agro Proc. Technol.* **2015**, *21*, 365–370.
107. Subedi, M.; Willenborg, C.J.; Vandenberg, A. Influence of harvest aid herbicides on seed germination, seedling vigor and milling quality traits of red lentil (*Lens culinaris* L.). *Front. Plant Sci.* **2017**, *8*, 311. [CrossRef] [PubMed]
108. Swanson, B.G. Pea and lentil protein extraction and functionality. *J. Am. Oil Chem. Soc.* **1990**, *67*, 276–279. [CrossRef]
109. Aryee, A.N.A.; Boye, J.I. Improving the digestibility of lentil flours and protein isolate and characterization of their enzymatically prepared hydrolysates. *Int. J. Food Prop.* **2016**, *19*, 2649–2665. [CrossRef]
110. Joshi, M.; Timilsena, Y.; Adhikari, B. Global production, processing and utilization. *J. Integr. Agric.* **2017**, *16*, 2898–2913. [CrossRef]
111. Vose, J.R. Production and functionality of starches and protein isolates form legume seeds (field peas and horsebeans). *Cereal Chem.* **1980**, *57*, 406–410.
112. Gueguen, J. Legume seed protein extraction, processing, and end product characteristics. *Plant Foods Hum. Nutr.* **1993**, *32*, 267–303. [CrossRef]
113. Tyler, R.T.; Youngs, C.G.; Sosulski, F.W. Air classification of legumes. 1. Separation efficiency, yield, and composition of the starch and protein fractions. *Cereal Chem.* **1981**, *58*, 144–148.
114. Reichert, R.D. Quantitative isolation and estimation of cell wall material from dehulled pea (*Pisum sativum*) flours and concentrates. *Cereal Chem.* **1981**, *58*, 266–270.

115. Lee, H.C.; Htoon, A.K.; Uthayakumaran, S.; Paterson, J.L. Chemical and functional quality of protein isolated from alkaline extraction of Australian lentil cultivars: Matilda and Digger. *Food Chem.* **2007**, *102*, 1199–1207. [CrossRef]
116. Fan, T.Y.; Sosulski, F.W. Dispersibility and isolation of proteins from legume flours. *Can. Inst. Food Sci. Technol. J.* **1974**, *7*, 256–259. [CrossRef]
117. Alsohaimy, S.A.; Sitohy, M.Z.; El-Masry, R.A. Isolation and partial characterization of chickpea, lupine and lentil seed proteins. *World J. Agric. Sci.* **2007**, *3*, 123–129.
118. Joshi, M.; Adhikari, B.; Aldred, P.; Panozzo, J.F.; Kasapis, S. Physicochemical and functional properties of lentil protein isolates prepared by different drying methods. *Food Chem.* **2011**, *129*, 1513–1522. [CrossRef]
119. Joshi, M.; Adhikari, B.; Aldred, P.; Panozzo, J.F.; Kasapis, S.; Barrow, C.J. Interfacial and emulsifying properties of lentil protein isolate. *Food Chem.* **2012**, *134*, 1343–1353. [CrossRef]
120. Kaur, M.; Singh, N.; Sandhu, K.S. Preparation and characterization of protein isolates from different lentil (*Lens culinaris*) cultivars. *J. Food Sci. Technol.* **2007**, *44*, 327–329.
121. Johnston, S.P.; Nickerson, M.T.; Nicholas, H.L. The physicochemical properties of legume protein isolates and their ability to stabilize oil-in-water emulsions with and without genipin. *J. Food Sci. Technol.* **2015**, *52*, 4135–4145. [CrossRef] [PubMed]
122. Jarpa-Parra, M.; Bamdad, F.; Wang, Y.; Tian, Z.; Temelli, F.; Han, J.; Chen, L. Optimization of lentil protein extraction and the influence of process pH on protein structure and functionality. *LWT Food Sci. Technol.* **2014**, *57*, 461–469. [CrossRef]
123. Matsusaka, S.; Maruyama, H.; Matsuyama, T.; Ghadiri, M. Triboelectric charging of powders: A review. *Chem. Eng. Sci.* **2010**, *65*, 5781–5807. [CrossRef]
124. Jafari, M.; Rajabzadeh, A.R.; Tabtabaei, S.; Marsolais, F.; Legge, R.L. Physicochemical characterization of a navy bean (*Phaseolus vulgaris*) protein fraction produced using a solvent-free method. *Food Chem.* **2016**, *208*, 35–41. [CrossRef]
125. Assatory, A.; Vitellia, M.; Rajabzdeh, A.R.; Legge, R.L. Dry fractionation methods for plant protein, starch and fiber enrichment: A review. *Trends Food Sci. Technol.* **2019**, *86*, 340–351. [CrossRef]
126. Tabtabaei, S.; Jafari, M.; Rajabzadeh, A.R.; Legge, R.L. Solvent-free production of protein-enriched fractions from navy bean flour using a triboelectrification-based approach. *J. Food Eng.* **2016**, *174*, 21–28. [CrossRef]
127. Tabtabaei, S.; Konakbayeva, D.; Rajabzadeh, A.R.; Legge, R.L. Analysis of protein enrichment during single- and multi-stage triboelectrostatic bioseparation processes for dry fractionation of legume flour. *Sep. Purif. Technol.* **2017**, *176*, 48–58. [CrossRef]
128. Tabtabaei, S.; Konakbayeva, D.; Rajabzadeh, A.R.; Legge, R.L. Functional properties of navy bean (*Phaseolus vulgaris*) protein concentrates obtained by pneumatic tribo-electrostatic separation. *Food Chem.* **2019**, *283*, 101–110. [CrossRef]
129. Udenigwe, C.C.; Aluko, R.E. Food protein-derived bioactive peptides: Production, processing, and potential health benefits. *J. Food Sci.* **2012**, *77*, R11–R24. [CrossRef]
130. Garcia-Mora, P.; Peñas, P.; Frias, J.; Martinez-Villaluenga, C. Savinase, the most suitable enzyme for releasing peptides from lentil (*Lens culinaris* var. castellana) protein concentrates with multifunctional properties. *J. Agric. Food Chem.* **2014**, *62*, 4166–4174. [CrossRef]
131. Barbana, C.; Boye, J.I. Angiotensin I-converting enzyme inhibitory properties of lentil protein hydrolysates: Determination of the kinetics of inhibition. *Food Chem.* **2011**, *127*, 94–101. [CrossRef]
132. Natesh, R.; Schwager, S.L.; Sturrock, E.D.; Acharya, K.R. Crystal structure of the human angiotensin-converting enzyme-Lysinopril complex. *Nature* **2003**, *421*, 551–554. [CrossRef] [PubMed]
133. García-Mora, P.; Martín-Martínez, M.; Bonache, M.A.; González-Múniz, R.; Peñas, E.; Frias, J.; Martinez-Villaluenga, C. Identification, functional gastrointestinal stability and molecular docking studies of lentil peptides with dual antioxidant and angiotensin I converting enzyme inhibitory activities. *Food Chem.* **2017**, *221*, 464–472. [CrossRef] [PubMed]
134. Wu, J.; Aluko, R.E.; Nakai, S. Structural requirements of angiotensin I converting enzyme inhibitory peptides: Quantitative structure-activity relationship study of di and tripeptides. *J. Agric. Food Chem.* **2006**, *54*, 732–738. [CrossRef]
135. Rizzello, C.B.; Verni, M.; Bordignon, S.; Gramaglia, V.; Gobbetti, M. Hydrolysate from a mixture of legume flours with antifungal activity as an ingredient for prolonging the shelf-life of wheat bread. *Food Microbiol.* **2017**, *64*, 72–82. [CrossRef]

136. Wang, H.X.; Ng, T.B. An antifungal peptide from red lentil seeds. *Peptides* **2007**, *28*, 547–552. [CrossRef]
137. Chalamaiah, M.; Ulug, S.K.; Hong, H.; Wu, J. Regulatory requirements of bioactive peptides (protein hydrolysates) from food proteins. *J. Funct. Foods* **2019**, *58*, 123–129. [CrossRef]
138. Mamilla, R.K.; Mishra, V.K. Effect of germination on antioxidant and ACE inhibitory activities of legumes. *LWT Food Sci. Technol.* **2017**, *75*, 51–58. [CrossRef]
139. Boschin, G.; Scigliuolo, G.M.; Resta, D.; Arnoldi, A. ACE-inhibitory activity of enzymatic protein hydrolysates from lupin and other legumes. *Food Chem.* **2014**, *145*, 34–40. [CrossRef] [PubMed]
140. Jakubczyk, A.; Baraniak, B. Activities and sequences of the angiotensin I-converting enzyme (ACE) inhibitory peptides obtained from the digested lentil (*Lens culinaris*) globulins. *Int. J. Food Sci. Technol.* **2013**, *48*, 2363–2369. [CrossRef]
141. Akıllıoğlu, H.G.; Karakaya, S. Effects of heat treatment and in vitro digestion on the Angiotensin converting enzyme inhibitory activity of some legume species. *Eur. Food Res. Technol.* **2009**, *229*, 915–921. [CrossRef]
142. Pelgrom, P.J.M.; Boom, R.M.; Schutyser, M.A.I. Method development to increase protein enrichment during dry fractionation of starch-rich legumes. *Food Bioprocess Technol.* **2015**, *8*, 1495–1502. [CrossRef]
143. Ladjal-Ettoumi, Y.; Boudries, H.; Chibane, M.; Romero, A. Pea, chickpea and lentil protein isolates: Physicochemical characterization and emulsifying properties. *Food Biophys.* **2016**, *11*, 43–51. [CrossRef]
144. Can Karaca, A.; Low, N.; Nickerson, M. Emulsifying properties of chickpea, faba bean, lentil and pea proteins produced by isoelectric precipitation and salt extraction. *Food Res. Int.* **2011**, *444*, 2742–2750. [CrossRef]
145. Boye, J.I.; Roufik, S.; Pesta, N.; Barbana, C. Angiotensin I-converting enzyme inhibitory properties and SDS-PAGE of red lentil protein hydrolysates. *Food Sci. Technol.* **2010**, *43*, 987–991. [CrossRef]
146. Chang, C.; Tu, S.; Ghosh, S.; Nickerson, M.T. Effects of pH on the inter-relationships between the physicochemical, interfacial and emulsifying properties for pea, soy, lentil and canola protein isolates. *Food Res. Int.* **2015**, *77*, 360–367. [CrossRef]
147. Primozic, M.; Duchek, A.; Nickerson, M.T.; Ghosh, S. Formation, stability and in vitro digestibility of nanoemulsions stabilized by high-pressure homogenized lentil protein isolates. *Food Hydrocoll.* **2018**, *77*, 126–141. [CrossRef]
148. Gumus, C.E.; Decker, E.A.; Mc Clements, D.J. Impact of legume protein type and location on lipid oxidation in fish oil-in-water emulsions: Lentil, peas and faba bean proteins. *Food Res. Int.* **2017**, *100*, 175–185. [CrossRef]
149. Aryee, A.N.A.; Boye, J.I. Comparative study of the effects of processing on the nutritional, physicochemical and functional properties of lentil. *J. Food Proc. Preserv.* **2017**, *41*, e12824. [CrossRef]
150. Toews, R.; Wang, N. Physicochemical and functional properties of protein concentrates from pulses. *Food Res. Int.* **2013**, *52*, 445–451. [CrossRef]
151. Stantiall, S.E.; Dale, K.J.; Calizo, F.S.; Serventi, L. Applications of pulses cooking water as functional ingredients: The foaming and gelling abilities. *Eur. Food Res. Technol.* **2018**, *244*, 97–104. [CrossRef]
152. Chang, C.; Stone, A.K.; Green, R.; Nickerson, M.T. Reduction of off-flavours and the impact on the functionalities of lentil protein isolate by acetone, ethanol, and isopropanol treatments. *Food Chem.* **2019**, *277*, 84–95. [CrossRef] [PubMed]
153. Shariati-Ievari, S.; Ryland, D.; Edel, A.; Nicholson, T.; Suh, M.; Aliani, M. Sensory and physicochemical studies of thermally micronized chickpea (*Cicer arietinum*) and green lentil (*Lens culinaris*) flours as binders in low-fat beef burgers. *J. Food Sci.* **2016**, *81*, S1230–S1242. [CrossRef] [PubMed]
154. Ma, Z.; Boye, J.I.; Azarnia, S.; Simpson, B.K. Volatile flavor profile of Saskatchewan grown pulses as affected by different thermal processing treatments. *Int. J. Food Prop.* **2016**, *19*, 2251–2271. [CrossRef]
155. Eckert, E.; Wismer, W.; Waduthanthiri, K.; Babii, O.; Yang, J.; Chen, L. Application of barley-and lentil-protein concentrates in the production of protein-enriched doughnuts. *J. Am. Oil Chem. Soc.* **2018**, *95*, 1027–1040. [CrossRef]
156. Jarpa-Parra, M.; Wong, L.; Wismer, W.; Temelli, F.; Han, J.; Huang, W.; Eckhart, E.; Tian, Z.; Shi, K.; Sun, T.; et al. Quality characteristics of angle3 food cake and muffin using lentil protein as egg/milk replacer. *Int. J. Food Sci. Technol.* **2017**, *52*, 1604–1613. [CrossRef]
157. Han, J.; Janz, J.A.M.; Gerlat, M. Development of gluten-free cracker snacks using pulse flours and fractions. *Food Res. Int.* **2010**, *43*, 627–633. [CrossRef]
158. Baugreet, S.; Kery, J.P.; Botinestean, C.; Allen, P.; Hamill, R.M. Development of novel fortified beef patties with added functional protein ingredients for elderly. *Meat Sci.* **2016**, *122*, 40–47. [CrossRef]

159. Baugreet, S.; Kerry, J.P.; Brodkorb, A.; Gomex, C.; Auty, M.; Allen, P.; Hamill, R.M. Optimisation of plant protein and transglutaminase content in novel beef restructured steaks for older adults by central composite design. *Meat Sci.* **2018**, *142*, 65–77. [CrossRef]
160. Ma, Z.; Boye, J.I.; Simpson, B.K. Preparation of salad dressing emulsions using lentil, chickpea and pea protein isolates: A response surface methodology study. *J. Food Qual.* **2016**, *39*, 274–291. [CrossRef]
161. Primozic, M.; Duchek, A.; Nickerson, M.T.; Ghosh, S. Effect of lentil protein isolate concentration on the formation, stability and rheological behavior of oil-in-water nanoemulsions. *Food Chem.* **2017**, *234*, 65–74. [CrossRef] [PubMed]
162. Chang, C.; Nickerson, M.T. Stability and in vitro release behaviour of encapsulated omega fatty acid-rich oils in lentil protein isolate-based microcapsules. *Int. J. Food Sci. Nutr.* **2018**, *69*, 12–23. [CrossRef] [PubMed]
163. Chang, C.; Varankovich, N.; Nickerson, M.T. Microencapsulation of canola oil by lentil protein isolate-based wall materials. *Food Chem.* **2016**, *212*, 264–273. [CrossRef] [PubMed]
164. Boyaci, D.; Yemenicioglu, A. Expanding horizons of active packaging: Design of consumer-controlled release systems helps risk management of susceptible individuals. *Food Hydrocoll.* **2018**, *79*, 291–300. [CrossRef]
165. Tam, N.; Oguz, S.; Aydogdu, A.; Sumnu, G.; Sahin, S. Influence of solution properties and pH on the fabrication of electro spun lentil flour/HPMC blend nanofibers. *Food Res. Int.* **2017**, *102*, 616–624. [CrossRef] [PubMed]

Article

Overall Nutritional and Sensory Profile of Different Species of Australian Wattle Seeds (*Acacia* spp.): Potential Food Sources in the Arid and Semi-Arid Regions

Kinnari J. Shelat [1,2], Oladipupo Q. Adiamo [1], Sandra M. Olarte Mantilla [1], Heather E. Smyth [1], Ujang Tinggi [3], Sarah Hickey [4], Broder Rühmann [5], Volker Sieber [5] and Yasmina Sultanbawa [1,*]

1 Queensland Alliance for Agriculture and Food Innovations, Health and Food Sciences Precinct, Cooper Plains, Brisbane 4108, QLD, Australia; k.shelat@uq.edu.au (K.J.S.); o.adiamo@uq.edu.au (O.Q.A.); s.olartemantilla@uq.edu.au (S.M.O.M.); h.smyth@uq.edu.au (H.E.S.)

2 Australian National Fabrication Facility–Queensland Node, Australian Institute of Bioengineering and Nanotechnology, The University of Queensland, St. Lucia, Brisbane 4067, QLD, Australia

3 Health Support Queensland, Queensland Health, Inorganic Chemistry, Forensic and Scientific Services, Coopers Plains, Brisbane 4108, QLD, Australia; ujang.tinggi@health.qld.gov.au

4 Karen Sheldon Catering, PO Box 2351, Parap 0812, NT, Australia; sarah@karensheldoncatering.com.au

5 Department of Chemistry of Biogenic Resources, Technical University of Munich, 94315 Straubing, Germany; broder.ruehmann@tum.de (B.R.); sieber@tum.de (V.S.)

* Correspondence: y.sultanbawa@uq.edu.au; Tel.: +61-7-3443-2474

Received: 24 August 2019; Accepted: 8 October 2019; Published: 11 October 2019

Abstract: Wattle seed (*Acacia* spp.) is a well-known staple food within indigenous communities in Australia. A detailed investigation of the overall nutritional and sensory profile of four abundant and underutilized *Acacia* species—*A. coriacea*, *A. cowleana*, *A. retinodes* and *A. sophorae*—were performed. Additionally, molecular weight of protein extracts from the wattle seeds (WS) was determined. The seeds are rich in protein (23–27%) and dietary fibre (33–41%). Relatively high fat content was found in *A. cowleana* (19.3%), *A. sophorae* (14.8%) and *A. retinodes* (16.4%) with oleic acid being the predominant fatty acid. The seeds contained high amounts of essential amino acids (histidine, lysine, valine, isoleucine and leucine). *A. coriacea* is rich in iron (43 mg/kg), potassium (10 g/kg) and magnesium (1.7 g/kg). Pentose (xylose/arabinose), glucose, galactose and galacturonic acids were the major sugars found in the four species. Raw seeds from *A. sophorae*, *A. retinodes* and *A. coriacea* have the highest protein molecular weight, between 50–90 kDa, 80 kDa and 50–55 kDa, respectively. There was variation in the sensory profile of the WS species. This study showed that the four WS species have good nutritional value and could be included in human diet or used in food formulations.

Keywords: wattle seed species; nutritional profile; sensory profile; gel electrophoresis

1. Introduction

As the world population increases and natural resources diminish, there has been a serious concern on available sustainable nutritious foods [1]. In addition, a majority of people from developing countries suffer from protein malnutrition, famine and different kinds of diseases due to inadequate food supply and poor quality food [2]. In order to meet these continued population growth and nutritional requirements, studies are required to examine and discover new sources of food. In the past few years, researchers have focused on the use of underutilized plant products as human food and animal feed [3–5].

The genus *Acacia*, commonly known as wattle, belongs to the family *Fabaceae* and it is a large group of woody species comprising of shrubs. *Acacia* subgenus *Phyllodineae* are naturally the most common *Acacia* species found in Australia and are among the most promising native leguminous plants [6,7]. These *Acacias* have been reported to exhibit significant potential to lower poverty in semi-arid regions of Africa [8,9]. Moreover, the seeds from various *Acacia* species, which were used traditionally as source of food by Australian Indigenous population, have been economically revived as food additives, such as emulsifying and flavouring agents [10–12]. *Acacia victoriae* Bentham is the most common species of *Acacia* with high water-soluble carbohydrates and protein contents and, thus, have been reported to have significant functional properties in food systems [13,14]. Additionally, *Acacia* plants have been frequently used to treat diseases, such as fever, leucorrhoea, throat infection, diarrhoea and haemoptysis [15].

However, several *Acacia* species which are also widely cultivated by indigenous people in different regions of Australia have not been fully utilized in food formulations or incorporated in human diets. These includes *A. coriacea* and *A. cowleana* which occurs throughout Northern Australia as well as *A. retinodes* and *A. sophorae* that are found in Southern and Southeastern Australia. These *Acacia* plants, particularly *A. retinodes*, are mainly used for gum production and ornamental purposes [16]. Nevertheless, information on the nutritive value of these *Acacia* species is sparse which may limit their use in foods. Therefore, this study investigated the overall nutritional value of seeds of these abundant and native Australian *Acacia* species. Furthermore, the seeds were roasted and the molecular weight profiles of protein extracts before and after treatment were examined. In addition, a preliminary sensory profiling of the four wattle seed species was carried out. This study will provide information on whether or not it is advisable to incorporate these seeds into the human diet and indicate the sensory characteristics of these species.

2. Materials and Methods

2.1. Materials

Mature seeds of four different species of Australian *Acacia* species (*A. coriacea*, *A. cowleana*, *A. retinodes* and *A. sophorae*) used in this study are shown in Figure 1. *A. coriacea* and *A. cowleana* were sourced from NATIF Australian Native Superfoods, Fruits Herbs Spices and Mixes, Victoria, Australia, and *A. retinodes* and *A. sophorae* were supplied by Valley Seeds Pty Ltd., Victoria, Australia. The seeds from each species were separately ground using a coffee grinder (power: 200 W, time: 30 s) and stored in the refrigerator until further analysis. Additionally, parts of the whole seeds were roasted at 180 °C for 5 min and used to determine the molecular weight profile of the seeds protein extracts for comparison with that obtained from raw seeds. All samples were analysed at least in duplicate and the average for each parameter was reported.

Figure 1. Four species of Australian wattle seeds—appearance and size comparison.

2.2. Proximate Analysis

The four different species of *Acacia* seeds were sent to Symbio Alliance Lab Pty Ltd., Eight Mile Plains, Queensland, Australia. A complete proximate analysis was performed at this accredited National Association of Testing Authorities (NATA) laboratory using AOAC [17] standard methods. The following analysis were measured: moisture (AOAC 925.10) by air oven with a measurement of uncertainty (MU) of ±15%, ash (AOAC 923.03), crude protein (AOAC 990.03) by Dumas combustion with a MU of ±10%, crude fat (AOAC 991.36) with a MU of ±15% and dietary fibre (985.29) with a MU of ±15%, carbohydrate and energy by calculation using information from the Food Standards Code.

2.3. Sugar Analysis

A combination of fast liquid chromatography coupled to UV and electrospray ionization trap detection (LC-UV-ESI-MS/MS) was used for the quantification of various sugars [18]. The hydrolysis was performed in duplicates by adding 6 mL of 2 M TFA to 12 mg of grounded *Acacia* seeds into 15 mL glass tubes. The tubes were incubated in a heating block (VLM GmbH, EC-Model, Heideblümchenweg, Bielefeld, Germany) for 90 min at 121°C. After cooling to room temperature, the hydrolysates were neutralized to pH ~8 by adding an aqueous solution of 3.2% NH_4OH, since light alkaline conditions are required for the subsequent derivatization of monosaccharides. A 25 µL of neutralized hydrolysate supernatant were derivatized via the high throughput 1 phenyl-3-methyl-5-pyrazolone (HT-PMP) method [18]. The calibration standards were diluted with neutralized TFA-matrix to compensate the influence on the derivatization process. Each sample was derivatized in triplicates and the carbohydrate fingerprint was analysed.

2.4. Fatty Acid Profiles

About 1 g of finely chopped seed samples were taken for initial extraction with chloroform and methanol (2:1) followed by agitation at room temperature for 1 h and centrifuged for 5 min at 3500× *g*. Fatty acid profiling was performed at the School of Agriculture and Food sciences, University of Queensland laboratory. The GC-MS (Shimadzu QP2010, Shimadzu Coporation, Tokyo, Japan) was used at oven temperature of 100 °C, injector temperature 250 °C, total program time was 39 min, and helium used as the carrier gas. The inlet pressure used for gas chromatography was 0.4 kPa, at linear gas velocity of 42.7 cm/s, column (Restek stabilwax capillary column; 30 m × 0.25 mm ID × 0.5 µm film thickness) flow 1.10 mL/min with a split ratio of 1:1 and injection volume of 0.2 µL. For mass spectrometry, the ion source temperature used was 200 °C, the interface temperature was 250 °C and the mass range was 35–500 atomic mass units. Identification of the compounds was done by comparing their retention times and mass spectra with corresponding data from a standard food industry FAME Mix (Restek Corporation, Bellefonte, PA, USA).

2.5. Amino Acid Analysis

The samples (100 mg per replicate) were first hydrolysed with 6 M HCl at 110 °C for 24 h. As asparagine is hydrolysed to aspartic acid and glutamine to glutamic acid, the reported amount of these acids is the sum of those respective components. After hydrolysis, all amino acids were analysed at the Department of Molecular Science, Australian Proteome Analysis Facility, Macquarie University, NSW, Australia, using the Waters AccQTag Ultra chemistry on a Waters Acquity UPLC. Samples were analysed in duplicate and results are expressed as an average. The coefficient of variation (CV) of the UPLC analysis of amino acids was less than 5%.

2.6. Mineral Analysis

A detailed description of method used for mineral analysis is outlined as described by Carter et al. [19]. The ground seed samples were accurately weighed (0.3 g) into digestion Teflon vessels and concentrated nitric acid (4 mL) was added. The samples were digested using a microwave digestion system (MarsXpress, CEM, Matthews, NC, USA) programmed to three steps: step 1 (400 W power, 85 °C, 14 min), step 2 (800 W power, 110 °C, 20 min), and step 3 (1600 W power, 160 °C, 10 min) and the analysis was performed using inductively coupled plasma mass spectrometry (ICP-MS 7500a, Agilent, Tokyo, Japan) and optical emission spectrometry (ICP-OES, Varian Australia, VIC, Australia).

2.7. Sodium Dodecyl Sulphate-Polyacrylamide Gel Electrophoresis (SDS-PAGE)

The SDS-PAGE analysis of lyophilized protein extracts from raw and roasted wattle seeds were conducted at an accredited Protein Expression Facility (PEF), University of Queensland, St Lucia, Queensland, Australia. The extracts were resuspended in PBS to a final concentration of 2 mg/mL. Samples were loaded onto a 4–12% Bis-Tris SDS-PAGE gel and run under denatured and reduced conditions, except where noted. Analysis was performed using a Bio-Rad Chemi-Doc™ XRS + imaging system.

2.8. Rapid Sensory Profiling

A sensory tasting session was carried out to develop sensory descriptors of four WS species: *A. coriacea*, *A. retinodes*, *A. sophorae* and *A. cowleana*. Twelve trained assessors (nine females; three males) with an average age of 50 years old participated in a two-hour session where they provided descriptors for aroma, flavour and aftertaste. For sensory evaluation the seeds of the four species were roasted in the oven for 7 min at 180 °C and ground when cooled. The wattle seeds were presented in two forms for sensory evaluation: as ground seeds on their own and the ground seeds mixed with semolina paste to make them more palatable for the assessors. One gram of ground seeds was presented for aroma assessment in a 30 mL plastic cup covered with a lid and labelled with a three digit blinding code. The ground seeds mixed with semolina where used to conduct a second assessment of aroma and to assess flavour and after taste. Each assessor was presented with 6 g of 1:20 ground wattle seeds and hydrated semolina mix in a 30 mL plastic cup covered with a lid and labelled with the same three digit blinding code used for the non-mixed sample. Green apple and water were used as palate cleansers. Upon completion of the session tasting component the panel leader facilitated session where sensory descriptors of each category for four wattle seed species were summarised and a consensus reached. This resulted in a preliminary sensory profile for each variety.

2.9. Statistical Analysis

The data were calculated using Microsoft Excel 2013 and the results are expressed as mean of the triplicate experiments unless otherwise stated. Statistical analysis of the data was done by one-way analysis of variance procedure (ANOVA) using SPSS software (Version 23.0 IBM Corporation, Armonk, NY, USA), and means comparison was done using Duncan's multiple range test at $p < 0.05$.

3. Results and Discussion

3.1. Proximate Composition

As shown in Table 1, all of the species had low moisture content (less than 9%) which are comparable with the 6.9% recorded for *A. victoriae* Bentham [20] and the range (6.3–8.0%) for *A. tumida* and *A. colei* [8]. The four species of wattle seeds (WS) showed high protein content, ranging from 22.5% in *A. coriacea* to 27.5% in *A. retinodes*. The values for crude protein obtained in this study were higher than that of *A. victoriae* Bentham [20] but lower than those found in different subspecies of *A. saligna* (28.6–32.6%) [3]. However, the protein contents were within the range (23.4–34.1%) reported for *A. tumida* and *A. colei* [8]. The results indicate that WS can serve as a source of protein in the diets of the Aboriginal population in Australia. Moreover, it can be included in food formulations as a source of protein. The crude fat content varied among the four species with relatively high values observed in *A. cowleana* followed by *A. retinodes* while *A. coriacea* had the least value (9.8%). These results showed that *A. cowleana* seed would be a good source of energy due to its relatively high crude fat content. All the four species studied had similar ash contents (3.4–3.9%) which were comparable to that found in *A. victoriae* Bentham [20] and *A. tortilis* [21]. All seeds also showed high amounts of dietary fibre ranging from 33.7% in *A. cowleana* to 41.4% in *A. coriacea*. These values were higher than that recorded for *A. tortilis* [21] and *A. victoriae* Bentham [20]. Therefore, the seeds from these *Acacia* species can be considered as a good source of dietary fibre. Dietary fibre has a vital function in human nutrition by maintaining the health of the gastrointestinal tract, however, in excess may bind to iron and zinc, thereby lowering their bioavailability [22]. Regarding the carbohydrate content, the seeds have a range of 12.8–15.6% carbohydrate. These levels of carbohydrate were lower than that reported for *A. victoriae* Bentham [20] and *A. tumida* [2], due to the greater amounts of crude fat, fibre and protein in the seeds. Traditionally, *A. coriacea* is known to lower postprandial glucose level [23,24], higher dietary fibre and protein content could be the reason for these potential health benefits. This suggest that using wattle seed flour can be a healthier option. Therefore, the chemical composition of these wattle seed species was found to be nutritious and adding these seeds into the human diets will enhance the nutrition status.

Table 1. Proximate composition of four different species of wattle seeds (%, dry weight).

Composition	*A. cowleana* *	*A. coriacea*	*A. sophorae*	*A. retinodes*	*A. victoriae* [19]
Moisture (%)	5.3	8.6	7.5	5.6	6.9
Crude protein (%)	23.0	22.5	22.7	27.5	17.5
Crude fat (%)	19.3	9.8	14.8	16.4	3.2
Ash (%)	3.4	3.9	3.5	3.7	3.5
Dietary fibre (%)	33.7	41.4	36.0	34.0	29.4
Non-fibre carbohydrate (%)	15.2	13.7	15.6	12.8	67.5
Energy (kJ)	1634.0	1310.0	1485.0	1563.0	1384.0

Values are the means of triplicate analyses. * Estimated measurement of uncertainty (MU) was calculated at 95% confidence interval. *A. cowleana*: *Acacia cowleana*.

3.2. Fatty Acid Composition

The fatty acid composition of the four species of wattle seed are presented in Table 2. The unsaturated fatty acids (UFA) constitute the bulk of the fatty acids, as in the case of certain edible legumes, such as peanut [25]. Palmitic acid (17.69–23.83%) was found to be the predominant saturated fatty acid (SFA) in all the seeds, followed by stearic acid (4.15–10.12%) with *A. coriacea* and *A. retinodes* having significantly ($p < 0.05$) higher values in the two fatty acids, respectively. Monounsaturated fatty acids (MUFA) contributed the larger percentage of the total UFA with values more than twice that of polyunsaturated fatty acids (PUFA) in *A. coriacea*, *A. sophorae* and *A. retinodes*, unlike in other legumes and wattle seed species [3,26]. The C18:1 was found to be the most abundant of all the MUFAs with similar values (50.82–57.57%) obtained in all the species except *A. cowleana* (36.24%) ($p < 0.05$). However, greatest ($p < 0.05$) amount of PUFA particularly linoleic acid was found in *A. cowleana* (34.77%) as compared to other species, with *A. sophorae* having significantly ($p < 0.05$) lowest PUFA value (7.17%). These four species have lower linoleic acids as compared to that of other legumes such as broad beans (48.3%), chickpeas (56.7%) and small lentils (52.3%) [26], as well as other WS species such as *A. saligna* [27] and *A. tortilis* [21]. Oils rich in linoleic acid can have an important role in lowering the levels of blood cholesterol [28], thus the oil obtained from *A. cowleana* is highly nutritious when consumed regularly as a part of diet. Moreover, only trace amount of *n*-3 fatty acids was identified in all the seeds. Overall, the high degree of unsaturation found in all the four WS species agrees with those of common vegetable oils indicating that WS composite flour can be a healthy option for human consumption.

Table 2. Fatty acid profile of four different species of wattle seeds expressed as percentage (± SD) of the total fatty acid profile as determined by FAME GC-MS analysis.

Fatty acid (%)	*A. cowleana*	*A. coriacea*	*A. sophorae*	*A. retinodes*
Lauric acid (C12:0)	0^{b}	0^{b}	0.56 ± 0.03^{a}	0^{b}
Myristic acid (14:0)	0^{b}	0^{b}	1.33 ± 0.07^{a}	0^{b}
Palmitic acid (C16:0)	21.3 ± 0.35^{b}	23.8 ± 0.35^{a}	21.6 ± 0.77^{b}	17.7 ± 0.11^{c}
Stearic acid (C18:0)	4.15 ± 0.05^{c}	4.36 ± 0.17^{c}	7.51 ± 0.31^{b}	10.1 ± 0.21^{a}
Arachidonic acid (C20:0)	0.99 ± 0.05^{a}	1.07 ± 0.10^{a}	0.86 ± 0.03^{b}	0.95 ± 0.06^{ab}
Behenic acid (C22:0)	2.49 ± 0.07^{a}	1.80 ± 0.13^{b}	0.34 ± 0.03^{c}	1.02 ± 0.04^{b}
Lignoceric acid (C24:0)	0^{b}	0^{b}	0.16 ± 0.01^{a}	0^{b}
Total SFA	28.9 ± 0.52^{c}	31.0 ± 0.75^{ab}	32.4 ± 1.25^{a}	29.8 ± 0.42^{bc}
Palmitic acid (C16:1)	3.40 ± 0.03^{b}	0.45 ± 0.01^{c}	3.84 ± 0.27^{a}	0.46 ± 0.03^{c}
Oleic acid (C18:1)	32.8 ± 0.43^{c}	50.8 ± 0.26^{b}	57.6 ± 0.49^{a}	50.1 ± 1.34^{b}
Ecosanoic acid (C20:1)	0^{c}	0^{c}	0.37 ± 0.01^{b}	1.57 ± 0.02^{a}
Erucic acid (C22:1)	0^{b}	0^{b}	0^{b}	0.52 ± 0.01^{a}
Total MUFA	36.2 ± 0.46^{c}	51.3 ± 0.27^{b}	61.8 ± 0.77^{a}	52.7 ± 1.4^{b}
Linoleic acid (C18:2)	34.3 ± 0.08^{a}	17.4 ± 0.33^{b}	6.76 ± 0.75^{c}	16.0 ± 1.64^{b}
Linolenic acid (C18:3)	0.49 ± 0.02^{b}	0.27 ± 0.04^{c}	0.41 ± 0.04^{b}	1.61 ± 0.08^{a}
Total PUFA	34.8 ± 0.10^{a}	17.7 ± 0.73^{b}	7.17 ± 0.79^{c}	17.6 ± 1.72^{b}

SFA: Saturated fatty acids; MUFA: Monounsaturated fatty acids; PUFA: polyunsaturated fatty acids. Means not sharing the same superscript's letters (a, b, c, d) in a row are significantly different at $p < 0.05$ as assessed by Duncan's multiple range tests.

3.3. Amino Acid Composition

The essential amino acid analysis of four different species of WS showed all four of them contained about 13–15% of glutamic acid, followed by about 9–11.5% of aspartic acid, 6.8–7.4% of lysine, 8.0–8.6% of leucine and about 7% of arginine, serine and alanine (Table 3). Moreover, methionine is the limiting amino acids found in all four seeds and this agrees with that reported in other *Acacia* seed species [21,22]. This is the first time that a complete detailed analysis of all essential amino acids of these four different species of wattle seeds have been investigated. Due to high amounts of protein in all these species, it is important to look at different types of amino acids available in these seeds. It is observed that the glutamic acid content is slightly lower than that of about 18% in mung bean flour [29]. The amount of lysine in all four seeds is comparable to that of soybean [30]. The amount of arginine is very similar to that of winged bean and soy beans [31]. Presence of high amount of arginine potentially can help to increase physiological pool of L-arginine and may have positive impact on cardiovascular health [32]. Comparing the amino acid profile of the seed proteins with FAO reference pattern [33] can be used to justify the potential food value of the proteins. The amino acid profile of the four WS species showed that histidine, lysine, valine, isoleucine and leucine had higher levels than those listed in FAO/WHO reference pattern.

Table 3. Amino acid (g/100 g dry weight) profile of four different species of wattle seeds.

Amino acid	*A. cowleana*	*A. coriacea*	*A. sophorae*	*A. retinodes*	FAO/WHO References [32]
			Essential amino acids		
Histidine	2.8	2.6	4.3	2.4	1.9
Threonine	4.1	4.0	4.2	4.3	3.4
Lysine	7.0	6.8	7.4	7.0	5.8
Tyrosine	2.0	2.1	2.1	2.2	6.3
Methionine	0.4	0.4	0.3	0.3	
Valine	5.9	5.8	6.0	6.2	3.5
Isoleucine	4.4	4.2	4.4	4.3	2.8
Leucine	8.4	8.0	8.6	8.5	6.6
Phenylalanine	3.6	3.3	3.5	3.5	
			Non-essential amino acids		
Serine	7.0	7.7	7.1	7.1	
Arginine	6.0	7.1	6.0	6.3	
Glycine	9.7	10.6	9.5	10	
Aspartic acid	10.5	11.5	9.4	10.7	
Glutamic acid	15.3	13.3	14.2	14.2	
Alanine	7.3	6.7	7.1	7.0	
Proline	5.7	5.8	6.0	6.0	

Results are expressed as the mean of duplicate experiments.

3.4. Mineral Composition

Table 4 outlines a detailed mineral analysis performed on four different species of Australian WS. As highlighted in the table, all four species are good sources of important essential minerals such as iron, potassium, magnesium, calcium and zinc. Potassium is the most abundant element in all the WS species with *A. coriacea* having significantly ($p < 0.05$) highest value (11,000 mg/kg dry weight (DW)). The results obtained agrees with that reported for different WS species [8,21,22]. A significantly ($p < 0.05$) higher iron content (195.0 mg/kg DW) was found in *A. sophorae* as compared to other species of WS, showing it is a good source of iron considering the Australian recommended dietary allowance (RDA) of iron is 7 and 12–16 mg/day for men and women during pregnancy, respectively [34,35]. Moreover, heavy metals, such as Hg, are less than 0.005 mg in all four WS species, indicating the use of all these four species of WS is safe for human consumption. As stated in the Food Standards Australia New Zealand (FSANZ), Standard 1.4.1 for contaminants and Natural Toxicants, the maximum level of Pb in legumes is set at 0.2 mg/kg and cadmium in rice is 0.1 mg/kg. The amount of Pb in all four WS species is within this range and in case of *A. sophorae* and *A. cowleana*, they have significantly ($p < 0.05$) higher values (0.1 and 0.175 mg/kg DW, respectively) as compared to other species, but the values still complied with the food standards code [36]. Based on the above results, the WS are a good source for minerals and thus can be incorporated into foods, such as commercial baked products, that are deficient in minerals to enhance their nutritional properties. Further studies should be carried out to investigate the bioavailability of the essentials minerals, particularly potassium in the WS species.

Table 4. Mineral analysis of four different species of wattle seeds.

Minerals (mg/kg DW)	*A. cowleana*	*A. coriacea*	*A. sophorae*	*A. retinodes*
Major				
Ca	2300 ± 0.0 [c]	4300 ± 141.4 [a]	2600 ± 424.3 [c]	3150 ± 212.1 [b]
K	8700 ± 0.0 [b]	11000 ± 0.0 [a]	7300 ± 141.4 [c]	9050 ± 495.0 [b]
Mg	1700 ± 0.0 [b]	2350 ± 70.7 [a]	1700 ± 0.0 [b]	2400 ± 141.4 [a]
Na	<20 [c]	<20 [c]	1100 ± 0.0 [a]	940 ± 70.7 [b]
P	1700 ± 0.0 [b]	2350 ± 70.7 [a]	2300 ± 0.0 [a]	2300 ± 141.4 [a]
Trace				
Co	0.1 ± 0.0 [c]	0.04 ± 0.0 [b]	0.365 ± 0.0 [a]	0.095 ± 0.0 [c]
Cr	2.1 ± 0.2 [a]	0.8 ± 0.1 [c]	1.95 ± 0.2 [a]	1.15 ± 0.2 [b]
Cu	4.8 ± 0.1 [c]	5.0 ± 0.0 [c]	8.65 ± 0.2 [a]	7.2 ± 0.1 [b]
Fe	75.0 ± 0.0 [b]	50.5 ± 0.7 [b]	195.0 ± 35.4 [a]	49.5 ± 0.7 [b]
Mn	14.0 ± 0.0 [c]	14.0 ± 0.0 [c]	46.5 ± 2.1 [b]	130.0 ± 0.0 [a]
Mo	0.9 ± 0.2 [c]	0.84 ± 0.1 [c]	1.75 ± 0.1 [b]	2.0 ± 0.0 [a]
Se	0.9 ± 0.0 [a]	0.68 ± 0.0 [b]	0.165 ± 0.0 [c]	0.34 ± 0.0 [d]
Zn	24.5 ± 0.7 [b]	23.0 ± 0.0 [c]	21.0 ± 0.0 [d]	34.0 ± 0.0 [a]
Other minerals				
Al	42.0 ± 1.4 [b]	16.5 ± 2.1 [c]	77.5 ± 3.5 [a]	3.75 ± 0.4 [d]
As	ND	<0.005 [b]	0.87 ± 0.2 [a]	<0.005 [b]
Ba	3.2 [c]	49.0 ± 1.4 [a]	1.0 ± 0.0 [d]	7.15 ± 0.1 [b]
Cd	<0.005 [c]	<0.005 [c]	0.054 ± 0.0 [a]	0.011 ± 0.0 [b]
Hg	<0.005 [a]	<0.005 [a]	<0.005 [a]	<0.005 [a]
Ni	2.7 ± 0.1 [c]	1.1 ± 0.0 [d]	3.8 ± 0.0 [a]	3.05 ± 0.1 [b]
Pb	0.1 ± 0.0 [ab]	0.074 ± 0.0 [b]	0.175 ± 0.1 [a]	0.0085 ± 0.0 [b]
Sb	ND	0.01 ± 0.0 [a]	<0.01 [a]	<0.01 [a]
Sn	<0.05 [b]	0.85 ± 0.1 [a]	<0.05 [b]	<0.05 [b]
Sr	19.0 ± 0.0 [b]	24.0 ± 1.4 [a]	5.9 ± 0.1 [d]	13.0 ± 0.0 [c]
V	0.1 ± 0.0 [b]	0.035 ± 0.0 [b]	1.65 ± 0.4[a]	0.01 ± 0.0[b]

ND: not detected. Means not sharing the same superscript's letters (a, b, c, d) in a row are significantly different at $p < 0.05$ as assessed by Duncan's multiple range tests.

3.5. Sugar Profile

The results in Table 5 show that the monosaccharides are the major kind of sugar present in all the four WS species. Pentose sugars (xylose/arabinose) were the predominant monosaccharides found in the seeds and these were followed by galactose and glucose, while a small amount of mannose and fucose were present. Among the species, *A. sophorae* has a significantly ($p < 0.05$) higher amount of pentose sugars (84.2%), while significantly ($p < 0.05$) higher glucose and galactose contents were found in both *A. sophorae* and *A. retinodes*. Moreover, the seeds are rich in galacturonic acid but contain lesser amounts of rhamnose. All these sugars were present in substantial amount in all the species, particularly in *A. retinodes*, *A. sophorae* and *A. coriacea* as compared to *A. cowleana*. To the best of our knowledge, this study was the first to report the sugar composition present in wattle seeds. Overall, the results revealed that these wattle seed species have high amount of reducing sugar and this may be due to action of endogenous enzymes involved in hydrolysing the stored carbohydrate during maturation and storage.

Table 5. Sugar profile of four different species of wattle seeds.

Sugar Composition (%)	*A. cowleana*	*A. coriacea*	*A. sophorae*	*A. retinodes*
Glucose	22.1 ± 1.30 [c]	37.8 ± 1.01 [b]	44.4 ± 1.60 [a]	43.9 ± 2.16 [a]
Mannose	4.17 ± 0.39 [b]	4.05 ± 0.31 [b]	4.47 ± 0.21 [b]	5.48 ± 0.57 [a]
Galactose	34.2 ± 1.30 [c]	57.6 ± 2.23 [b]	65.5 ± 3.35 [a]	67.3 ± 2.91 [a]
Galacturonic acid	30.3 ± 1.28 [c]	51.7 ± 3.68 [a]	44.1 ± 0.55 [b]	51.3 ± 1.50 [a]
Rhamnose	4.98 ± 0.32 [c]	7.90 ± 0.53 [a]	5.40 ± 0.32 [c]	5.82 ± 0.19 [b]
Fucose	2.77 ± 0.29 [a]	2.43 ± 0.35 [a]	2.63 ± 0.28 [a]	2.75 ± 0.38 [a]
Xylose/Arabinose	79.7 ± 3.43 [b]	71.5 ± 4.84 [c]	84.2 ± 2.24 [a]	78.4 ± 1.08 [b]

Means not sharing the same superscript's letters (a, b, c, d) in a row are significantly different at $p < 0.05$ as assessed by Duncan's multiple range tests.

3.6. SDS-PAGE Profile of Soluble Proteins Extracted from Raw and Roasted Wattle Seed Species

Figure 2 shows the SDS-PAGE gel electrophoretograms of extracts from four species of raw and roasted WS obtained under non-reducing and reducing conditions. For raw WS extracts in the four species, similar protein bands appeared under reducing and non-reducing conditions but protein molecules smaller than 18 kDa were broken down under reducing condition probably due to reduction in disulphide bonds. Moreover, more protein bands were found in raw *A. sophorae* (lanes 10 and 11) followed by *A. coriacea* (lanes 6 and 7) and *A. retinodes* (lanes 16 and 17) while *A. cowleana* (lanes 1 and 2) showed faint protein bands. Overall, the results suggest that water soluble proteins from raw WS species were mainly polypeptides with less than 80 kDa molecular weights for *A. sophorae* and *A. retinodes* and 55 kDa for *A. coriacea*. This differs from previous study on different wattle seed species where the molecular weight of the protein was found to be lower than 66 kDa [13,27]. After roasting, no protein band was visible in the four wattle seed species with the exception of only two faint protein bands observed in *A. coriacea* (lanes 8 and 9) at 55 kDa. The results showed that the heat treatment had caused the soluble proteins in WS to be degraded into fragments with molecular weight smaller than 10 kDa. A similar effect of thermal treatment on protein had been reported in previous studies on wattle seeds [27] and peanuts [37].

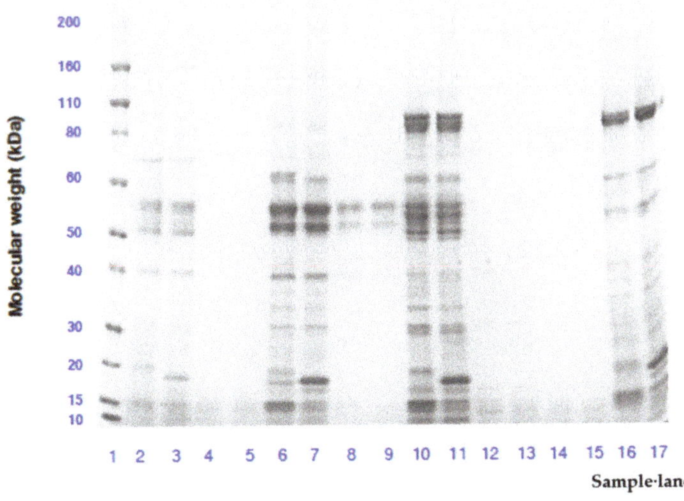

Figure 2. Molecular weight analysis by SDS-PAGE of protein extracts from raw and roasted wattle seeds species. Lanes 1: Novex Sharp Pre-stained molecular weight ladder; Lanes 2, 3, 4 and 5: Raw and reducing, raw and non-reducing, roasted and reducing, roasted and non-reducing *A. cowleana*, respectively; Lanes 6, 7, 8 and 9: Raw and reducing, raw and non-reducing, roasted and reducing,

roasted and non-reducing *A. coriacea*, respectively; Lanes 10, 11, 12 and 13: Raw and reducing, raw and non-reducing, roasted and reducing, roasted and non-reducing *A. sophorae*, respectively; Lanes 14, 15, 16 and 17: Roasted and reducing, roasted and non-reducing, raw and reducing, raw and non-reducing *A. retinodes*, respectively.

3.7. Rapid Sensory Profiling

The results of the rapid sensory profiling presented in Table 6 showed a wide diversity of sensory characteristics for the four WS species. The summary of the descriptors provided by the trained panel showed that, while all four species had savoury notes, certain descriptors characterised each species. *A. cowleana*, for instance, was perceived as a complex cultivar where the broth, popcorn and savoury notes where more intense for aroma and flavour. *A. coriacea* was distinguished by some confectionary and tropical fruit notes. *A. sophorae* was characterised by more savoury and grassy notes but not very intense. Cultivar *A. retinodes* was characterised by gravy, lemony earthy and nutty notes. Previous preliminary studies in *A. victoriae* have shown that this wattle seed cultivar has a savoury note as the four WS species in this study but also its own nutty and wheat biscuit notes [38]. The informal sensory component of this study has shown diversity in sensory profiles between WS seed species, opening the door for different applications of WS species in the food industry. Future work could investigate more rigorously the sensory differences between these four species and other species of the WS, as well as investigate effects of roasting and grinding on flavour.

Table 6. Rapid sensory profiling of four different species of wattle seeds.

Characteristics	*A. retinodes*	*A. cowleana*	*A. coriacea*	*A. sophorae*
Aroma ground powder	Mushroom, gravy, onion, coffee, chocolate, roasted nuts, lemon	Spicy, curried, savoury, popcorn, burnt toast, savoury jam, chemical fish stock.	Savoury, spinach, chocolate, sweet, fairy floss, tropical fruit, water of boiled vegetables	Yeasty (vegemite), meaty, burnt tyres, roasted onion
Aroma ground powder in semolina	Earthy, grassy, musty, pepper, oily peanut butter	Smoky, broth, lentil, turmeric, chemical	Cinnamon, dairy	Buttery, popcorn, big change from peppery, meaty to earthy.
Flavour	Bitter, nutty, leaves, earthy, peanut, sesame paste	Intense, bitter, burnt, peppery, chives, onion salt, roasted chicken stuffing, dried legumes.	Fruity, almost coffee, tahini, mashed vegetables, cereal, not very intense	Vegemite, nutty, savoury, slight grassy, onion
After taste	Grass, barnyard, smoky, savoury, cucumber.	Bitter, pepper, very lingering	Spinach, not very lingering	Mild, grassy, bitter, savoury

4. Conclusions

This study shows that the four species of wattle seeds are good source of protein and dietary fibre. In addition, the seeds can be considered as a potential dietary source of major and minor minerals, essential amino acids and oleic acids, thus, the seeds have promising nutritional profiles. Nevertheless, further studies should be carried out to investigate the impact of processing on these nutrients and also to evaluate the anti-nutritional components present in these species before and after processing.

Author Contributions: K.J.S., B.R., U.T., S.M.O.M., O.Q.A. performed the experiments, collected, analysed and interpreted the data and drafted the manuscript. Y.S., O.Q.A., H.E.S., S.H. and V.S. conceived and designed the experiments, checked and approved the results and critically revised the manuscript. All authors read and approved the final version of the manuscript.

Funding: This project was funded by the University of Queensland, Australia, The Australian Research Council Transformational Training Centre for Uniquely Australian Foods Grant number IC180100045 and the Innovation Connections Grant number RC52539.

Acknowledgments: The authors acknowledge the Traditional Owners of the lands on which the wattle seed species were harvested, and respect the knowledge and experience the Traditional Owners hold regarding the care, harvest and use of these plants. We thank NATIF Australian Native Superfoods, Fruits Herbs Spices and Mixes, Victoria, Australia and Valley Seeds Pty Ltd., Victoria, Australia for supplying the wattle seeds. The authors thank the Australian Government via the Innovation Connections Grant Scheme for funding this research.

Conflicts of Interest: The authors declare no conflict of interest.

References

1. Karhagomba, I.B.; Mirindi, T.A.; Mushagalusa, T.B.; Nabino, V.B.; Koh, K.; Kim, H.S. The cultivation of wild food and medicinal plants for improving community livelihood: The case of the Buhozi site, DR Congo. *Nutr. Res. Pract.* **2013**, *7*, 510–518. [CrossRef] [PubMed]
2. Falade, M.S.; Owoyomi, O.; Harwood, C.E.; Adewusi, S.R.A. Chemical composition and starch hydrolysis of *Acacia colei* and *Acacia tumida* seeds. *Cereal Chem.* **2005**, *82*, 479–484. [CrossRef]
3. Ee, K.Y.; Yates, P. Nutritional and antinutritional evaluation of raw and processed Australian wattle (*Acacia saligna*) seeds. *Food Chem.* **2013**, *138*, 762–769. [CrossRef]
4. El-Adawy, T.A.; Khalil, A.H. Characteristics of roselle seeds as a new source of protein and lipid. *J. Agric. Food Chem.* **1994**, *42*, 1896–1900. [CrossRef]
5. Embaby, H.E.S.; Mokhtar, S.M. Chemical composition and nutritive value of lantana and sweet pepper seeds and nabak seed kernels. *J. Food Sci.* **2011**, *76*, 736–741. [CrossRef] [PubMed]
6. Ahmed, A.K.; Johnson, K.A. Horticultural development of Australian native edible plants. *Aus. J. Bot.* **2000**, *48*, 417–426. [CrossRef]
7. Seigler, D.S. Economic potential of Western Australian Acacia species: Secondary plant products. *Conserv. Sci. West. Aust.* **2002**, *4*, 109–116.
8. Adewusi, S.R.A.; Falade, O.S.; Harwood, C. Chemical composition of *Acacia colei* and *Acacia tumida* seeds—Potential food sources in the semi-arid tropics. *Food Chem.* **2003**, *80*, 187–195. [CrossRef]
9. Rinaudo, A.; Cunningham, P. Australian Acacias as multi-purpose agro-forestry species for semi-arid regions of Africa. *Muelleria* **2008**, *26*, 79–85.
10. Maslin, R.; McDonald, M.W. *AcaciaSearch: Evaluation of Acacia as a Woody Crop Option for Southern Australia: Canberra;* Rural Industries Research and Development Corporation: Macquaire Street, Barton, ACT, Australia, 2004.
11. Zhao, J.; Agboola, S. *A Report for the Rural Industries Research and Development Corporation: Functional Properties of AUSTRALIAN Bush Food;* Canberra, Rural Industries Research and Development Corporation: Canberra, Australia, 2007.
12. Maslin, B.R.; Thomson, L.A.J.; McDonald, M.W.; Hamilton-Brown, S. *Edible Wattle Seeds of Southern Australia: A Review of Species for Use in Semi-Arid Regions;* CSIRO Publishing: Collingwood, Australia, 1998.
13. Agboola, S.; Ee, K.Y.; Mallon, L.; Zhao, J. Isolation, characterization, and emulsifying properties of wattle seed (*Acacia victoriae* Bentham) extracts. *J. Agric. Food Chem.* **2007**, *55*, 5858–5863. [CrossRef]
14. Ee, K.Y.; Zhao, J.; Rehman, A.; Agboola, S. Characterisation of trypsin and α-chymotrypsin inhibitors in Australian wattle seed (*Acacia victoriae* Bentham). *Food Chem.* **2008**, *107*, 337–343. [CrossRef]
15. Agrawal, N.K.; Gupta, U. Evaluation of hypoglycemic and antihyperglycemic effects of *Acacia tortilis* seed extract in normal and diabetic rats. *Int. J. Pharm. Technol. Res.* **2013**, *5*, 330–336.
16. ILDIS. International Legume Database & Information Service: LegumeWeb. Available online: https://ildis.org/cgi-bin/Araneus.pl?version~{}10.01&LegumeWeb&tno~{}5839&genus~{}Acacia&species~{}retinodes (accessed on 16 June 2018).
17. AOAC. *Official Methods of Analysis;* Association of Official Analytical Chemists: Washington, DC, USA, 2000.
18. Rühmann, B.; Schmid, J.; Sieber, V. Fast carbohydrate analysis via liquid chromatography coupled with ultra violet and electrospray ionization ion trap detection in 96-well format. *J. Chromatogr. A* **2014**, *1350*, 44–50. [CrossRef] [PubMed]
19. Carter, J.F.; Tinggi, U.; Yang, X.; Fry, B. Stable isotope and trace metal compositions of Australian prawns as a guide to authenticity and wholesomeness. *Food Chem.* **2015**, *170*, 241–248. [CrossRef] [PubMed]

20. Miller, J.B.; James, K.W.; Maggiore, P.M.A. *Tables of Composition of Australian Aboriginal Foods*; Aboriginal Studies Press: Canberra, ACT, Australia, 1993.
21. Embaby, H.E.; Rayan, A.M. Chemical composition and nutritional evaluation of the seeds of *Acacia tortilis* (Forssk.) Hayne ssp. *raddiana*. *Food Chem.* **2016**, *200*, 62–68. [CrossRef]
22. Siddhuraju, P.; Vijayakumari, K.; Janardhanan, K. Chemical composition and nutritional evaluation of an underexploited legume, *Acacia nilotica* (L.). *Del. Food Chem.* **1996**, *57*, 385–391. [CrossRef]
23. Thorburn, A.W.; Brand, J.C.; Cherikoff, V.; Truswell, A.S. Lower postprandial plasma glucose and insulin after addition of *Acacia coriacea* flour to wheat bread. *Aust. N. Z. J. Med.* **1987**, *17*, 24–26. [CrossRef]
24. Thorburn, A.W.; Brand, J.C.; Truswell, A.S. Slowly digested and absorbed carbohydrate in traditional bushfoods: A protective factor against diabetes? *Am. J. Clin. Nutr.* **1987**, *45*, 98–106. [CrossRef]
25. Mukhtar, A. Oil and fatty acid composition of peanut species grown in Pakistan. *Pak. J. Bot.* **2012**, *44*, 627–630.
26. Kalogeropoulos, N.; Chiou, A.; Ioannou, M.; Karathanos, V.T.; Hassapidou, M.; Andrikopoulos, N.K. Nutritional evaluation and bioactive microconstituents (phytosterols, tocopherols, polyphenols, triterpenic acids) in cooked dry legumes usually consumed in the Mediterranean countries. *Food Chem.* **2010**, *121*, 682–690. [CrossRef]
27. Ee, K.Y.; Zhao, J.; Rehman, A.U.; Agboola, S. Effects of roasting on the characteristics of Australian wattle (*Acacia victoriae* Bentham) seed and extracts. *Int. J. Food Prop.* **2013**, *16*, 1135–1147. [CrossRef]
28. El Adawy, T.A.; Taha, K.M. Characteristics and composition of different seed oils and flours. *Food Chem.* **2001**, *74*, 47–54. [CrossRef]
29. Coffmann, C.W.; Garciaj, V.V. Functional properties and amino acid content of a protein isolate from mung bean flour. *Int. J. Food Sci. Technol.* **1977**, *12*, 473–484. [CrossRef]
30. Deshpande, S.S. Food legumes in human nutrition: A personal perspective. *Crit. Rev. Food Sci. Nutr.* **1992**, *32*, 333–363. [CrossRef] [PubMed]
31. Ekpenyong, T.E.; Borchers, R.L. Amino acid profile of the seed and other parts of the winged bean. *Food Chem.* **1982**, *9*, 175–182. [CrossRef]
32. Chivandi, E.; Davidson, B.C.; Erlwanger, K.H. Proximate, mineral, fibre, phytate-phosphate, vitamin E, amino acid and fatty acid composition of *Terminalia sericea*. *S. Afr. J. Bot.* **2013**, *88*, 96–100. [CrossRef]
33. FAO/WHO. Protein quality evaluation. In *Report of a Joint FAO/WHO Expert Consultation*; Viale delle Terme di Caracalla: Rome, Italy, 1990.
34. Lehrfeld, J.; Wu, Y.V. Distribution of phytic acid in milled fractions of scout-66 hard red winter-wheat. *J. Agric. Food Chem.* **1991**, *39*, 1820–1824. [CrossRef]
35. National Health and Medical Research Council (NHMRC). *Recommended Dietary Intakes for Use in Australia*; Australian Government Publishing Service: Canberra, Australia, 1991.
36. Arise, A.K.; Alashi, A.M.; Nwachukwu, I.D.; Ijabadeniyi, O.A.; Aluko, R.E.; Amonsou, E.O. Antioxidant activities of bambara groundnut (*Vigna subterranea*) protein hydrolysates and their membrane ultrafiltration fractions. *Food Funct.* **2016**, *7*, 2431–2437. [CrossRef]
37. Ejigui, J.; Savoie, L.; Marin, J.; Desrosiers, T. Influence of traditional processing methods on the nutritional composition and antinutritional factors of red peanuts (*Arachis hypogyea*) and small red kidney beans (*Phaseolus vulgaris*). *J. Biol. Sci.* **2005**, *5*, 597–605. [CrossRef]
38. Smyth, H.E.; Sanderson, J.E.; Sultanbawa, Y. Lexicon for the sensory description of Australian native plant foods and ingredients. *J. Sens. Stud.* **2012**, *27*, 471–481. [CrossRef]

Article

Technofunctional and Sensory Properties of Fermented Lupin Protein Isolates

Katharina Schlegel [1,2], Anika Leidigkeit [2], Peter Eisner [2,3] and Ute Schweiggert-Weisz [2,*]

1 Chair of Aroma and Smell Research, Department of Chemistry and Pharmacy, Emil Fischer Center, Friedrich-Alexander-Universität Erlangen-Nürnberg, 91054 Erlangen, Germany; katharina.schlegel@ivv.fraunhofer.de

2 Department Food Process Development, Fraunhofer Institute for Process Engineering and Packaging (IVV), 85354 Freising, Germany; anika.leidigkeit@ivv.fraunhofer.de (A.L.); peter.eisner@ivv.fraunhofer.de (P.E.)

3 ZIEL-Institute for Food & Health, TUM School of Life Sciences Weihenstephan, Technical University of Munich, 85354 Freising, Germany

* Correspondence: ute.weisz@ivv.fraunhofer.de; Tel.: +49-8161-491-483

Received: 28 October 2019; Accepted: 11 December 2019; Published: 13 December 2019

Abstract: Lupin protein isolate was fermented with eight different microorganisms to evaluate the influence on sensory profile, techno-functional properties and protein integrity. All investigated microorganisms were able to grow in lupin protein isolate. The results showed that the foaming activity in the range of 1646–1703% and the emulsifying capacity in the range of 347–595 mL of the fermented lupin protein isolates were similar to those of the unfermented ones. Protein solubility at pH 4 showed no significant changes compared to unfermented lupin protein isolate, whereas the solubility at pH 7 decreased significantly from 63.59% for lupin protein isolate to solubilities lower than 42.35% for fermented lupin protein isolate. Fermentation with all microorganisms showed the tendency to decrease bitterness from 2.3 for lupin protein isolate (LPI) to 1.0–2.0 for the fermented ones. The most promising microorganisms for the improvement of the sensory properties of lupin protein isolates were *Lactobacillus brevis* as it reduced the intensity of characteristic aroma impression (*pea-like, green bell pepper-like*) from 4.5 to 1.0. The SDS-PAGE results showed the fermentation treatment appeared not to be sufficiently effective to destruct the protein integrity and thus, deplete the allergen potential of lupin proteins. Fermentation allows the development of food ingredients with good functional properties in foam formation and emulsifying capacity, with a well-balanced aroma and taste profile.

Keywords: fermentation; lupin; plant protein; aroma profile; techno-functional properties; lactobacteria; foam; SDS-PAGE; solubility; emulsifying capacity

---

1. Introduction

The human organism relies on the nutritional intake of protein to maintain its health. The supply of animal protein has drastic environmental impacts on land use, air and water quality, and greenhouse gases [1]. The partial replacement of animal proteins by plant proteins could be a promising strategy to reduce the environmental impact of nutrition [2]. Soy protein is currently one of the most common plant proteins, but has some disadvantages including deforestation and genetic modification. Therefore, the search for alternative high-quality plant protein sources is a persistent challenge.

Lupins of the family *Fabaceae* are widely cultivated in Europe and are rich in seed proteins with valuable functional properties and a well-balanced amino acid profile [3,4]. However, like other leguminous plant proteins, lupin protein preparations exhibit a distinct aroma profile. Schlegel et al. [5] described the aroma profile of lupin protein isolate with oatmeal-like and fatty, cardboard-like impressions followed by earthy, moldy, beetroot-like; grassy; metallic; cooked potato-like and pea-like impressions. In addition, taste perceptions of bitter, salty, and astringent were identified in the taste pattern [5]. This characteristic aroma profile makes lupin proteins less suitable for some food applications as, for example, plain non-dairy based products.

The lactic fermentation of legume proteins might be a promising approach to influence the sensory profile of these ingredients as various studies have shown that lactic fermentation of plant proteins results in a reduction or masking of off-flavors in legumes [6,7]. The influence of the fermentation on the aroma profile of the plant proteins strongly depends on the microorganisms used [6–8]. For the development of ingredients with specific aroma and taste properties (e.g., plant-based dairy alternatives or meat substitutes), different microorganisms must be investigated for their ability to improve the aroma profile of lupin proteins.

Besides the sensory profile, appropriate techno-functional properties of the proteins, such as solubility, emulsifying capacity and foam properties, are also a prerequisite for their use as food ingredients. Several authors have shown that the lactic fermentation of plant proteins affected their techno-functional properties in particular the protein solubility as shown by Lampart-Szczapa et al. [9] and Meinlschmidt et al. [10]. *Lactobacilli* produce metabolites such as organic acids, which alter the conformation of proteins and thus their physicochemical properties as relevant factors for their techno-functional behavior.

In addition, several studies have shown that the fermentation of plant proteins can reduce their allergenic potential [10,11] due to proteolytic enzyme activities. In particular for soy proteins, the impact of fermentation on the reduction of the allergenic potential has been investigated extensively [10–14]. It could be shown that the molecular weight distribution of the soy proteins could give a first indication of the degradation of the main allergens [10]. Data for lupins are still missing.

The objective of this study was to investigate the impact of lactic fermentation on the changes of the sensory profile of lupin protein isolates to improve the sensory properties of this promising food ingredient. Furthermore, the functional properties of the fermented isolates were also examined. To get a first indication of the reduction of the allergenic potential of the lupin proteins, the molecular weight distribution of the individually fermented protein samples were compared to the unfermented isolates.

2. Materials and Methods

2.1. Lupin Seeds

Lupin (*Lupinus angustifolius* cultivar Boregine) seeds were purchased from Saatzucht Steinach GmbH & Co KG (Steinach, Germany).

2.2. Preparation of Lupin Protein Isolate

Lupin protein isolate (LPI) was prepared from *Lupinus angustifolius* L. cultivar Boregine. Seeds were dehulled with an under runner disc sheller and separated and classified using an air-lift system. The dehulled seeds were passed through a counter-rotating roller mill and the resulting flakes were de-oiled in *n*-hexane. The processed flakes were suspended in 0.5 M HCL (pH 4.5) at a 1:8 (*w/w*) ratio and stirred for 1 h. After this, flakes were recovered in a decanter centrifuge with 5600× *g* at 4 °C for 1 h. Pre-extracted flakes were dispersed in 0.5 M NaOH (pH 8) at a 1:8 ratio for 1 h. After extraction, the suspension was centrifuged at 5600× *g* at 20 °C for 1 h and then neutralized (0.5 M NaOH), pasteurized (70 °C, 10 min), and spray dried.

2.3. Microbiological Compositions

2.3.1. Nutrient Media

Liquid growth media and agar were purchased from Carl Roth (Karlsruhe, Germany) and Merck KGaA (Darmstadt, Germany).

2.3.2. Bacteria Strains and Culture Conditions

Microorganisms were purchased from Deutsche Sammlung von Mikroorganismen und Zellkulturen (Braunschweig, Germany), Lehrstuhl für Technische Mikrobiologie, Technische Universität München (Freising, Germany) and Prof. Werner Back (Table 1). The microorganisms were stored as a cryo-culture in our strain collection and were recovered on MRS (De Man, Rogosa, & Sharpe) and TSYE (Trypticase soy yeast extract) agar. The selection of the microorganisms based on the results of previous experiments in which 26 microorganisms were tested for growth in LPI (data not shown). The microorganisms investigated in this study were the eight most promising microorganisms and were further processed.

Table 1. Bacteria Strains.

Bacteria Strain	
Lactobacillus reuteri	DSM 20016
Lactobacillus brevis	TMW 1.1326
Lactobacillus amylolyticus	TL 5
Lactobacillus parabuchneri	DSM 5987
Lactobacillus sakei subsp. *carnosus*	DSM 15831
Staphylococcus xylosus	DSM 20266
Lactobacillus helveticus	DSM 20075
Lactobacillus delbrueckii	DSM 20081

Liquid cultures were incubated at aerobic (*Lactobacillus reuteri, Lactobacillus brevis, Lactobacillus amylolyticus, Lactobacillus sakei* subsp. *carnosus, Staphylococcus xylosus, Lactobacillus parabuchneri*) and anaerobic (*Lactobacillus helveticus, Lactobacillus delbrueckii*) conditions in sealed tubes (Sarstedt AG & Co, Nümbrecht, Germany) without shaking. Liquid pre-cultures (15 mL) were prepared from single colonies on agar plates and incubated for 36–48 h at 30 °C (*L. parabuchneri, L. brevis*), 37 °C (*L. helveticus, L. delbrueckii, L. sakei* subsp. *carnosus, L. reuteri, S. xylosus*) and 42 °C (*L. amylolyticus*), respectively.

2.3.3. Determining Growth Conditions of Microorganisms

Growth curves were recorded with a microplate reader (Infinite M1000 Pro, Tecan Group Ltd., Männedorf, Switzerland) measuring with an OD of 600 nm each 15 min with previous shaking at various temperatures. Aliquots of 200 µL liquid medium in 96 well micro test plates were inoculated from pre-cultures. For anaerobic conditions, cultures were covered with 100 µL sterile paraffin oil.

2.3.4. Determination of pH and Viable Cell Counts

The pH course during lupin fermentation was recorded for 24 h with one measurement point each 30 min with a wtw pH 3310 pH electrode (Xylem Analytics Germany GmbH, Weilheim, Germany). The viable cell counts were determined on nutrients agar from 1 mL of diluted (0.9% NaOH) culture aliquots and expressed as $\log_{10}$ of colony forming units per milliliter per sample (CFU/mL).

2.4. *Fermentation of Lupin Protein Isolates*

Fermentation of LPI was carried out in a 5 L glass reaction vessel in an incubator under aerobic conditions and in a 5 L glass reaction vessel with a fermenter (Satorius) under anaerobic conditions, respectively. A 5% LPI (*w/w*) solution with 0.5% glucose (*w/w*) was pasteurized separately at 80 °C for 20 min and mixed under sterile conditions. LPI solution was inoculated with the activated culture in the late exponential growth phase (10^7 CFU/mL). Anaerobic conditions were achieved by flushing the reactor with N_2. The inoculated lupin protein isolate was incubated for 24 h without stirring and sampled at 0, 4, 18, and 24 h. Fermentation was stopped by heat treatment at 90 °C for 20 min. All samples were neutralized (pH 7) with 1 M NaOH and spray-dried with a Niro Atomizer 2238 (GEA, Düsseldorf, Germany). The whole experiment was repeated in duplicate.

2.5. *Analysis of* D*-Glucose*

For determination of D-glucose, 100 µL of the sample was mixed with 450 µL zinc sulphate (10%) and 450 µL NaOH (0.5 M) and incubated for 20 min at room temperature. After incubation, the sample was centrifuged at 12,045× *g* for 10 min and the supernatant was filtrated using a 0.45 µm nylon filter. D-Glucose was analyzed using the enzymatic D-glucose test from R-Biopharm (Darmstadt, Germany) following the manufacturer's instructions.

2.6. *Chemical Composition*

The protein content was estimated based on the nitrogen content according to the Dumas combustion method (AOAC 968.06) with a factor of N × 5.8 [15] using a Nitrogen Analyzer FP 528 (Leco Corporation, St. Joseph, MI, USA). The dry matter was identified according to AOAC methods 925.10 in a TGA 601 thermogravimetric system (Leco Corporation) at 105 °C.

2.7. *Sensory Analysis*

2.7.1. Panelists

The panel consisted of trained volunteers recruited from Fraunhofer IVV (Freising, Germany), exhibiting no known illness at the time of examination and with normal olfactory function. The panel consisted of 10 panelists (eight female and two male). All panelists were trained in weekly training sessions with selected, super-threshold aroma solutions to correctly identify and name fragrances.

2.7.2. Sensory Evaluation

For sample evaluation, 2% (*w/w*) solutions of the LPI and fermented LPI, respectively, and tap water were prepared by stirring. The samples were evaluated in two sessions on one day. Each panelist received five samples of 20 mL aliquots (room temperature) in the first session and four samples in the second session in covered glass vessels, with tap water and flavorless crackers used to neutralize between each sample. To obtain the retronasal aroma and taste attributes, panelists were required to open the lid of the beaker and record the retronasal aroma and taste attributes. After the discussion regarding the most prominent aroma and taste attributes, panelists were asked to select and rate attributes on a scale from 0 (no perception) to 10 (strong perception) that were perceived by at least half of the panel. In addition, the panelists were asked to rate the overall intensity of the aroma of the sample also on a scale from 0 (no perception) to 10 (strong perception), and a hedonic scaling was performed on a scale from 0 (strong dislike) to 5 (neutral) to 10 (strong like). We used supra-threshold aroma solutions for orthonasal perception as a reference for each selected aroma attribute.

2.8. *Techno-Functional Properties*

2.8.1. Protein Solubility

The solubility (%) of LPI and fermented LPI was measured in duplicate at pH 4 and 9 according to Morr et al. [16]. For each measurement, 1.5 g of protein was suspended in 50 mL 0.1 M NaCl and the pH was adjusted with 0.1 M NaOH or 0.1 M HCl. After stirring for 1 h at room temperature, the non-dissolved fractions of the samples were separated by centrifugation (20,000× *g*, 15 min, room temperature) and the supernatants were passed through Whatman No. 1 filter paper to remove any remaining particulates. The protein content of the supernatant was determined by Lowry et al. [17] using the DC Protein Assay (Bio-Rad Laboratories, Hercules, CA, USA). The absorbance was read at 750 nm and the protein concentration was calculated using the generated BSA standard curve. The resulting amount of dissolved proteins was related to the total amount of protein and the protein solubility (%) was determined.

2.8.2. Foam Properties

For the determination of the foaming activity, 50 mL of a 5% (*w/w*) protein solution (pH 7) was whipped at room temperature for 8 min in a Hobart 50-N device (Hobart GmbH, Offenburg, Germany) according to the method described by Phillips et al. [18]. The increase of foam volume was used to determine the foam activity. The percentage leftover of foam volume after 1 h was defined as the foaming stability (%).

2.8.3. Emulsifying Capacity

Emulsifying capacity (EC) was identified according to Wang and Johnson [19]. Samples were dispersed in deionized water (1%, *w/w*), adjusted to pH 7 and stirred with an Ultraturrax at 18 °C. Rapeseed oil was continuously added using a Titrino 702 SM titration system (Metrohm GmbH & Co. KG, Hertisau, Switzerland) at a rate of 10 mL/min until phase inversion was detected by means of an LF 521 meter fitted with a KLE1/T electrode (Wissenschaftlich-technische Werkstätten GmbH, Weilheim, Germany). For EC calculation, volume of oil needed to achieve the phase inversion was used (mL oil per g sample). Measurement was repeated in duplicate.

2.9. Analysis of the Molecular Weight Profiles of LPI and Fermented LPI

The molecular weight distribution of the untreated and fermented LPI was determined by sodium dodecyl sulfate-polyacrylamide gel electrophoresis (SDS-PAGE) as described by Laemmli [20] with modification under reducing conditions. Untreated LPI, fermented LPI and control samples were re-suspended in 1 mL loading buffer (0.125 mol/l Tris-HCl, 4% SDS (*w/v*), 20% glycerol (*v/v*), 0.2 mol/l DDT, 0.02% bromophenol blue, pH 6.8), dissolved for 30 min at 30 °C in an ultrasonic bath and boiled for 5 min at 95 °C in an Eppendorf ThermoMixer C (Eppendorf AG, Hamburg, Germany). Following centrifugation at 12,045× *g* for 10 min (Mini Spin, Eppendorf AG), an aliquot of the supernatant was transferred to a fresh tube and supplemented in a ratio of 1:10 with loading buffer (see above). An aliquot of 10 µL of each sample was transferred into the wells of pre-cast Criterion TGX stain-free 12% polyacrylamide gels (Bio-Rad Laboratories, Hercules, CA, USA). The samples were separated for 36 min at 200 V (60 mA, 100 W) (Amersham Biosciences Europe GmbH, Freiburg, Germany) at room temperature in a vertical electrophoresis cell (Bio-Rad Laboratories) with a 10–250 kDa Precision Plus Protein Unstained Standard (Bio-Rad Laboratories) alongside as size markers. Protein subunits were visualized using a Gel Doc™ EZ Imager system (Bio-Rad Laboratories). The molecular weight distribution was determined using Image Lab software (Bio-Rad Laboratories).

2.10. Statistical Analysis

Results are expressed as means ± standard deviations and for sensory evaluation (aroma profile) as median ± standard deviations. Data were analyzed using pairwise *t*-test to determine the significance of differences between a sample and the unfermented LPI, with a threshold of $p < 0.05$. For the microbial growth (CFU, pH, glucose), data were analyzed using one-way analysis of variances (ANOVA) and means were generated and adjusted with Tukey's honestly significant difference post hoc test to determine the significance of differences between all samples, with a threshold of $p < 0.05$. Statistical analysis was performed with SigmaPlot 12.5 for Windows (Systat Software GmbH, Erkrath, Germany).

3. Results and Discussion

3.1. Chemical Composition

Dry matter and protein content of LPI and fermented LPI are given in Table 2.

Table 2. Dry matter and protein content of unfermented lupin protein isolate (LPI) and fermented (24 h) LPI.

	Dry Matter	Protein Content
LPI (unfermented)	95.4 ± 0.0%	89.6 ± 0.0%
L. reuteri	94.6 ± 0.0% *	82.4 ± 1.3%
L. brevis	94.9 ± 0.9%	80.1 ± 0.5% *
L. amylolyticus	94.7 ± 0.2%	79.7 ±0.7% *
L. parabuchneri	94.7 ± 0.0% *	81.7 ± 0.2% *
L. sakei subsp. *carnosus*	94.2 ± 0.4%	80.6 ± 1.5%
S. xylosus	95.3 ± 1.1%	80.1 ± 1.3%
L. helveticus	94.7 ± 0.2%	78.5 ± 0.4% *
L. delbrueckii	92.8 ± 1.4%	78.5 ± 1.7%

The data are expressed as mean ± standard deviation ($n = 4$). Means marked with an asterisk (*) within a column indicate significant differences between the individual sample and the unfermented LPI ($p < 0.05$) following pairwise *t*-test.

Dry matter of LPI and fermented LPI ranged from 92.8% for *L. delbrueckii* to 95.4% for unfermented LPI. Unfermented LPI contained the highest protein content with 89.6%.

3.2. Comparison of Microbial Growth on Lupin Protein Isolate Solutions

The growing parameters consisting of CFU, pH and glucose for all eight microorganisms investigated are shown in Table 3. In addition, growth curves of four of these microorganisms were selected in Figure 1 in order to highlight the temporally different transitions into the exponential phase.

The results showed that all microorganisms were able to grow in LPI solution. The minimum increase in CFU/mL (ΔE_{CFU}) was recorded for *L. reuteri* with 1.36×10^7 CFU/mL and the maximum for *S. xylosus* with 6.01×10^8 CFU/mL. The results of the pH curve showed that *L. amylolyticus* and *L. helveticus* appear to have the best metabolism and were adapted most rapidly to the lupine solution. The pH curves showed a direct and constant pH acidification over 24 h for *L. helveticus* (exemplarily shown in Figure 1a) and *L. amylolyticus* (similar to Figure 1a). After 24 h of fermentation, a change into the stationary phase could not be observed for both microorganisms. *S. xylosus* (exemplarily shown in Figure 1b) and *L. delbrueckii* (similar to Figure 1b) showed a lag phase and a transition into the log phase after approximately 8 h. *S. xylosus* reached the stationary phase after 14 h. *L. delbrueckii* did not show a clear transition into the stationary phase after 24 h. *L. sakei* subsp. *carnosus* (exemplarily shown in Figure 1c) and *L. reuteri* (similar to Figure 1c) changed from the lag phase to the log phase after 10 h and reached the stationary phase after 18 h. The largest lag phases were recorded for *L. parabuchneri* (exemplarily shown in Figure 1d) and *L. brevis* (similar to Figure 1d). Both microorganisms reached the log phase after 14 h and changed to the stationary phase after 20 h. The results of glucose concentrations showed that the added carbon source of 5 g/kg glucose was metabolized by all microorganisms. After 24 h fermentation, residues of glucose were present in all fermented samples. The degradation of glucose ($\Delta E_{Glucose}$) ranged from 3.4 g/kg with remaining 1.7 g/kg glucose after 24 h for *L. amylolyticus* to 4.8 g/kg with remaining 0.2 g/kg glucose after 24 h for *L. reuteri*, respectively. Fritsch et al. [21] and Lampart-Szczapa, Konieczny, Nogala-Kałucka, Walczak, Kossowska and Malinowska [9] showed similar results and confirmed the suitability of lupine flour and lupine protein, respectively, for lactic fermentation.

Table 3. Colony forming units (CFU) (**a**) and pH, and glucose amount (**b**) after 0 h, 4 h, 18 h and 24 h of fermentation.

(**a**)

	CFU (CFU/mL)				
	0 h	**4 h**	**18 h**	**24 h**	**ΔE_{CFU} [1]**
L. reuteri	$2.68 \times 10^6 \pm 2.40 \times 10^5$ a	$6.52 \times 10^6 \pm 2.02 \times 10^6$ a	$5.77 \times 10^7 \pm 5.16 \times 10^6$ a	$1.63 \times 10^7 \pm 2.83 \times 10^6$ a	$1.36 \times 10^7 \pm 3.07 \times 10^6$ a
L. brevis	$1.83 \times 10^7 \pm 2.33 \times 10^6$ a	$4.20 \times 10^7 \pm 8.91 \times 10^6$ a	$1.24 \times 10^8 \pm 3.75 \times 10^7$ a	$1.31 \times 10^8 \pm 1.16 \times 10^8$ a	$1.13 \times 10^8 \pm 1.19 \times 10^8$ a
L. amylolyticus	$1.38 \times 10^7 \pm 7.38 \times 10^6$ a	$4.39 \times 10^6 \pm 1.11 \times 10^6$ a	$5.77 \times 10^7 \pm 2.33 \times 10^6$ a	$5.88 \times 10^7 \pm 4.60 \times 10^6$ a	$4.50 \times 10^7 \pm 1.20 \times 10^7$ a
L. parabuchneri	$1.21 \times 10^7 \pm 3.39 \times 10^6$ a	$2.30 \times 10^7 \pm 1.00 \times 10^7$ a	$3.43 \times 10^7 \pm 7.07 \times 10^5$ a	$5.91 \times 10^7 \pm 4.62 \times 10^7$ a	$4.70 \times 10^7 \pm 4.29 \times 10^7$ a
L. sakei subsp. *carnosus*	$1.95 \times 10^7 \pm 9.19 \times 10^6$ a	$1.74 \times 10^7 \pm 5.09 \times 10^6$ a	$1.33 \times 10^7 \pm 3.32 \times 10^6$ a	$3.99 \times 10^7 \pm 1.56 \times 10^6$ a	$2.04 \times 10^7 \pm 7.64 \times 10^6$ a
S. xylosus	$1.24 \times 10^7 \pm 1.54 \times 10^7$ a	$3.13 \times 10^7 \pm 1.27 \times 10^7$ a	$1.86 \times 10^8 \pm 1.91 \times 10^8$ a	$6.13 \times 10^8 \pm 7.59 \times 10^8$ a	$6.01 \times 10^8 \pm 7.75 \times 10^8$ a
L. helveticus	$3.08 \times 10^7 \pm 1.25 \times 10^7$ a	$3.67 \times 10^7 \pm 2.03 \times 10^7$ a	$5.09 \times 10^7 \pm 1.12 \times 10^7$ a	$7.44 \times 10^7 \pm 1.94 \times 10^7$ a	$4.37 \times 10^7 \pm 3.19 \times 10^7$ a
L. delbrueckii	$3.10 \times 10^6 \pm 9.97 \times 10^5$ a	$3.01 \times 10^6 \pm 1.77 \times 10^5$ a	$4.34 \times 10^7 \pm 1.51 \times 10^7$ a	$1.20 \times 10^8 \pm 1.84 \times 10^7$ a	$1.17 \times 10^8 \pm 1.74 \times 10^7$ a

(**b**)

	pH				Glucose (g/kg)				
	0 h	**4 h**	**18 h**	**24 h**	**0 h**	**4 h**	**18 h**	**24 h**	**$\Delta E_{Glucose}$ [2]**
L. reuteri	6.5 ± 0.0 b	6.4 ± 0.0 c,d,e	4.9 ± 0.0 b,c,d	4.8 ± 0.0 b,c	5.0 ± 0.0 a	4.1 ± 0.2 a,b,c	0.9 ± 0.1 a	0.2 ± 0.2 a	4.8 ± 0.0 d
L. brevis	6.6 ± 0.0 b,c	6.5 ± 0.1 d,e	5.3 ± 0.4 c,d,e	5.0 ± 0.3 b,c	5.0 ± 0.0 a	4.7 ± 0.3 d	3.7 ± 1.1 e	0.9 ± 0.3 d	4.1 ± 0.2 b
L. amylolyticus	6.6 ± 0.0 c	6.2 ± 0.0 c	5.6 ± 0.3 e	5.2 ± 0.1 c	5.0 ± 0.0 a	4.1 ± 0.1 a,b,c	2.5 ± 0.0 c,d	1.7 ± 0.3 e	3.4 ± 0.1 a
L. parabuchneri	6.6 ± 0.0 c	6.5 ± 0.0 d,e	5.5 ± 0.0 d,e	4.9 ± 0.0 b,c	5.0 ± 0.0 a	4.3 ± 0.2 b,c,d	3.3 ± 0.3 d,e	0.7 ± 0.1 c,d	4.3 ± 0.1 b,c
L. sakei subsp. *carnosus*	6.6 ± 0.0 c	6.3 ± 0.0 c,d	4.7 ± 0.0 a,b,c	4.7 ± 0.0 b	5.0 ± 0.0 a	4.4 ± 0.4 c,d	1.4 ± 0.0 a,b	0.4 ± 0.1 a,b	4.6 ± 0.1 c,d
S. xylosus	6.6 ± 0.0 c	6.5 ± 0.0 e	4.8 ± 0.1 b,c	4.8 ± 0.0 b,c	5.0 ± 0.0 a	4.3 ± 0.2 b,c,d	1.4 ± 0.3 a,b	0.5 ± 0.4 a,b,c	4.5 ± 0.1 c,d
L. helveticus	6.0 ± 0.0 a	5.4 ± 0.1 a	4.1 ± 0.2 a	3.9 ± 0.2 a	5.0 ± 0.0 a	3.8 ± 0.1 a	1.8 ± 0.2 b	0.6 ± 0.1 b,c,d	4.4 ± 0.0 b,c
L. delbrueckii	6.0 ± 0.1 a	6.0 ± 0.0 b	4.2 ± 0.1 a,b	4.4 ± 0.0 a	5.0 ± 0.0 a	4.0 ± 0.4 a,b	2.1 ± 0.2 b,c	0.8 ± 0.2 c,d	4.2 ± 0.1 b,c

The data are expressed as mean ± standard deviation from duplicates. Values followed by different letter in a column indicate significant differences between groups ($p < 0.05$) following one-way ANOVA (Tukey). [1] ΔE_{CFU} = Growth rate over 24 h fermentation; [2] $\Delta E_{Glucose}$ = Difference of the start and end glucose content of the fermentation.

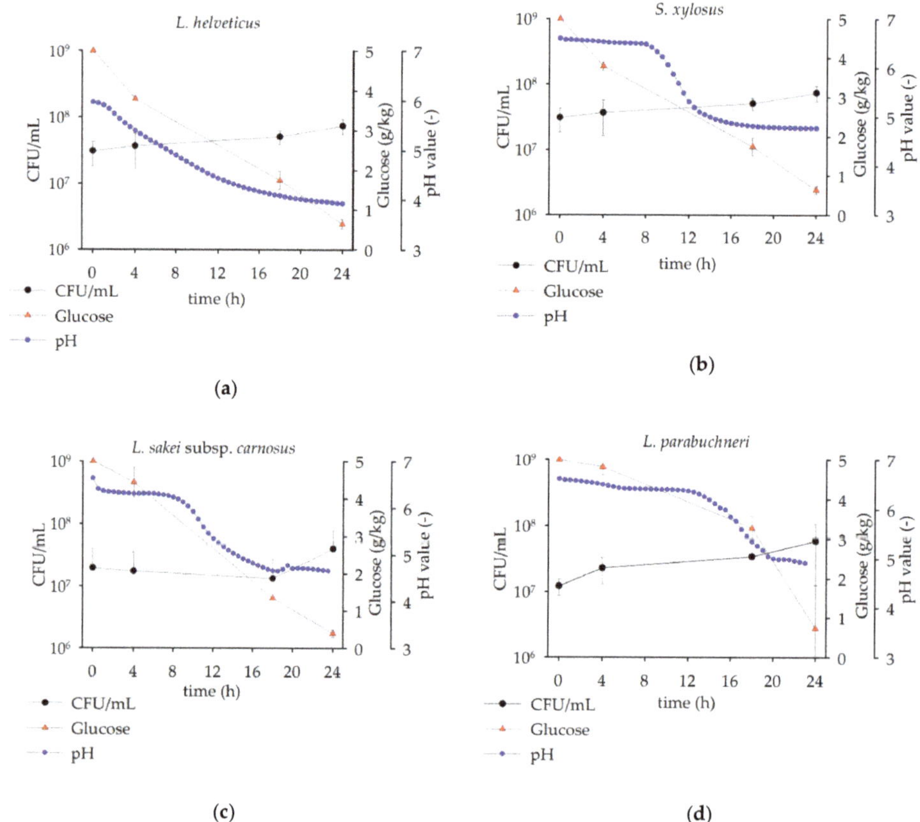

Figure 1. CFU/mL, glucose amount, and pH after 0 h, 4 h, 18 h, and 24 h for *L. helveticus* (**a**), *S. xylosus* (**b**), *L. sakei* subsp. *carnosus* (**c**), *and L. parabuchneri* (**d**). The data are expressed as mean ± standard deviation from duplicates.

3.3. Sensory Anaylsis

In the retronasal sensory evaluation, the following six aroma qualities and corresponding references were selected by the panelists for the description of LPI and fermented LPI: *cheesy* (butanoic acid); *popcorn-like, roasty* (2-acetylpyrazine); *earthy, moldy, beetroot-like* (geosmin); *pea-like, green bell pepper-like* (2-isopropyl-3-methoxypyrazine); *cooked potato-like* (3-(methylthio-)propanal); and *oatmeal-like, fatty* (oatmeal).

Comparative aroma profile analyses (Figure 2) of LPI, *L. brevis*, *L. amylolyticus* and *S. xylosus* were emphasized to highlight the increase of aroma perception in *cheesy* and *roasty, popcorn-like* as well as the reduction of the mean aroma perceptions in comparison to LPI, across the panel. The primary aroma attributes in the LPI were *pea-like, green bell pepper-like*, and *oatmeal-like, fatty* with median values of 4.5 and 3.0, respectively. Otherwise, the aroma impression was evaluated with low intensities for *earthy, moldy, beetroot-like*; *cooked potato-like*, and *popcorn-like, roasty* with median values of 2.5, 2.5, and 1.0, respectively, while the attribute *cheesy* was imperceptible (median value of 0). The dominant aroma attributes in the LPI samples obtained after fermentation with *L. reuteri* were described as *cheesy* (median of 4.0) and *oatmeal-like, fatty* (median value of 3.0). The aroma perception was otherwise evaluated with low intensities for *earthy, moldy, beetroot-like*; *pea-like, green bell pepper-like*, and *cooked potato-like* with values of 2.0 and *popcorn-like, roasty* with a value of 1.0. The aroma perception of the samples obtained after *L. brevis* fermentation was evaluated with a maximum intensity of 3.0 for *oatmeal-like, fatty*. Followed by low intensities of *earthy, moldy, beetroot-like; cooked potato-like*, and *popcorn-like, roasty* with values of 2.0 and *cheesy* and *pea-like, green bell pepper-like* with values of 1.0. Fermentation of LPI with *L. amylolyticus* was described with a dominant aroma impression of *popcorn-like, roasty* with a value of 5.0, followed by *oatmeal-like, fatty* with a value of 3.5. Low intensities were judged for *pea-like, green bell pepper-like* with a value of 2.0, *cooked potato-like* with a value of 1.5, and *earthy, moldy, beetroot-like* with 1.0. Attribute *cheesy* was not perceptible (median value of 0). The aroma profile of the *L. parabuchneri* fermented samples showed the aroma impressions of *pea-like, green bell pepper-like; cheesy;* and *oatmeal-like, fatty* with intensities of 4.0, 3.0, and 3.0 respectively. Less dominant were the attributes *popcorn-like, roasty* and *cooked potato-like*, both with median values of 2.0, and *earthy, moldy, beetroot-like* with an intensity median value of 1.5. *L. sakei* subsp. *carnosus* fermentation was described with *popcorn-like, roasty* as main aroma impression with an intensity of 4.0, followed by *oatmeal-like, fatty* with an intensity of 3.0. The attributes *earthy, moldy, beetroot-like; pea-like, green bell pepper-like*, and *cooked potato-like* were described equally less intensely with values of 2.0. The attribute *cheesy* was imperceptible in the samples obtained after fermentation with *L. sakei* subsp. *carnosus*. Samples fermented with *S. xylosus* exhibited a *cheesy* intensity of 5.0, followed by *oatmeal-like, fatty* with a value of 3.0. The other attributes were less intense with *pea-like, green bell pepper-like* (2.5), *earthy, moldy, beetroot-like* (2.0), *popcorn-like, roasty* (2.0), and *cooked potato-like* (1.0). Similar to *L. amylolyticus* and *L. sakei* subsp. *Carnosus* fermentation, the attribute *popcorn-like, roasty* was described as a dominant aroma impression for the *L. helveticus* fermented samples with an intensity of 4.0. The aroma profile of *L. helveticus* was otherwise described with aroma impressions *pea-like, green bell pepper-like* with a value of 3.0. The attributes *oatmeal-like, fatty; cooked potato-like; earthy, moldy, beetroot-like*, and *cheesy* were rated less intensively with values of 2.5, 2.0, 1.0, and 0.5 respectively. The main aroma impression of *L. delbrueckii* was assessed as *oatmeal-like, fatty* with an intensity of 4.0. The aroma impressions *popcorn-like, roasty; earthy, moldy, beetroot-like; pea-like, green bell pepper-like;* and *cooked potato-like* were rated with an intensity of 2.0. The attribute *cheesy* could not be observed by the panelists.

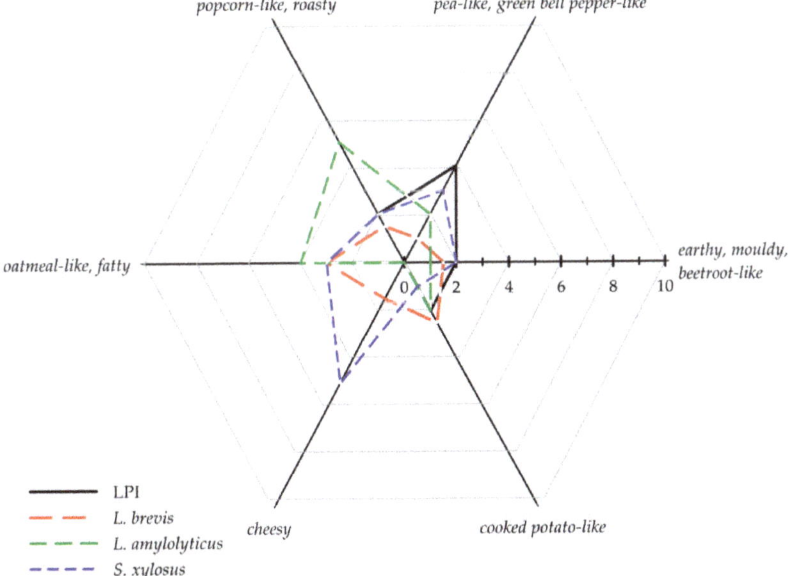

Figure 2. Retronasal aroma profile analyses of LPI, *L. brevis*, *L. amylolyticus*, and *S. xylosus* fermented samples on a scale from (no perception) to 10 (strong perception). The data are displayed as median values of the sensory evaluations ($n = 10$).

The aroma profile showed, with the exception of the *L. parabuchneri* fermented samples, that the aroma perception *pea-like, green bell pepper-like* decreased in intensity due to the fermentation. The maximum decreases were determined for the samples obtained after *L. brevis* fermentation with an intensity of 1.0 compared to unfermented LPI with a value of 4.5. Furthermore, fermentation increased the intensity of the aroma impressions *popcorn-like, roasty* and *cheesy*. In the samples fermented with *L. amylolyticus*, *L. sakei* subsp. *carnosus* and *L. helveticus* the aroma impressions *popcorn-like, roasty* increased from 1.0 for unfermented LPI to 5.0, 4.0 and 4.0 respectively. *S. xylosus* and *L. reuteri* showed an increase in intensity of the attribute *cheesy* from 0 for unfermented LPI to 5.5 and 4.0 respectively. Several authors confirm for lupin, soy and pea protein, respectively, the significant modification of aroma profile by fermentation [6–8,10]. The authors described the reduction of *n*-hexanal content, which contributes most to the *green* and *beany* off-flavor of pea, by fermentation with lactic acid extract in pea protein extract and soy, respectively [6–8]. Further studies showed that the fermentation of soy protein isolate with lactic acid bacteria significantly reduced the aroma impression of *beany* [10]. A statement about the reduction of *n*-hexanal content due fermentation cannot be obtained in this study. A reduced perception of the *pea-like, green bell pepper-like* aroma may also be caused by masking effects. The total aroma intensities of all samples differed only slightly with median values of 5.5 for the samples fermented with *L. amylolyticus*, 5.0 for LPI, *L. parabuchneri*, *S. xylosus*, *L. reuteri*, and *L. sakei* subsp. *carnosus* fermented samples, 4.5 for *L. helveticus* and *L. delbrueckii*, and 4.0 for *L. brevis* fermented samples.

The taste impressions *bitter* and *salty* were analyzed by the panel in the sensory evaluation and displayed in Figure 3. The *bitter* intensity of unfermented LPI was described with a mean value of 2.3. All fermented samples did not differ significantly ($p < 0.05$) in bitterness compared to LPI, with the exception of *L. sakei* subsp. *carnosus* (1.0). However, the trend of the mean values of the fermented samples showed slightly lower intensities of bitterness compared to LPI. The intensity of *salty* was described for LPI with a mean value of 1.9. In comparison, no significant differences ($p < 0.05$) in the intensity of the fermented samples were found. However, the samples obtained after fermentation with *L. amylolyticus*, *S. xylosus*, and *L helveticus* tended to be more salty.

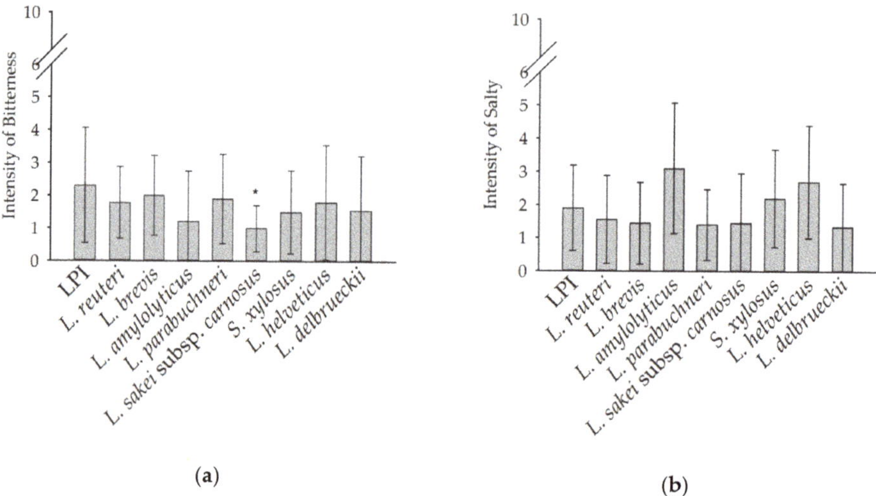

Figure 3. Intensity of bitter (**a**) and salty (**b**) taste perception of LPI and fermented LPI rate on a scale from 0 (no perception) to 10 (strong perception). Means marked with an asterisk (*) indicate significant differences between the individual sample and the unfermented LPI ($p < 0.05$) following pairwise *t*-test.

The hedonic evaluation was performed for a first indication of the prevalence of the samples and was not performed according to ISO standards. The evaluation (Figure 4) of the panel resulted in a rating of 4.2 for unfermented LPI. The sample, which was most popular with the panelists (6.4), was fermented with *L. sakei* subsp. *carnosus*, followed by the samples fermented with *L. helveticus* (5.5) and *L. amylolyticus* (5.4). The most unpopular samples were the ones fermented with *S. xylosus* and *L. reuteri* with values of 2.8 and 3.0, respectively. The results showed that both saltiness and bitterness do not have a considerable effect on the acceptance of the samples. The acceptance seems to be influenced by the differences in the aroma profile. It was found that the samples with the aroma attributes popcorn like were rated as popular by all subjects. The samples with the maximum evaluation had the attribute *popcorn-like, roasty* in their aroma profile as dominant aroma impression and also the highest intensity in this attribute compared to LPI and all other fermented samples. Meanwhile, the aroma impression *cheesy* dominated in *S. xylosus* and *L. reuteri* fermented samples—the samples with the minimal hedonic rating. In addition, both of these samples had the highest intensity of this attribute compared to LPI and the other samples.

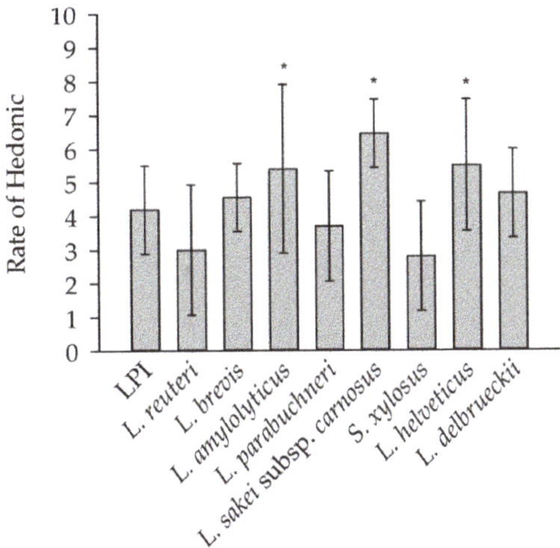

Figure 4. Rate of hedonic evaluation of LPI and unfermented LPI; scale from 0 (strong dislike) to 5 (neutral) to 10 (strong like). Means marked with an asterisk (*) indicate significant differences between the individual sample and the unfermented LPI ($p < 0.05$) following pairwise t-test.

3.4. Techno-Functional Properties

3.4.1. Protein Solubility

The protein solubility of LPI and fermented LPI was determined at pH 4 and pH 7 and given in Table 4. The protein solubility of all samples was higher under neutral condition (pH 7) than under acidic condition (pH 4). Usually, protein solubility is minimal at the isoelectric point (pH 4.5) [4,22–24]. Unfermented LPI showed a significantly ($p < 0.05$) higher protein solubility of 63.6% at pH 7 than all fermented samples. The minimum solubility was measured for the samples obtained after *L. helveticus* and *L. delbrueckii* fermentation with 23.57% and 27.48%, respectively, and the maximum solubility of the fermented samples with *L. reuteri* with 42.35%. Similar results have been found in other studies. Lampart-Szczapa, Konieczny, Nogala-Kałucka, Walczak, Kossowska, and Malinowska [9] determined also lower solubility of lupin proteins after fermentation than non-fermented samples and other authors observed this for fermented soy [10,11,25,26]. *Lactobacilli* produce organic acids during fermentation, which might have induced an irreversible coagulation of proteins and thus a reduced solubility. Further, heat treatment for the fermentation stop (90 °C, 20 min) might have promoted aggregation and cross-linking of partially hydrolyzed lupin proteins [10]. In contrast, the protein solubility at pH 4 after fermentation was not significantly different ($p < 0.05$) to unfermented LPI

(7.31%), with the exception of the samples fermented with *L. delbrueckii* (5.55%) and *L. helveticus* (5.92%). The highest protein solubility at pH 4 was measured for *S. xylosus* fermented samples with 8.11%.

Table 4. Protein solubility (%) at pH 4 and pH 7 of unfermented und fermented (24 h) LPI.

Samples	Protein Solubility	
	pH 4	pH 7
LPI (unfermented)	7.31 ± 0.26	63.59 ± 3.04
L. reuteri	7.40 ± 0.69	42.35 ± 3.76 *
L. brevis	7.41 ± 0.97	38.45 ± 2.87 *
L. amylolyticus	7.13 ± 0.31	35.47 ± 3.16 *
L. parabuchneri	8.01 ± 0.90	27.37 ± 4.00 *
L. sakei subsp. *carnosus*	7.17 ± 1.01	37.40 ± 4.53 *
S. xylosus	8.11 ± 0.77 *	28.04 ± 3.01 *
L. helveticus	5.92 ± 0.92 *	23.57 ± 2.99 *
L. delbrueckii	5.55 ± 0.41 *	27.48 ± 2.51 *

The data are expressed as mean ± standard deviation ($n = 4$). Means marked with an asterisk (*) within a column indicate significant differences between sample and unfermented LPI ($p < 0.05$) following pairwise t-test.

3.4.2. Foam Properties

The foam properties (foam activity and stability) of LPI and fermented LPI are shown in Table 5. All fermented samples showed a significant higher ($p < 0.05$) foam activity compared to unfermented LPI, with the exception of the samples fermented with *L. brevis* and *L. delbrueckii*. *L. brevis* and *L. delbrueckii* fermented samples showed an increase in foam activity compared to unfermented LPI, although the differences were not significant ($p < 0.05$). Similar results were obtained by Klupsaite et al. [27] for lupin proteins and Meinlschmidt, Ueberham, Lehmann, Schweiggert-Weisz, and Eisner [10] for soy protein isolates.

Table 5. Foam activity (%) and foam stability (%) of unfermented and fermented (24 h) LPI.

Samples	Foam Activity (%)	Foam Stability (%)
LPI (unfermented)	1613 ± 11	89 ± 3
L. reuteri	1646 ± 20 *	87 ± 3
L. brevis	1683 ± 57	86 ± 6
L. amylolyticus	1688 ± 52 *	94 ± 2
L. parabuchneri	1703 ± 25 *	16 ± 5 *
L. sakei subsp. *carnosus*	1670 ± 32 *	83 ± 3
S. xylosus	1678 ± 23 *	91 ± 1
L. helveticus	1698 ± 17 *	20 ± 0 *
L. delbrueckii	1652 ± 36	80 ± 0 *

The data are expressed as mean ± standard deviation ($n = 4$). Means marked with an asterisk (*) within a column indicate significant differences between sample and unfermented LPI ($p < 0.05$) following pairwise t-test.

Unfermented LPI showed a foam stability of 89%. The foam stability of all fermented samples, with exception of *L. parabuchneri* and *L. helveticus*, was above 80%. *L. parabuchneri* and *L. helveticus* showed significantly ($p < 0.05$) lower stability of 16% and 20%, respectively, compare to unfermented LPI.

3.4.3. Emulsifying Capacity

Unfermented LPI showed an emulsifying capacity of 552.9 mg/mL (Table 6). The treatments with the various microorganisms resulted in emulsifying capacities in the range of 347.7 mg/mL for *L. parabuchneri* up to 595.6 mg/mL for *S. xylosus*. Samples fermented with *L. parabuchneri* (347.7 mg/mL), *L. delbrueckii* (370.3 mg/mL), *L. sakei* subsp. *carnosus* (407 mg/mL) and *L. brevis* (447 mg/mL), showed a significantly ($p < 0.05$) lower emulsifying capacity than the unfermented LPI sample. The residual samples did not show a significantly ($p < 0.05$) lower emulsifying capacity compared to unfermented

LPI, but a tendency was observed. El-Adawy et al. [28] and Qi et al. [29] described a positive correlation between the emulsifying capacity and protein solubility of a protein. A reduced protein solubility was observed in fermented LPI compared to unfermented LPI. It could be an indication of the tendency decrease in emulsifying capacity of fermented LPI. The literature offers controversial data regarding the emulsifying capacity after fermentation. Lampart-Szczapa, Konieczny, Nogala-Kałucka, Walczak, Kossowska, and Malinowska [9] observed a decrease in emulsifying capacity after the fermentation of lupine protein while Klupsaite, Juodeikiene, Zadeike, Bartkiene, Maknickiene, and Liutkute [27] reported a significant increase in capacity after 24 h of fermentation. The fermentation of soy protein isolate resulted in a decrease in emulsifying capacity [10].

Table 6. Emulsifying capacity (mg/mL) of unfermented and fermented (24 h) LPI.

Samples	Emulsifying Capacity (mg/mL)
LPI (unfermented)	552.9 ± 9.8
L. reuteri	471.5 ± 10.7 *
L. brevis	447.2 ± 27.4 *
L. amylolyticus	564.1 ± 18.7
L. parabuchneri	347.7 ± 5.2 *
L. sakei subsp. *carnosus*	407.2 ± 7.6 *
S. xylosus	595.6 ± 55.6
L. helveticus	455.2 ± 51.9
L. delbrueckii	370.3 ± 6.8 *

The data are expressed as mean ± standard deviation ($n = 4$). Values marked with an asterisk (*) indicate significant differences between the individual sample and the unfermented LPI ($p < 0.05$) following pairwise t-test.

3.5. Analysis of the SDS-PAGE Profiles of Unfermented and Fermented (24 h) LPI

The molecular weight distribution of LPI and fermented LPI was analyzed to obtain an indication of the integrity of the proteins. The results of the SDS-PAGE showed that the microorganisms were able to change the SDS-PAGE profile of the lupin protein isolates. It could be observed that the polypeptides of lupin protein isolates were partially decomposed. All fermented samples showed a degradation of low molecular weight fractions (<23 kDa) associated with the increase in small polypeptides with a molecular weight of <15 kDa compared to unfermented LPI (data not presented). In particular, polypeptides with a molecular weight of 17 kDa appear to be decomposed in all samples and exemplarily shown for *L. reuteri* and *L. parabuchneri* in Figure 5. The amount of polypeptides within the medium molecular weight range (25–36 kDa) increased due to the breakage of larger polypeptides. It appears that all used microorganisms are proteolytic active to decompose medium molecular weight and low molecular weight polypeptides from lupin protein isolate into small fragments.

Meinlschmidt, Ueberham, Lehmann, Schweiggert-Weisz, and Eisner [10] also investigated the effect of fermentation with *L. helveticus* on the integrity of proteins in soy protein isolate. The number of subunits in the 25–37 kDa range was also increased, as were the number of low molecular weight bands <20 kDa.

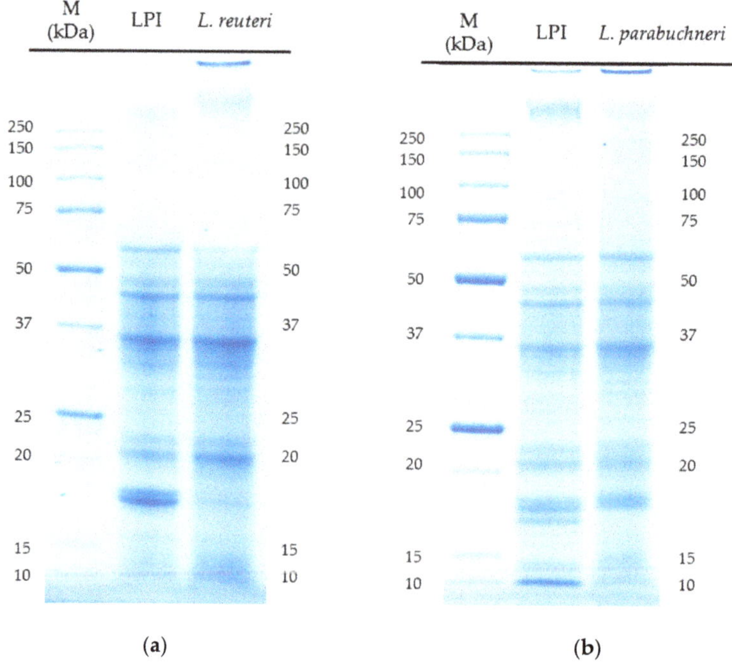

Figure 5. Peptide band profiles of fermented LPI produced by fermentation (24 h) with *L. reuteri* (**a**) and *L. parabuchneri* (**b**) as determined by SDS-PAGE under reducing conditions.

4. Conclusions

Fermentation maintains the functional properties and improves the sensory properties of lupin protein isolate. Following the determination of the aroma profile using a trained panel, the formulation of the new aroma components should be qualitatively and quantitatively analyzed. Based on the results of SDS-PAGE, the microorganisms mainly decomposed polypeptides with low and medium molecular weight. As the protein integrity provides a first indication of the allergenic potential, in vitro and in vivo studies should be performed to investigate the allergenicity of these fermented lupin proteins. In addition, it is necessary to validate the allergen structure of LPI and fermented LPI and to develop more reliable methods for the quantification of allergens. The use of fermented plant proteins in food products such as meat and dairy alternatives must be examined in practice by subsequent application trails.

Author Contributions: Conceptualization, K.S. and U.S.-W.; methodology, K.S. and U.S.-W.; formal analysis, K.S.; investigation, K.S. and A.L.; resources, P.E.; writing—Original draft preparation, K.S.; writing—Review and editing, K.S., U.S.-W., and P.E.; visualization, K.S.

Funding: This research received no external funding.

Acknowledgments: This work was supported by the Fraunhofer-Zukunftsstiftung.

Conflicts of Interest: The authors declare no conflict of interest.

References

1. Eshel, G.; Shepon, A.; Makov, T.; Milo, R. Land, irrigation water, greenhouse gas, and reactive nitrogen burdens of meat, eggs, and dairy production in the United States. *Proc. Natl. Acad. Sci. USA* **2014**, *111*, 11996–12001. [CrossRef] [PubMed]
2. Westhoek, H.; Lesschen, J.P.; Rood, T.; Wagner, S.; De Marco, A.; Murphy-Bokern, D.; Leip, A.; van Grinsven, H.; Sutton, M.A.; Oenema, O. Food choices, health and environment: Effects of cutting Europe's meat and dairy intake. *Glob. Environ. Chang.* **2014**, *26*, 196–205. [CrossRef]
3. Arnoldi, A.; Boschin, G.; Zanoni, C.; Lammi, C. The health benefits of sweet lupin seed flours and isolated proteins. *J. Funct. Foods* **2015**, *18*, 550–563. [CrossRef]
4. Bader, S.; Bez, J.; Eisner, P. Can protein functionalities be enhanced by high-pressure homogenization?—A study on functional properties of lupin proteins. *Procedia Food Sci.* **2011**, *1*, 1359–1366. [CrossRef]
5. Schlegel, K.; Sontheimer, K.; Hickisch, A.; Wani, A.A.; Eisner, P.; Schweiggert-Weisz, U. Enzymatic hydrolysis of lupin protein isolates—Changes in the molecular weight distribution, technofunctional characteristics and sensory attributes. *Food Sci. Nutr.* **2019**, *7*, 2747–2759. [CrossRef] [PubMed]
6. Schindler, S.; Zelena, K.; Krings, U.; Bez, J.; Eisner, P.; Berger, R.G. Improvement of the Aroma of Pea (*Pisum sativum*) Protein Extracts by Lactic Acid Fermentation. *Food Biotechnol.* **2012**, *26*, 58–74. [CrossRef]
7. Schindler, S.; Wittig, M.; Zelena, K.; Krings, U.; Bez, J.; Eisner, P.; Berger, R.G. Lactic fermentation to improve the aroma of protein extracts of sweet lupin (*Lupinus angustifolius*). *Food Chem.* **2011**, *128*, 330–337. [CrossRef]
8. Blagden, T.D.; Gilliland, S.E. Reduction of Levels of Volatile Components Associated with the "Beany" Flavor in Soymilk by Lactobacilli and Streptococci. *J. Food Sci.* **2005**, *70*, M186–M189. [CrossRef]
9. Lampart-Szczapa, E.; Konieczny, P.; Nogala-Kałucka, M.; Walczak, S.; Kossowska, I.; Malinowska, M. Some functional properties of lupin proteins modified by lactic fermentation and extrusion. *Food Chem.* **2006**, *96*, 290–296. [CrossRef]
10. Meinlschmidt, P.; Ueberham, E.; Lehmann, J.; Schweiggert-Weisz, U.; Eisner, P. Immunoreactivity, sensory and physicochemical properties of fermented soy protein isolate. *Food Chem.* **2016**, *205*, 229–238. [CrossRef]
11. Song, Y.-S.; Frías, J.; Martinez-Villaluenga, C.; Vidal-Valdeverde, C.; de Mejia, E.G. Immunoreactivity reduction of soybean meal by fermentation, effect on amino acid composition and antigenicity of commercial soy products. *Food Chem.* **2008**, *108*, 571–581. [CrossRef] [PubMed]
12. Song, Y.S.; Pérez, V.G.; Pettigrew, J.E.; Martinez-Villaluenga, C.; de Mejia, E.G. Fermentation of soybean meal and its inclusion in diets for newly weaned pigs reduced diarrhea and measures of immunoreactivity in the plasma. *Anim. Feed Sci. Technol.* **2010**, *159*, 41–49. [CrossRef]
13. Frias, J.; Song, Y.S.; Martínez-Villaluenga, C.; De Mejia, E.G.; Vidal-Valverde, C. Immunoreactivity and Amino Acid Content of Fermented Soybean Products. *J. Agric. Food Chem.* **2008**, *56*, 99–105. [CrossRef] [PubMed]
14. Kobayashi, M. Immunological functions of soy sauce: Hypoallergenicity and antiallergic activity of soy sauce. *J. Biosci. Bioeng.* **2005**, *100*, 144–151. [CrossRef] [PubMed]
15. Mosse, J.; Huet, J.-C.; Baudet, J. Relationships between nitrogen, amino acids and storage proteins in *Lupinus albus* seeds. *Phytochemistry* **1987**, *26*, 2453–2458. [CrossRef]
16. Morr, C.V.; German, B.; Kinsella, J.E.; Regenstein, J.M.; Vanburen, J.P.; Kilara, A.; Lewis, B.A.; Mangino, M.E. A Collaborative Study to Develop a Standardized Food Protein Solubility Procedure. *J. Food Sci.* **1985**, *50*, 1715–1718. [CrossRef]
17. Lowry, O.H.; Rosebrough, N.J.; Farr, A.L.; Randall, R.J. Protein measurement with the Folin phenol reagent. *J. Biol. Chem.* **1951**, *193*, 265–275.
18. Phillips, L.G.; Haque, Z.; Kinsella, J.E. A Method for the Measurement of Foam Formation and Stability. *J. Food Sci.* **1987**, *52*, 1074–1077. [CrossRef]
19. Wang, C.Y.; Johnson, L.A. Functional properties of hydrothermally cooked soy protein products. *J. Am. Oil Chem. Soc.* **2001**, *78*, 189–195. [CrossRef]
20. Laemmli, U.K. Cleavage of structural proteins during the assembly of the head of bacteriophage T4. *Nature* **1970**, *227*, 680–685. [CrossRef]
21. Fritsch, C.; Vogel, R.F.; Toelstede, S. Fermentation performance of lactic acid bacteria in different lupin substrates—Influence and degradation ability of antinutritives and secondary plant metabolites. *J. Appl. Microbiol.* **2015**, *119*, 1075–1088. [CrossRef] [PubMed]

22. Lqari, H.; Pedroche, J.; Girón-Calle, J.; Vioque, J.; Millán, F. Production of *Lupinus angustifolius* protein hydrolysates with improved functional properties. *Grasas Aceites* **2005**, *56*, 135–140. [CrossRef]
23. Piornos, J.A.; Burgos-Díaz, C.; Ogura, T.; Morales, E.; Rubilar, M.; Maureira-Butler, I.; Salvo-Garrido, H. Functional and physicochemical properties of a protein isolate from AluProt-CGNA: A novel protein-rich lupin variety (*Lupinus luteus*). *Food Res. Int.* **2015**, *76*, 719–724. [CrossRef] [PubMed]
24. Rodríguez-Ambriz, S.L.; Martínez-Ayala, A.L.; Millán, F.; Dávila-Ortíz, G. Composition and Functional Properties of *Lupinus campestris* Protein Isolates. *Plant Foods Hum. Nutr.* **2005**, *60*, 99–107. [CrossRef]
25. Weng, T.M.; Chen, M.T. Changes of Protein in Natto (a fermented soybean food) Affected by Fermenting Time. *Food Sci. Technol. Res.* **2010**, *16*, 537–542. [CrossRef]
26. Murekatete, N.; Hua, Y.; Kong, X.; Zhang, C. Effects of Fermentation on Nutritional and Functional Properties of Soybean, Maize, and Germinated Sorghum Composite Flour. *Int. J. Food Eng.* **2012**, *8*, 1–15. [CrossRef]
27. Klupsaite, D.; Juodeikiene, G.; Zadeike, D.; Bartkiene, E.; Maknickiene, Z.; Liutkute, G. The influence of lactic acid fermentation on functional properties of narrow-leaved lupine protein as functional additive for higher value wheat bread. *LWT-Food Sci. Technol.* **2017**, *75*, 180–186. [CrossRef]
28. El-Adawy, T.A.; Rahma, E.H.; El-Bedawey, A.A.; Gafar, A.F. Nutritional potential and functional properties of sweet and bitter lupin seed protein isolates. *Food Chem.* **2001**, *74*, 455–462. [CrossRef]
29. Qi, M.; Hettiarachchy, N.S.; Kalapathy, U. Solubility and Emulsifying Properties of Soy Protein Isolates Modified by Pancreatin. *J. Food Sci.* **1997**, *62*, 1110–1115. [CrossRef]

Article

Sensory Improvement of a Pea Protein-Based Product Using Microbial Co-Cultures of Lactic Acid Bacteria and Yeasts

Cynthia El Youssef [1], Pascal Bonnarme [2], Sébastien Fraud [1], Anne-Claire Péron [2], Sandra Helinck [2,*,†] and Sophie Landaud [2,†]

1 General Mills Yoplait, Vienne Technical Center, 38205 Vienne, France; Cynthia.Elyoussef@genmills.com (C.E.Y.); Sebastien.Fraud@genmills.com (S.F.)

2 Université Paris-Saclay, INRAE, AgroParisTech, UMR SayFood, 78850 Thiverval-Grignon, France; pascal.bonnarme@inrae.fr (P.B.); anne-claire.peron@inrae.fr (A.-C.P.); sophie.landaud@agroparistech.fr (S.L.)

* Correspondence: sandra.helinck@agroparistech.fr; Tel.: +33-1-30-81-54-85

† These authors jointly supervised the study.

Received: 3 February 2020; Accepted: 11 March 2020; Published: 17 March 2020

Abstract: Consumer demands for plant-based products have increased in recent years. However, their consumption is still limited due to the presence of off-flavor compounds, primarily beany and green notes, which are mainly associated with the presence of aldehydes, ketones, furans, and alcohols. To overcome this problem, fermentation is used as a lever to reduce off-flavors. A starter culture of lactic acid bacteria (LAB) was tested in a 4% pea protein solution with one of the following yeasts: *Kluyveromyces lactis*, *Kluyveromyces marxianus*, or *Torulaspora delbrueckii*. The fermented samples were evaluated by a sensory panel. Non-fermented and fermented matrices were analyzed by gas chromatography coupled with mass spectrometry to identify and quantify the volatile compounds. The sensory evaluation showed a significant reduction in the green/leguminous attributes of pea proteins and the generation of new descriptors in the presence of yeasts. Compared to the non-fermented matrix, fermentations with LAB or LAB and yeasts led to the degradation of many off-flavor compounds. Moreover, the presence of yeasts triggered the generation of esters. Thus, fermentation by a co-culture of LAB and yeasts can be used as a powerful tool for the improvement of the sensory perception of a pea protein-based product.

Keywords: pea protein; fermentation; lactic acid bacteria; yeast; beany; green

1. Introduction

The demand for plant-based products is on the rise as consumers become increasingly health-conscious and seek out non-dairy alternatives for dairy products. Reasons for the plant-based movement stem from three different factors: environmental, health, and concerns about animal welfare [1,2]. Consequently, the food industries are undoubtedly experiencing disruptive pressures in favor of plant-based products and are evolving rapidly to meet consumer needs [3].

Due to their high protein content ranging from 20% to 25%, pulses present an important economic and nutritional interest [4]. These dried seeds belong to the family of legumes, Fabaceae, which includes *Pisum sativum* (peas), *Cicer arietinum* (chickpeas), *Lupinus* (lupins), *Phaseolus vulgaris* (beans), *Glycine* max (soybeans), and *Arachis hypogaea* (peanuts) [5]. As a result of their limited amount of sulfur-containing amino acids (methionine, cystine, and tryptophan), they are often consumed along with cereal grains that have a complementary amino acid profile, i.e., that are rich in sulfur-containing amino acids but deficient in lysine. In addition to their high nutritive value, the consumption of pulses can offer many health benefits to human beings, e.g., the reduction of cardiovascular diseases, diabetes, and cancer risk [6,7].

Thanks to its functional properties and high protein content, pea (*Pisum sativum*) offers a great opportunity to develop an alternative plant protein source that will meet nutritional needs in an affordable and sustainable way [8]. This plant-based protein can be used in food applications as an isolate (protein content higher than 80%) or in concentrate form (protein content lower than 80%). However, the presence of off-flavors described as "beany," "green," "vegetal," "hay-like", and "rancid" is a limiting factor for its consumption [9,10]. These undesirable notes are partially inherent in peas or produced during harvesting, processing, and storage.

The typical beany flavor of pea is generally associated with aldehyde molecules, mainly hexanal, which is the most frequently reported compound responsible for the undesirable aroma of pea proteins. In addition to aldehyde molecules, other compounds such as ketones, alcohols, pyrazines, and furans also contribute to the off-flavor profile of pea proteins [5]. Thus, the perception of the beany flavor could arise from the interactions between different molecules and would, therefore, not be the result of the perception of one single compound [11].

Fermentation, one of the oldest stabilization processes, has been used for centuries as a method to preserve agricultural raw materials and to improve their organoleptic properties. In that respect, the preparation of many traditional cereal- and pulse-based food products relies on the use of fermentation [12]. The diversity of fermented foods depends on the types of substrates and microorganisms involved in the process. Some examples of cereal-based fermented products are beer, bread, and a less common product, boza [13]. Alcoholic beverages are made from different substrates such as malted barley in Western countries (in the case of beer), as well as from sorghum (burukutu) and maize/sorghum (pito) in Africa. These alcoholic beverages are mainly produced using yeasts (*Saccharomyces* sp.) but also use co-cultures of yeasts and lactic and acetic acid bacteria, e.g., *Brettanomyces*, *Saccharomyces*, *Lactobacillus*, *Leuconostoc*, and *Acetobacter* for lambic beers; *Geotrichum candidum* and *Lactobacillus* for pito; and *S. cerevisae*, *S. chavelieri*, *Leuconostoc mesenteroides*, *Candida*, and *Acetobacter* for burukutu [13,14]. On the other hand, boza, a beverage made from wheat, rye, millet, maize, and other cereals, is mainly fermented with lactic acid bacteria (LAB) such as *Lactobacillus* and *Leuconostoc* sp., and sometimes associated with yeasts (*S. cerevisiae*) [13]. Bread, which is also known as naan, bhatura, and kulcha in Asian countries, is a dough obtained mainly by yeast fermentation (*S. cerevisiae*) but can also be produced via a co-culture of LAB and yeasts [15].

In addition to fermented cereal-based beverages, fermented food products made from pulses are generally found in traditional Asian food products. One of the major types of pulses used is soybean, which contributes to the formation of a wide range of products such as sufu, tempeh, miso, and natto [16]. As in the case of cereal-based beverages, these fermented products can be made with an association of LAB (*Pediococcus*), *Bacillus*, fungi (*Rhizopus*, *Actinomucor*, and *Mucor*), and yeasts (*Saccharomyces*, *Candida*) [17–19]. It should be noted that several species used in these fermented products, such as *Rhizopus* spp., *Mucor* spp., and *Candida* spp., are not on the Qualified Presumption of Safety (QPS) list or are not recommended for QPS status in European countries [20].

Consequently, most plant-based fermented foods are produced using microbial communities due to their potential synergistic effects that help to improve the qualities of fermented products [21]. However, the use of fermentation for pea-based matrices using microbial communities with variable complexities is very rare.

Previous studies have reported the impact of lactic acid fermentation on the modification of the green/beany perception of pea. They showed that lactic acid fermentation had a limited effect on the reduction of the negative descriptors associated with pea proteins [10,22,23]. Moreover, the addition of yeasts such as *Candida catenulate* and *Geotrichum candidum* triggered the formation of banana and apricot aroma in a cheese-like pea-based product [24].

Thus, a promising line of research would be to co-ferment pea protein with LAB and yeasts. The reason that these microorganisms were combined in our study is based on the need to use LAB to obtain a pea protein "yogurt-like" product with a final pH value of 4.55 that ensures both its sanitary quality and gelation properties. Moreover, the genera *Saccharomyces* and *Kluyveromyces* display enzymatic activities such as alcohol and aldehyde dehydrogenase, enzymes that have been reported to decrease green beany flavor in soybeans through the degradation of aldehydes and alcohols [25]. Yeasts are known to improve the aroma quality of fermented beverages via the formation of ester compounds with fruity and floral notes such as ethyl acetate and 2-methylbutyl acetate [26]. The aim of this study is, therefore, to investigate the impact of fermentation with yeasts in co-culture with LAB on the volatile profile associated with off-flavor in peas.

The specific objectives of this study are: (1) to evaluate the impact of the addition of yeasts on the acidifying activity of LAB; and (2) to compare the flavor profile of the samples fermented with a pure culture of LAB and the ones fermented with LAB in co-culture with yeasts.

2. Materials and Methods

2.1. Raw Materials, Ingredients, and Strains

Pea protein isolates of *Pisum sativum L.* (Purispea Tm 870, batch1708TL1) were supplied by Cargill (Chicago, IL, USA). The sample had the following characteristics: pH 7.0, moisture 4.2%, and protein content 82.1%. Sucrose from sugar cane was provided by Tereos (France). The raw materials were stored at room temperature.

VEGE 047 LYO was obtained from DuPont Danisco (Dangé-Saint-Romain, France) and consisted of freeze-dried defined strains of lactic acid bacteria: *Lactobacillus acidophilus* NCFM®, *Streptococcus thermophilus, Lactobacillus delbrueckii subsp. bulgaricus*, and *Bifidobacterium lactis* HN019™. Using API 50 CHL medium (Biomérieux SA; Marcy l'Etoile, France), this mixture of lactic acid bacteria and *Bifidobacterium* was found to be glucose (+), fructose (+), and sucrose (+). *Torulaspora delbrueckii* TD 291 (freeze-dried BIODIVA™) was provided by LALLEMAND S.A.S, France. *Kluyveromyces lactis* Clib 196 and *Kluyveromyces marxianus* 3810 were obtained from the INRA collection (UMR GMPA, Grignon, France).

2.2. Fermentation of Pea Protein Isolates

2.2.1. Inoculum Preparation

Kluyveromyces lactis Clib 196 and *Kluyveromyces marxianus* 3810 were incubated from frozen glycerol stocks (−80 °C) in potato dextrose broth (PDB, Becton, Dickinson & Company, Sparks, MD, USA) for 21 h at 30 °C with an agitation of 200 rpm using an incubator (INFORS HT). When the stationary phase of growth was reached, cells were harvested by centrifugation at 4000 rpm (1699× *g*) for 15 min at 10 °C (Eppendorf, 5804R). The pellet was then washed and suspended in sterile physiological water (9 g of NaCl in 1 L of osmotic water) to obtain 10^7 colony forming unit/mL (CFU/mL) and used as the inoculum.

For the rehydration of VEGE 047 LYO, we followed the supplier's instructions. The freeze-dried culture was rehydrated in a 500-mL of a 4% thermally treated pea protein solution using Purispea Tm 870. The mixture was left to settle for 15 min and then agitated gently before the preparation of the cryotubes. The latter was then stored at −80 °C.

As for *Torulaspora delbrueckii*, 0.25 g/L was inoculated in a pea protein solution and then transferred to cryotubes at −80 °C until the fermentation process.

Upon fermentation, the samples were inoculated with 10^7 CFU/mL of VEGE and *Torulaspora delbrueckii.*

2.2.2. Preparation of Fermented Pea Protein Isolate

Osmotic water was used to prepare a 4% pea protein solution with 3% sucrose. The mixture was stirred using a magnetic stirrer at room temperature for 10 min until a homogeneous solution was obtained. The solution was then thermally treated at 110 °C for 15 min. This process was required to eliminate the endogenous microflora of pea proteins before fermentation. Prior to inoculation, the solution was cooled down to 30 °C. The initial pH of the solution was 7.1 ± 0.1.

Four different fermented samples were inoculated with the following strains: VEGE 047 (VEGE), VEGE 047 + *Kluyveromyces lactis* Clib 196 (VEGE + KL), VEGE 047 + *Kluyveromyces marxianus* 3810 (VEGE + KM), and VEGE 047 + *Torulaspora delbrueckii* (VEGE + TOR). The experiments were carried out in triplicate.

Fermentations were stopped at pH 4.55 (an optimal pH to ensure sanitary and textural qualities) by rapid cooling in an ice bath until the temperature reached 4 °C. Microbial enumerations were performed at this point.

All fermented samples were stored in 150-mL glass vials at 4 °C for 7 days. At day 7, the products were sent to an expert panel for sensory evaluation and the remaining samples were frozen at −80 °C until analysis.

2.3. Fermentation Monitoring

2.3.1. Acidification Activity Measurement

The Cinac system 4 (AMS, Frépillon France) [27] was used to measure the acidification activity of the microbial strains at 30 °C. The pH of the inoculated pea protein samples was continuously measured and automatically recorded at 3-min intervals. The time to reach pH 4.55 (tpH 4.55 in min) was used as a descriptor for the acidification activity. It was calculated using Cinac 4 (version 4, release 0.4.4). Measurements were made in triplicate.

2.3.2. Microbial analyses

LAB and yeast populations were determined at the beginning and at the end of the fermentation process when a pH of 4.55 was reached. The samples were diluted 1:10 with sterile physiological water (9 g of NaCl/L) and then homogenized using an Ultra Turrax®device (Labortechnik, Germany) at 8000 rpm for 1 min. The yeast population was determined by surface plating in triplicate using yeast–glucose–chloramphenicol agar (YGCA, BIOKAR, Beauvais, France) after three days of incubation at 30 °C. Lactic acid bacteria were counted by spread plating technique in triplicate on non-acidified Man Rogosa and Sharpe agar (MRS, BIOKAR, Beauvais, France) after 3 days at 42 °C under anaerobic conditions (Bugbox Anaerobic System, Ruskinn, Bridgend, United Kingdom).

2.3.3. Biochemical Analysis

Analyses Using HPLC–MS to Determine Sugar Content

To obtain an accurate determination of the amount of sucrose, fructose, and glucose in a complex matrix, a sugar analysis was done using liquid chromatography coupled with mass spectrometry.

Sugars were extracted as previously described [28]. After thawing, the samples were diluted in 50/50 (*v/v*) LC/MS water/acetonitrile and were quantified using high-performance liquid chromatography coupled with mass spectrometry (Waters, Beaver Dam, WI, USA).

Metabolites were separated on an XBridge BEH Amide column (length: 150 mm; internal diameter: 4.6 mm; particle size: 3.5 μm; WATERS). The column temperature was set at 75 °C. The flow was 0.4 mL/min and the solvent were acetonitrile with 0.1% formic acid (D) and ultra-pure water (B) + 0.1% formic acid. The elution gradient was as follows: 0 min at 80% D + 20% B, then 50% D + 50% B for 23 min, level at 80% D and 20% D for 2 min. The injection volume was 5 μL, and the injector temperature was 7 °C. Each analysis took 25 min.

Mass spectrometric detection was performed with an ISQ™ EC-LC Quadrupole with a heated electrospray source (HESI–II) operated in the negative ionization mode (Thermofisher Scientific). Metabolites were identified and quantified (ng/g wet weight) using Chromeleon 7.2.10 software (Thermofisher scientific, Waltham, MA, USA).

Analyses Using HPLC to Determine Ethanol and Lactic Acid Concentrations

The concentrations of ethanol and lactic acid were determined by high performance liquid chromatography (HPLC). Similar to the preparation of sugar extracts, ethanol and lactic acid were extracted as previously described [28].

The analysis was performed using a Waters Associates chromatographic system (Alliance) equipped with a pump, an automatic injector (Waters e2695) and two columns, a pre-column of 30 × 4.6 mm (Bio-Rad Labs; Richmond, CA, USA), and an HPX-87H columns (300 × 7.8 mm; Bio-Rad Labs; Richmond, CA, USA) connected in series. The columns were operated at 35 °C. The samples were eluted with 0.01 N sulfuric acid at a flow rate of 0.6 mL/min. The eluting compounds were detected by a UV detector (Model 2489). This detector was connected in series to an RI detector (Model 2414); Empower ^{TM} 3 chromatography data software (Waters Corporation) was used to integrate peak areas using calibration by an external standard solution.

2.4. Sensory Evaluation

The sensory evaluation of thé four products was performed using descriptive analysis. A panel of 15 trained panelists was recruited for their familiarity with plant-based products.

Sensory analysis was carried out in an air-conditioned room (20 °C), in individual booths, under daylight. Samples of the fermented products (80 g) were presented in plastic cups labeled with randomly selected three-digit numbers. The sample evaluation order was balanced over the panel following a Williams Latin square design to account for potential order and carry-over effects. Panelists were asked to rinse their mouths with water and crackers between samples.

A one-hour session was dedicated to the generation of attributes, followed by training in the use of these attributes to obtain a quantitative description of the products. The 15 panelists generated a vocabulary of sensory attributes that covered the odor, texture, aroma, and taste of the samples. During the second session, the panelists had to rate the intensity of the 13 attributes (global intensity, sour, bitter, astringent, tangy, sparkling, green flavor/vegetal, leguminous plant, citrus fruit, nut, beer/yeast, sourdough, cultured apple cider) generated for each product on an interval scale ranging from 0 to 15 (from nonexistent to marked). Samples were presented in a monadic sequence. The panel performances were validated in a third session in terms of repeatability, using different means of analysis of variance (ANOVA)

2.5. Aroma Compound Analysis

To identify the aroma compounds present in the non-fermented and fermented pea protein solutions, GC/MS analysis was performed. All analyses were performed in triplicate.

Volatile compounds were extracted using the purge and trap method by means of a Gerstel Dynamic Headspace System (DHS) coupled with a Gerstel Multipurpose Sampler (MPS) Autosampler (Mulheim an der Ruhr, Denmark). Five grams of the fermented or non-fermented samples were weighed in a vial. The DHS system heated the samples to 40 °C for 3 min with an agitation speed of 500 rpm. The samples were purged with a helium flow at 30 mL/min for 10 min and analytes (volatile molecules) were collected on sorbent material at 30 °C. The sorbent material used for volatile molecule collection was Tenax TA (2, 6-diphenylene oxide polymer) (Gerstel). The sorbent material was dried to remove residual water vapor at 30 °C with a helium flow of 50 mL/min for 6 min.

GCMS was performed using a 7890 Agilent GC system coupled to an Agilent 5977B quadruple mass spectrometer (Agilent, Santa Clara, CA, USA). A non-polar Agilent column DB-5MS (60 m × 0.32 mm × 1 µm) was used. The injection was performed in splitless mode using helium at a flow rate of 1.6 mL/min. The oven temperature of the column was programmed as follows: temperature increase from 40 to 155 °C at 4 °C/min, followed by 155 to 250 °C at 20 °C/min. The oven temperature was then maintained at 250 °C for 5 min. The gas chromatogram was recorded and analyzed for volatile retention time. Volatile compounds identified by comparison with a mass spectra library (NIST database) were chosen based on their percentage of identity.

For the quantification of the volatile compounds, two different standard solutions were prepared: Solution A containing the molecules responsible for the off-notes, and solution B comprising the ester molecules. The choice of the volatile compounds responsible for the off-flavor perception in pea was based on their occurrence in the literature.

For Solution A, a mixture of 21 selected compounds was prepared. Pure commercial volatile standards belonging to different families of compounds were purchased from Sigma Aldrich (Milwaukee, USA). These standards were as follows: 2-methylpropanal, trans-2-methyl-2-butenal, hexanal, (E)-2- hexenal, heptanal, (E)-2-octenal, nonanal, butanal, (E)-2-heptenal, decanal, 1-penten-3-ol, 1-octen-3-ol, 1-hexanol, 1-octanol, 6-methyl-5-hepten-2-one, 2-octanone, 2-nonanone, 2-n-heptylfuran, 2-ethylfuran, and 2-pentylfuran.

Solution B consisted of ester compounds, major aromatic molecules resulting from yeast fermentation: isoamyl acetate, 2-methylbutyl acetate, 2-phenylethyl acetate, isobutyl acetate, ethyl octanoate, ethyl hexanoate, hexyl acetate, ethyl isobutyrate, ethyl propionate, propyl acetate, and ethyl acetate.

To obtain the calibration curves, four concentrations were prepared for Solution A or B, made from two mother solutions in water and injected three times.

To consider the interactions between aroma compounds and proteins, dilutions were carried out in a 4% sodium caseinate solution (Sigma Aldrich, St. Louis, MO, USA). The choice of the dairy protein is justified by the intrinsic contents of off-flavors in pea proteins, the neutral taste, and the negligible volatile profile of dairy proteins.

2.6. Statistical Analysis

One-way analysis of variance (ANOVA) was performed using Xlstat sensory software (version 2019.4.1) (Addinsoft, New York, NY, USA). All tests were performed at $p = 0.05$. The heat map was generated using Euclidean distance and the complete linkage algorithm (90) implemented in the gplots package (version 3.0.1.1) (https://CRAN.Rproject.org/package=gplots) of R software (version 3.6.1) (http://www.r-project.org/).

3. Results and Discussion

3.1. The Impact of the Addition of Yeasts on Pea Fermentation by LAB

Table 1 summarizes the main parameters that characterize the fermentations performed by the different cultures: the bacterial and yeast biomass at the initial and/or final time (t0 and tf, respectively, where tf is the time necessary to reach pH 4.55), the residual sugar level at tf, and finally the lactic acid and ethanol concentration at tf.

To evaluate the impact of the addition of yeasts on the acidification rate of VEGE, the parameter tf was analyzed. As shown in Table 1, the addition of yeasts to VEGE did not have a significant impact on the fermentation time. This is a positive result in the aim of producing commercial products. However, we can observe a difference in the tf between two yeast species: there was an increase in the time necessary to reach pH 4.55 for VEGE + KM compared to VEGE + TOR. Nevertheless, this cannot be directly linked to the final bacteria biomass since higher concentrations were found in VEGE + TOR.

Table 1. Fermentation characteristics of the different cultures.

	Bacteria Biomass ($\times 10^8$ CFU/mL)	Yeast Biomass ($\times 10^7$ CFU/mL)		Kinetic Parameters (g/L)			
	at tf	at t0	at tf *	tf (h)	Total Residual Sugar at tf	Lactic Acid at tf	Ethanol at tf
VEGE047	2.9 ± 0.2 [b]	-	-	13.1 ± 0.5 [ab]	25.4 ± 0.3 [a]	3.5 ± 0.03 [b]	-
VEGE047 + *K. marxianus*	4.2 ± 0.6 [ab]	1.4 ± 0.1 [b]	4.4 ± 0.5 [a]	14.9 ± 0.9 [a]	13.2 ± 0.1 [d]	3.0 ± 0.05 [d]	4.8 ± 0.007 [a]
VEGE047 + *K . lactis*	6.0 ± 1.6 [a]	3.4 ± 0.6 [a]	4.1 ± 0.7 [a]	13.1 ± 0.4 [ab]	20.8 ± 0.6 [b]	3.6 ± 0.03 [a]	1.7 ± 0.004 [c]
VEGE047 + *T. delbrueckii*	5.4 ± 0.3 [a]	0.4 ± 0.1 [b]	0.7 ± 0.1 [a]	12.3 ± 0.01 [b]	16 ± 0.3 [c]	3.3 ± 0.02 [c]	4.2 ± 0.02 [b]

Each mean is based on three independent replicates. The values with letters of the same color were compared with each other. Values with the same letters are not significantly different ($p > 0.05$). * tf is the time needed to reach pH 4.55.

On the other hand, differences in the final concentration of lactic acid were observed in the presence of yeasts compared to VEGE alone: a lower concentration of lactic acid was observed for VEGE + KM and VEGE + TOR. These differences could be explained by an acidifying metabolite produced by yeasts. In fact, these two co-cultures displayed a higher production of ethanol compared to VEGE alone (in which no production was identified) or in VEGE + KL. Ethanol production paralleled carbon dioxide generation, a gas that, by dissolution, acidifies the medium. Thus, the production of CO_2 could explain the lower concentration of lactic acid generated to reach pH 4.55 in the two former conditions. As for VEGE + KL, the low final ethanol concentration could explain that the lactic acid concentration is nearly the same in this co-culture compared to VEGE alone. Since *Kluyveromyces lactis* is an aerobic-respiring yeast, it could be limited by the oxygen availability in our static fermentation, leading to limited growth and ethanol production compared to *Kluyveromyces marxianus* [29,30].

Considering the sucrose metabolism, the total residual sugars were lower in the presence of yeasts, compared to VEGE. In fact, VEGE alone consumed 6 g/L of sucrose, and no residual monosaccharides (fructose or glucose) were detected (Supplementary Table S1). The yield of lactic acid/sucrose calculated for VEGE was approximately 0.8 g/g, which is close to a homofermentative yield. In fact, VEGE contains three homofermentative species (*Lactobacillus acidophilus*, *Streptococcus thermophilus*, and *Lactobacillus delbrueckii* subsp. *bulgaricus*).

In co-cultures, nearly 100% of the initial sucrose (30 g/L) was hydrolyzed into glucose and fructose, and these monosaccharides were partially metabolized (Supplementary Table S1). The consumed sugars were 9 g/L for (VEGE+KL), 14 g/L for (VEGE + TOR), and 17 g/L for (VEGE + KM). These results are in agreement with the final concentration of ethanol obtained in each condition. As a conclusion, the addition of yeasts had a weak impact on the behavior of VEGE, although slight differences were identified due to probable negative and/or positive interactions between VEGE and the yeast used.

3.2. Modification of the Sensory Perception in the Presence of Yeasts

In order to evaluate the impact of yeasts, a sensory analysis was performed only on the fermented products. It was previously shown that VEGE cultures improved the sensory perception of a pea protein yogurt-like product, but not enough for consumer acceptability (General Mills, personal communication).

The panelists generated typical descriptors for a plant matrix: green flavor/vegetal, leguminous plant, bitter, astringent, nut, and other descriptors such as sparkling, tangy, sour, citrus fruit, beer/yeast, and cultured apple cider. Figure 1 shows the characteristics of each fermented product as the average intensity of the individual panelists' scores, and the detailed data are shown in Table 2.

Considering the main defects detected in the pea matrix, the intensities of leguminous plant and green flavor/vegetal were significantly reduced in the presence of yeasts. There is a lack of information in the literature about the impact of yeasts on the improvement of the sensory characteristics of pea proteins. However, one recent study revealed a significant decrease in the beany odor using analytical methods after the fermentation of soybean residue, okara, by *K. marxianus* [31].

The global intensity of aroma in the presence of yeasts was significantly higher compared to VEGE culture. This increase could be linked to the presence of trigeminal sensations such as sparkling, tangy, and sour attributes, which were significantly higher in the samples with yeasts. The sparkling attribute could be directly linked to the presence of CO_2 [32]. Considering the sour attribute, the yeasts can produce pyruvic and acetic acid, which could explain this perception [33]. In addition, it was previously shown that high levels of carbonation significantly enhanced the sourness and astringency in flavored milk beverages [34].

Finally, the fermented products obtained with yeasts were characterized by a significantly higher "beer/yeast" attribute than the products obtained with VEGE. In fermented beverages such as beer, esters are the most important set of yeast-derived aroma-active compounds, and they are responsible for their fruity character [35].

The modifications of the sensory perception in the presence of yeasts could be attributed to a reduction in the concentration of pea off-notes or the generation of new notes that could modify the perception of sensory defects. Thus, to obtain a better understanding of the sensory modifications, the volatile profiles were analyzed.

Table 2. Average intensity of sensory attributes determined for different fermented products using a scale ranging from 0 to 15.

	VEGE	VEGE + *K. marxianus*	VEGE + *K. lactis*	VEGE + *T. delbrueckii*
Global intensity	4.660 b	7.033 a	6.390 a	6.900 a
Sour	2.257 b	5.000 a	4.250 a	5.400 a
Bitter	1.837 a	3.200 a	2.840 a	2.867 a
Astringent	3.730 a	4.733 a	4.243 a	4.933 a
Tangy	0.417 c	5.267 a	3.500 b	5.200 a
Sparkling	0.050 d	7.467 b	4.817 c	9.367 a
Green flavor/Vegetal	2.637 a	0.733 b	1.533 ab	0.867 b
Leguminous plant	3.723 a	1.147 b	1.500 b	0.800 b
Citrus fruit	0.400 a	0.300 a	0.850 a	0.200 a
Nut	0.267 a	0.167 a	0.333 a	0.200 a
Beer/yeast	0.183 c	5.000 a	3.150 b	2.800 b
Sourdough	0.243 b	1.260 ab	1.380 ab	1.800 a
Cultured apple cider	0.200 a	1.340 a	1.067 a	1.067 a

Mean values in the same row that are not followed by the same letter are significantly different ($p < 0.05$).

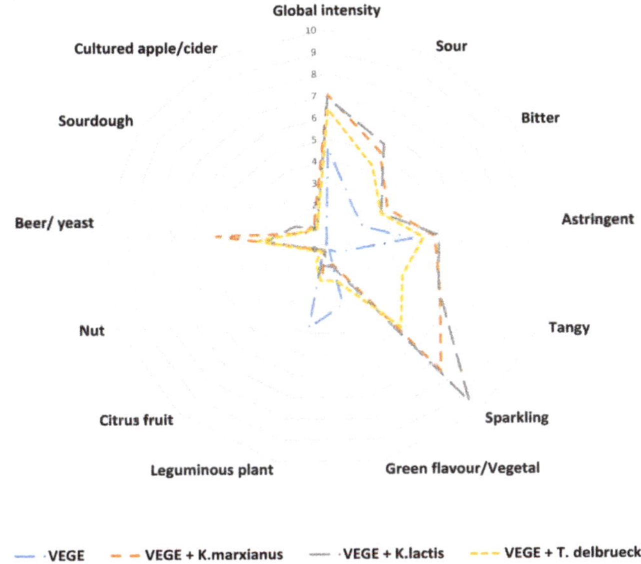

Figure 1. Aroma profile analyses of the four fermented pea protein isolate products. Data are displayed as mean numerical values of the sensory evaluations.

3.3. Characterization of Volatile Compounds Identified Using GC–MS Analysis

To obtain a better insight into the impact of fermentation on the volatile profile, GC–MS analyses were performed on the uninoculated and fermented samples (VEGE, VEGE + KM, VEGE + KL and VEGE + TOR).

3.3.1. Volatile profiles of Uninoculated and Fermented Samples

A total of 87 volatile molecules were detected. These compounds were grouped into five families, including aldehydes, alcohols, ketones, furans, and esters. A heat map (Figure 2) was drawn up using the proportions of each molecule (proportions calculated using the surface areas of the peaks) among the samples (Supplementary Table S2).

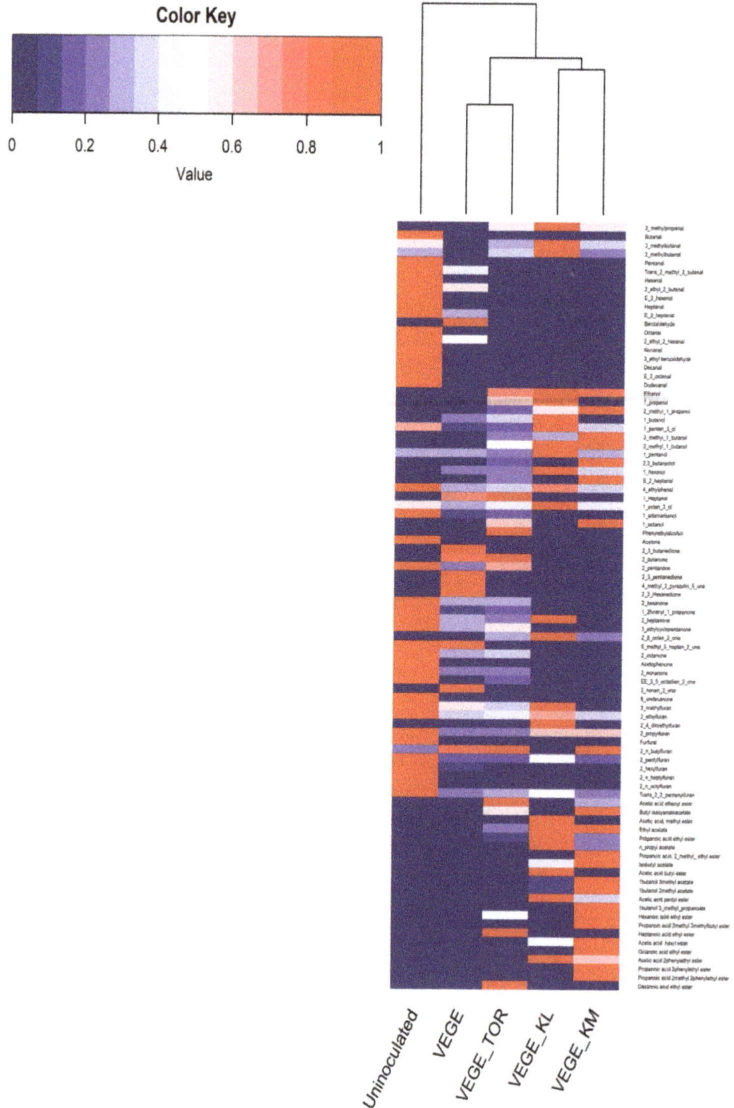

Figure 2. A hierarchically clustered heat map showing the patterns of the different samples for the identified volatile compounds.

First of all, it was observed that the uninoculated sample contained most of the aldehyde, ketone, and furan compounds, which are major families of molecules responsible for pea off-flavor. Upon fermentation with VEGE or VEGE + yeasts, the volatile profiles were strikingly modified compared to the uninoculated sample. As shown in Figure 2, most of the aldehyde, furan, and ketone molecules were degraded. Moreover, the esters were only present in the fermented products with VEGE and yeasts. The concentrations of alcohol were also generally higher in these products compared to VEGE.

The distribution of the proportions of the molecules revealed the presence of two main groups for the fermented cluster: VEGE and VEGE + TOR vs. VEGE + KL and VEGE + KM. These two clusters were not in agreement with the sensory results, which revealed differences in the sensory perception of products fermented with VEGE or VEGE and yeasts. In fact, there is not a direct link between the volatile compounds identified by GC–MS and the sensory descriptors.

Consequently, to acquire a better understanding of the sensory perceptions, we focused on the off-flavor molecules.

3.3.2. Degradation of Off-Flavor Molecules in the Fermented Samples

Twenty molecules among the 87 aroma compounds detected were reported in the literature as being responsible for the off-flavor perception in peas [9,36,37]. The concentrations of these 20 molecules are presented in Table 3. The perception thresholds were also reported for information purposes because they were determined by orthonasal olfaction in water [38].

First, it should be emphasized that most of the molecules responsible for the pea off-flavors were present in the uninoculated sample in which some of them were detected at high concentrations, including hexanal, butanal, 2-pentylfuran, and 2-ethylfuran. Hexanal has already been reported to be the major molecule responsible for the "green" and "herbal" perception in pea protein isolates [39]. Moreover, hexanal, as well as heptanal and nonanal, were identified as the main volatile compounds of soymilk flavor [40].

In the four fermented samples, there was a significant elimination of off-flavor molecules, mainly aldehydes, ketones, and furans. Previous studies showed that *Lactobacilli* and *Streptococci* could eliminate compounds related to the beany flavor of soymilk during fermentation, such as hexanal and 2-pentylfuran [40,41]. Moreover, lactic acid fermentation with *L. plantarum* or with *P. pentosaceus* had the potential to decrease the concentration of hexanal in lupin protein extracts [22].

Other compounds such as (E)-2-heptenal, 6-methyl-5-hepten-2-one, and trans-2-methyl-2-butenal were present in the products fermented with VEGE, but were not detectable in those fermented with VEGE and yeasts. These differences might explain the fact that the products fermented with VEGE were perceived as being greener and more leguminous by the panelists compared to the products fermented with VEGE and yeasts. This hypothesis should be supported by the determination of the odor perception thresholds by retronasal olfaction in pea.

In the fermented products with VEGE or with VEGE and yeasts, the incomplete reduction or the increase in off-flavor concentrations were notable. For example, 2-pentyl-furan and 2-ethyl-furan were reduced but remained at high levels in all the products. As for 2-methylpropanal, 1-hexanol, and 1-octanol, the concentrations increased when yeasts were added. These compounds could be the result of yeast metabolism under anaerobic conditions. Previous studies showed that the level of 2-methylpropanal increased in fermented pea gels using a microbial consortium, including yeasts [42]. This compound, which contributes to malty and chocolate-like notes, is the result of valine degradation by yeasts through the Ehrlich degradation pathway [43]. As for the two other alcohols, they could have been produced by the reduction of hexanal and octanal through the action of alcohol dehydrogenase activities. It was previously suggested that fermentation of soybean residue by *K. lactis* drastically decreased the amount of hexanal to trace levels, with a corresponding increase in hexanoic acid and/or hexanol [25–44]. The products with yeasts were perceived as being less "leguminous plant and green flavor/vegetal" than the products with VEGE; it is possible that the concentrations of 2-methylpropanal, 1-hexanol, and 1-octanol could be under their perception threshold. As previously mentioned, it will be necessary to determine the odor threshold of the molecules responsible for the sensory defects in peas.

Table 3. Concentrations of the off-flavor molecules in the non-fermented and fermented samples (µg/L).

Volatile Compounds	Descriptors	Uninoculated	VEGE	VEGE + *K. marxianus*	VEGE + *K. lactis*	VEGE + *T. delbrueckii*	Detection Threshold [a]
2-pentylfuran	Musty/earthy, mushroom, floral, buttery, rancid, green	304.3	46.4	56.7	148	55	6
Hexanal	Green, grass	181.3	<DL	<DL	<DL	<DL	4.5
2-ethylfuran	Beany, earthy, malty, sweet	77.7	28.1	29.7	56.7	32.8	
Butanal	Pungent, green, malty, chocolate, cocoa	54.9	<DL	<DL	<DL	<DL	
1-penten-3-ol	Green, vegetable, fruity	19.6	3.7	11	28.1	5.6	
(E)-2-octenal	Green, cucumber, musty/earthy, waxy, fatty, grass, banana, sweet	10.3	<DL	<DL	<DL	<DL	3
Nonanal	Aldehydic, fatty, green, geranium, floral, soapy, citrus, waxy	8.1	<DL	<DL	<DL	<DL	1
Heptanal	Green, fresh, fatty	5.7	<DL	<DL	<DL	<DL	3
2-nonanone	Green, earthy, grassy, fruity, sweet,	5.0	<DL	<DL	<DL	1.2	
2-methylpropanal	Aldehylic, grass, green, floral	4.0	0.7	40.9	70	39.6	6
1-octen-3-ol	Mushroom, earthy, burnt, green, vegetable, stale	3.9	2.2	2.9	6.7	2.9	1
(E)-2-heptenal	Pungent green, fatty	2.7	0.8	<DL	<DL	<DL	13
2-octanone	Green, floral, soapy, fruity, fatty	2.5	0.7	<DL	0.9	0.8	
Octanal	Aldehylic, green, soapy, citrus-like, sweet, waxy, fruity	2.5	<DL	<DL	<DL	<DL	0.7
(E)-2-Hexenal	Tea-like, green grass, almond, cherry, juicy, rancid	2.4	<DL	<DL	<DL	<DL	17
Decanal	Fresh, marine, aldehydic, iodized, soapy, grapefruit, bitter, sweet	1.04	<DL	<DL	<DL	<DL	0.1
6-methyl-5-hepten-2-one	Nutty, moldy, green, vegetable, citrus	0.5	0.5	<DL	<DL	<DL	2
Trans-2-methyl-2-butenal	Strong green-type odor and a fruity flavor	0.3	0.1	<DL	<DL	<DL	
1-hexanol	Green, musty/earthy, peanut hull, chemical-like, fruity, grassy	<DL	126.5	175.4	500.1	116.7	500
1-octanol	Moss, mushroom, green, vegetable, fatty, waxy, citrus, floral	<DL	<DL	4.4	<DL	2.9	

Concentrations are classified as per decreasing concentrations in the initial matrix. [a] Determined in water by orthonasal olfaction [38]. <DL: value inferior to the detection limit.

In addition, the perception of the above—mentioned compounds could have been modified by the presence of other molecules generated by yeasts such as esters. It was already suggested that the reduction and/or masking effect of off-flavors led to a more pleasant odor in the fermented lupin protein extracts compared to the unfermented protein extracts [22].

3.4. Understanding the Sensory Perception Generated by Yeasts

Esters were only detected in the products fermented with (Supplementary Table S2). Their concentrations, as well as their perception thresholds, are reported in Table 4.

As shown in Table 4, two main groups of esters were identified. The first group includes acetate esters (ethyl acetate, isobutyl acetate, 3-methylbutyl acetate, and 2-phenylethyl acetate), and the second group is composed of the ethyl esters (ethyl hexanoate and ethyl octanoate). The acetate esters in these two groups were found to be produced at higher levels compared to ethyl esters. These esters and their proportions are typical of fermented beverages such as beer [36]. This could explain the beer/yeast descriptor used in sensory analysis, which was shown to be significantly different between products with VEGE and those with VEGE and yeasts.

Ethyl acetate, an undesirable compound, was found to be present in higher concentrations in the three fermented samples compared to the other ester compounds. This molecule was also reported as the most abundant ester in wine and beers [50,51]. However, ethyl acetate could have a limited impact on sensory perception due to its high odor perception threshold [46]. As shown in Table 4, KM and KL produced more ethyl acetate compared to TOR.

Acetate esters are generated by yeasts as a result of the reaction between higher alcohols and acetyl co-A [36]. Higher alcohols are probably generated from the amino acid of pea proteins by the Ehrlich pathway [52].

As shown in Table 4, 2-phenylethyl acetate was produced in great quantity in the samples fermented with *Kluyveromyces* spp. It was shown that *K. marxianus* could produce this compound [53] and that *K. lactis* increased the amounts of this compound in the fermented soybean, okara [44].

In addition, hexyl acetate was identified in our fermented samples. This ester is not a common molecule found in beer. It is predicted that hexanol could be a direct precursor for this ester compound through the action of an alcohol acetyl transferase enzyme [53].

Other volatile compounds such as ethyl hexanoate and ethyl octanoate are produced by the lipid metabolism in yeasts and involve a reaction between ethanol and a fatty acid chain [48].

The presence of esters can add more complexity to pea products. In fact, esters can act in synergy with other molecules, thus influencing the final beer flavor at concentrations below the perception threshold [49].

To sum up, we can clearly see that most of the ester compounds were present at higher levels in co-cultures with *Kluyveromyces* sp. compared to *Torulaspora*. This highlights the fact that the different yeasts evaluated exert different metabolic activities. Moreover, the presence of these ester molecules might have generated a masking effect on the sensory defects in peas. Thus, further investigations are needed to identify the underlying phenomena that explain the modification of sensory perception in the presence of the following yeast species: *Kluyveromyces marxianus*, *Kluyveromyces lactis*, and *Torulaspora delbrueckii*.

Table 4. Concentrations of esters in the fermented samples with yeasts (μg/L).

Esters Compounds	Odor Descriptor *	Threshold Level [a]	Threshold Level in Beer [b]	VEGE + *K. marxianus*	VEGE + *K. lactis*	VEGE + *T. delbrueckii*
Ethyl acetate	Ethereal, fruity, sweet, grape and rum-like	12,000	25,000	2095	3040	633
Propyl acetate	Solvent-like pungency, lifting, fusel, amyl alcohol, sweet and fruity	4700		9.2	39.1	>DL
Hexyl acetate	Green, fruity, sweet, fatty, fresh, apple and pear	670		73	5.7	>DL
Isobutyl acetate	Sweet, fruity, ethereal with an apple banana nuance	1600	500	14	6.5	>DL
2-methyl butyl acetate	Sweet, banana, fruity, ripe, estery and tropical with a juicy, fruit-like note			78	0.84	>DL
3-methyl butyl acetate	Sweet, banana, fruity with a ripe estery nuance	160	2000	40	4.9	6.8
2-phenylethyl acetate	Sweet, honey, floral rosy, with a slight yeasty honey note with a cocoa and balsamic nuance	1800	200	354	372	2.7
Ethyl propanoate	Sweet, ethereal, rummy, grape, winey and fermented with an eggnog nuance	1800		51	240	30
Ethyl isobutyrate	Citrus, fruity, sweet			45	16	1
Ethyl hexanoate	Sweet, fruity, pineapple, waxy, fatty and estery with a green banana nuance	80	200	5	>DL	10
Ethyl octanoate	Waxy, sweet, musty, pineapple and fruity with a creamy, dairy nuance	580	1000	2	>DL	3

[a] Determined by orthonasal olfaction in 10% (*v/v*) ethanol solution adjusted to pH 3.5 with tartaric acid [45]. [b] Determined by orthonasal olfaction in beer [46]. * Descriptors presented in [47–49]

4. Conclusions

A microbial co-culture of VEGE and yeasts modified the sensory perception of a pea protein-based product. The intensity of the leguminous and green perception was decreased compared to VEGE alone. Moreover, new descriptors were generated in the presence of yeasts. A hedonic evaluation should be carried out to evaluate the acceptability of the fermented products.

Analytical results have shown that most of the molecules responsible for the leguminous and green off-notes were degraded by VEGE and by VEGE and yeasts, which is not in agreement with the sensory results. Nevertheless, three molecules, (E)-2-heptenal, trans-2-methyl-2-butenal, and 6-methyl-5-hepten-2-one, were still found in the products with VEGE but absent in those with VEGE and yeasts. Thus, it will be necessary to increment the list of molecules potentially responsible for the off-notes and to determine their threshold perception in the pea matrix.

The presence of yeasts modified the sensory perception of a pea "yogurt-like" product. In particular, the beer/yeast descriptor, which is probably due to esters, could be responsible for the decrease of the leguminous and green off-notes. This potential masking effect should be further investigated.

In conclusion, the fermentation process using a co-culture of LAB and yeasts is of major industrial interest because this process can compensate for the sensory defects in a plant matrix without adding any artificial aroma compounds.

Supplementary Materials: The following are available online at http://www.mdpi.com/2304-8158/9/3/349/s1. Table S1: Amount of residual sugars in the fermented samples (g/L), Table S2: Volatile compounds identified in the non-fermented and fermented pea samples (expressed as the surface area of the peaks).

Author Contributions: Planning of the work and supervision: S.H., S.L., C.E.Y.; writing—original draft paper: S.H., S.L., C.E.Y.; project administration: S.F., S.H., and S.L.; funding acquisition: S.F., P.B., S.H., and S.L.; writing, review, and editing: S.H., S.L., C.E.Y., P.B., S.F., and A.-C.P. All authors have read and agreed to the published version of the manuscript.

Funding: This research was supported by General Mills Yoplait.

Acknowledgments: Brigitte Pollet and Pascale Lieben are gratefully acknowledged for their excellent assistance in GC–MS and LC–MS assays.

Conflicts of Interest: The authors declare no conflict of interest. The sponsors had no role in the design, execution, interpretation, or writing of the study.

References

1. Graça, J.; Godinho, C.A.; Truninger, M. Reducing meat consumption and following plant-based diets: Current evidence and future directions to inform integrated transitions. *Trends Food Sci. Technol.* **2019**, *91*, 380–390. [CrossRef]
2. Silva, A.R.A.; Silva, M.M.N.; Ribeiro, B.D. Health Issues and Technological Aspects of Plant-based Alternative Milk. *Food Res. Int.* **2020**, 108972. [CrossRef]
3. McClements, D.J.; Newman, E.; McClements, I.F. Plant-based Milks: A Review of the Science Underpinning Their Design, Fabrication, and Performance. *Compr. Rev. Food Sci. Food Saf.* **2019**, *18*, 2047–2067. [CrossRef]
4. Singh, N. Pulses: An overview. *J. Food Sci. Technol.* **2017**, *54*, 853–857. [CrossRef]
5. Roland, W.S.; Pouvreau, L.; Curran, J.; van de Velde, F.; De Kok, P. Flavor Aspects of Pulse Ingredients. *Cereal Chem.* **2017**, *94*, 58–65. [CrossRef]
6. Dahl, W.J.; Foster, L.M.; Tyler, R.T. Review of the health benefits of peas (*Pisum sativum* L.). *Br. J. Nutr.* **2012**, *108*, S3–S10. [CrossRef]
7. Iqbal, A.; Khalil, I.A.; Ateeq, N.; Sayyar Khan, M. Nutritional quality of important food legumes. *Food Chem.* **2006**, *97*, 331–335. [CrossRef]
8. Lu, Z.X.; He, J.F.; Zhang, Y.C.; Bing, D.J. Composition, physicochemical properties of pea protein and its application in functional foods. *Crit. Rev. Food Sci. Nutr.* **2019**, *0*, 1–13. [CrossRef] [PubMed]
9. Murat, C.; Bard, M.-H.; Dhalleine, C.; Cayot, N. Characterisation of odour active compounds along extraction process from pea flour to pea protein extract. *Food Res. Int.* **2013**, *53*, 31–41. [CrossRef]
10. Schindler, S.; Zelena, K.; Krings, U.; Bez, J.; Eisner, P.; Berger, R.G. Improvement of the Aroma of Pea (*Pisum sativum*) Protein Extracts by Lactic Acid Fermentation. *Food Biotechnol.* **2012**, *26*, 58–74. [CrossRef]

11. Bott, L.; Chambers, E. Sensory Characteristics of Combinations of Chemicals Potentially Associated with Beany Aroma in Foods. *J. Sens. Stud.* **2006**, *21*, 308–321. [CrossRef]
12. Tangyu, M.; Muller, J.; Bolten, C.J.; Wittmann, C. Fermentation of plant-based milk alternatives for improved flavour and nutritional value. *Appl. Microbiol. Biotechnol.* **2019**, *103*, 9263–9275. [CrossRef] [PubMed]
13. Blandino, A.; Al-Aseeri, M.E.; Pandiella, S.S.; Cantero, D.; Webb, C. Cereal-based fermented foods and beverages. *Food Res. Int.* **2003**, *36*, 527–543. [CrossRef]
14. Spitaels, F.; Wieme, A.D.; Janssens, M.; Aerts, M.; Daniel, H.-M.; Van Landschoot, A.; De Vuyst, L.; Vandamme, P. The Microbial Diversity of Traditional Spontaneously Fermented Lambic Beer. *PLoS ONE* **2014**, *9*, e95384. [CrossRef]
15. Aidoo, K.E.; Nout, M.J.R.; Sarkar, P.K. Occurrence and function of yeasts in Asian indigenous fermented foods. *FEMS Yeast Res.* **2006**, *6*, 30–39. [CrossRef]
16. Park, Y.K.; Lee, J.H.; Mah, J.-H. Occurrence and reduction of biogenic amines in traditional Asian fermented soybean foods: A review. *Food Chem.* **2019**, *278*, 1–9. [CrossRef]
17. Onda, T.; Yanagida, F.; Uchimura, T.; Tsuji, M.; Ogino, S.; Shinohara, T.; Yokotsuka, K. Analysis of Lactic Acid Bacterial Flora during Miso Fermentation. *Food Sci. Technol. Res.* **2003**, *9*, 17–24. [CrossRef]
18. Feng, X.M.; Ostenfeld Larsen, T.; Schnürer, J. Production of volatile compounds by Rhizopus oligosporus during soybean and barley tempeh fermentation. *Int. J. Food Microbiol.* **2007**, *113*, 133–141. [CrossRef] [PubMed]
19. Xu, D.; Wang, P.; Zhang, X.; Zhang, J.; Sun, Y.; Gao, L.; Wang, W. High-throughput sequencing approach to characterize dynamic changes of the fungal and bacterial communities during the production of sufu, a traditional Chinese fermented soybean food. *Food Microbiol.* **2020**, *86*, 103340. [CrossRef]
20. Koutsoumanis, K.; Allende, A.; Alvarez-Ordóñez, A.; Bolton, D.; Bover-Cid, S.; Chemaly, M.; Davies, R.; Cesare, A.D.; Hilbert, F.; Lindqvist, R.; et al. Update of the list of QPS-recommended biological agents intentionally added to food or feed as notified to EFSA 10: Suitability of taxonomic units notified to EFSA until March 2019. *EFSA J.* **2019**, *17*, e05753. [CrossRef]
21. Sieuwerts, S.; de Bok, F.A.M.; Hugenholtz, J.; van Hylckama Vlieg, J.E.T. Unraveling Microbial Interactions in Food Fermentations: From Classical to Genomics Approaches. *Appl. Environ. Microbiol.* **2008**, *74*, 4997–5007. [CrossRef] [PubMed]
22. Schindler, S.; Wittig, M.; Zelena, K.; Krings, U.; Bez, J.; Eisner, P.; Berger, R.G. Lactic fermentation to improve the aroma of protein extracts of sweet lupin (*Lupinus angustifolius*). *Food Chem.* **2011**, *128*, 330–337. [CrossRef] [PubMed]
23. Yousseef, M.; Lafarge, C.; Valentin, D.; Lubbers, S.; Husson, F. Fermentation of cow milk and/or pea milk mixtures by different starter cultures: Physico-chemical and sensorial properties. *LWT Food Sci. Technol.* **2016**, *69*, 430–437. [CrossRef]
24. Ben-Harb, S.; Saint-Eve, A.; Panouillé, M.; Souchon, I.; Bonnarme, P.; Dugat-Bony, E.; Irlinger, F. Design of microbial consortia for the fermentation of pea-protein-enriched emulsions. *Int. J. Food Microbiol.* **2019**, *293*, 124–136. [CrossRef]
25. Chiba, H.; Takahashi, N.; Sasaki, R. Enzymatic Improvement of Food Flavor II. Removal of Beany Flavor from Soybean Products by Aldehyde Dehydrogenase. *Agric. Biol. Chem.* **1979**, *43*, 1883–1889. [CrossRef]
26. Hirst, M.B.; Richter, C.L. Review of Aroma Formation through Metabolic Pathways of Saccharomyces cerevisiae in Beverage Fermentations. *Am. J. Enol. Vitic.* **2016**, *67*, 361–370. [CrossRef]
27. Picque, D.; Corrieu, G. Characterization and classification of lactic acid bacteria based on their acidification kinetics. *Food Sci. Technol. Lebensm. Wiss. Technol.* **1992**, *25*, 181–186.
28. Le Boucher, C.; Courant, F.; Jeanson, S.; Chereau, S.; Maillard, M.-B.; Royer, A.-L.; Thierry, A.; Dervilly-Pinel, G.; Le Bizec, B.; Lortal, S. First mass spectrometry metabolic fingerprinting of bacterial metabolism in a model cheese. *Food Chem.* **2013**, *141*, 1032–1040. [CrossRef]
29. Merico, A.; Galafassi, S.; PiÅ¡kur, J.; Compagno, C. The oxygen level determines the fermentation pattern in *Kluyveromyces lactis*. *FEMS Yeast Res.* **2009**, *9*, 749–756. [CrossRef]
30. Lane, M.M.; Morrissey, J.P. *Kluyveromyces marxianus:* A yeast emerging from its sister's shadow. *Fungal Biol. Rev.* **2010**, *24*, 17–26. [CrossRef]
31. Hu, Y.; Piao, C.; Chen, Y.; Zhou, Y.; Wang, D.; Yu, H.; Xu, B. Soybean residue (okara) fermentation with the yeast *Kluyveromyces marxianus*. *Food Biosci.* **2019**, *31*, 100439. [CrossRef]

32. McMahon, K.M.; Culver, C.; Castura, J.C.; Ross, C.F. Perception of carbonation in sparkling wines using descriptive analysis (DA) and temporal check-all-that-apply (TCATA). *Food Qual. Prefer.* **2017**, *59*, 14–26. [CrossRef]
33. Fonseca, G.G.; Heinzle, E.; Wittmann, C.; Gombert, A.K. The yeast Kluyveromyces marxianus and its biotechnological potential. *Appl. Microbiol. Biotechnol.* **2008**, *79*, 339–354. [CrossRef] [PubMed]
34. Lederer, C.L.; Bodyfelt, F.W.; McDaniel, M.R. The Effect of Carbonation Level on the Sensory Properties of Flavored Milk Beverages. *J. Dairy Sci.* **1991**, *74*, 2100–2108. [CrossRef]
35. Verstrepen, K.J.; Derdelinckx, G.; Dufour, J.-P.; Winderickx, J.; Thevelein, J.M.; Pretorius, I.S.; Delvaux, F.R. Flavor-active esters: Adding fruitiness to beer. *J. Biosci. Bioeng.* **2003**, *96*, 110–118. [CrossRef]
36. Azarnia, S.; Boye, J.I.; Warkentin, T.; Malcolmson, L. Changes in volatile flavour compounds in field pea cultivars as affected by storage conditions. *Int. J. Food Sci. Technol.* **2011**, *46*, 2408–2419. [CrossRef]
37. Murat Chloe Gourrat, K.; Jerosch, H.; Cayot, N. Analytical comparison and sensory representativity of SAFE, SPME, and Purge and Trap extracts of volatile compounds from pea flour. *Food Chem.* **2012**, *135*, 913–920. [CrossRef]
38. Buttery, R.G.; Orts, W.J.; Takeoka, G.R.; Nam, Y. Volatile Flavor Components of Rice Cakes. *J. Agric. Food Chem.* **1999**, *47*, 4353–4356. [CrossRef]
39. Heng, L.; Vincken, J.-P.; Koningsveld G van Legger, A.; Gruppen, H.; Boekel T van Roozen, J.; Voragen, F. Bitterness of saponins and their content in dry peas. *J. Sci. Food Agric.* **2006**, *86*, 1225–1231. [CrossRef]
40. Achouri, A.; Boye, J.; Zamani, Y. Identification of volatile compounds in soymilk using solid-phase microextraction-gas chromatography. *Food Chem.* **2006**, *99*, 759–766. [CrossRef]
41. Blagden, T.D.; Gilliland, S.E. Reduction of Levels of Volatile Components Associated with the "Beany" Flavor in Soymilk by Lactobacilli and Streptococci. *J. Food Sci.* **2005**, *70*, M186–M189. [CrossRef]
42. Ben-Harb, S.; Irlinger, F.; Saint-Eve, A.; Panouillé, M.; Souchon, I.; Bonnarme, P. Versatility of microbial consortia and sensory properties induced by the composition of different milk and pea protein-based gels. *LWT* **2020**, *118*, 108720. [CrossRef]
43. Smit, B.A.; Engels, W.J.M.; Smit, G. Branched chain aldehydes: Production and breakdown pathways and relevance for flavour in foods. *Appl. Microbiol. Biotechnol.* **2009**, *81*, 987–999. [CrossRef] [PubMed]
44. Vong, W.C.; Liu, S.-Q. Bioconversion of green volatiles in okara (soybean residue) into esters by coupling enzyme catalysis and yeast (Lindnera saturnus) fermentation. *Appl. Microbiol. Biotechnol.* **2018**, *102*, 10017–10026. [CrossRef]
45. Peinado, R.A.; Moreno, J.; Bueno, J.E.; Moreno, J.A.; Mauricio, J.C. Comparative study of aromatic compounds in two young white wines subjected to pre-fermentative cryomaceration. *Food Chem.* **2004**, *84*, 585–590. [CrossRef]
46. Engan, S. Organoleptic threshold values of some alcohols and esters in beer. *J. Inst. Brew.* **1972**, *78*, 33–36. [CrossRef]
47. Dennis, E.G.; Keyzers, R.A.; Kalua, C.M.; Maffei, S.M.; Nicholson, E.L.; Boss, P.K. Grape Contribution to Wine Aroma: Production of Hexyl Acetate, Octyl Acetate, and Benzyl Acetate during Yeast Fermentation Is Dependent upon Precursors in the Must. *J. Agric. Food Chem.* **2012**, *60*, 2638–2646. [CrossRef]
48. Pires, E.J.; Teixeira, J.A.; Brányik, T.; Vicente, A.A. Yeast: The soul of beer's aroma—A review of flavour-active esters and higher alcohols produced by the brewing yeast. *Appl. Microbiol. Biotechnol.* **2014**, *98*, 1937–1949. [CrossRef]
49. Meilgaard, M.C.; Dalgliesh, C.E.; Clapperton, J.F. Beer flavour terminology[1]. *J. Inst. Brew.* **1979**, *85*, 38–42. [CrossRef]
50. Plata, C.; Millán, C.; Mauricio, J.C.; Ortega, J.M. Formation of ethyl acetate and isoamyl acetate by various species of wine yeasts. *Food Microbiol.* **2003**, *20*, 217–224. [CrossRef]
51. Lyumugabe, F.; Bajyana Songa, E.; Wathelet, J.P.; Thonart, P. Volatile compounds of the traditional sorghum beers "ikigage" brewed with Vernonia amygdalina "umubirizi". *Cerevisia* **2013**, *37*, 89–96. [CrossRef]
52. Hazelwood, L.A.; Daran, J.-M.; van Maris, A.J.A.; Pronk, J.T.; Dickinson, J.R. The Ehrlich Pathway for Fusel Alcohol Production: A Century of Research on Saccharomyces cerevisiae Metabolism. *Appl. Environ. Microbiol.* **2008**, *74*, 2259–2266. [CrossRef] [PubMed]
53. Liu, S.-Q.; Holland, R.; Crow, V.L. Esters and their biosynthesis in fermented dairy products: A review. *Int. Dairy J.* **2004**, *14*, 923–945. [CrossRef]

Article

A Comprehensive Characterisation of Volatile and Fatty Acid Profiles of Legume Seeds

Prit Khrisanapant [1,2], Biniam Kebede [1], Sze Ying Leong [1,2] and Indrawati Oey [1,2,*]

1 Department of Food Science, University of Otago, PO Box 56, Dunedin 9054, New Zealand; prit.khrisanapant@postgrad.otago.ac.nz (P.K.); biniam.kebede@otago.ac.nz (B.K.); sze.leong@otago.ac.nz (S.Y.L.)

2 Riddet Institute, Private Bag 11 222, Palmerston North 4442, New Zealand

* Correspondence: indrawati.oey@otago.ac.nz; Tel.: +64-3-479-8735

Received: 30 October 2019; Accepted: 4 December 2019; Published: 6 December 2019

Abstract: Legumes are rich in unsaturated fatty acids, which make them susceptible to (non) enzymatic oxidations leading to undesirable odour formation. This study aimed to characterise the volatile and fatty acid profiles of eleven types of legumes using headspace solid-phase microextraction gas chromatography–mass spectrometry (HS-SPME-GC-MS) and GC coupled with a flame ionisation detector (GC-FID), respectively. Volatile aldehydes, alcohols, ketones, esters, terpenes and hydrocarbons were the chemical groups identified across all the legumes. The lipids comprised palmitic, stearic, oleic, linoleic and α-linolenic acids, with unsaturated fatty acids comprising at least 66.1% to 85.3% of the total lipids for the legumes studied. Multivariate data analysis was used to compare volatile and fatty acid profiles between legumes, which allow discriminant compounds pertinent to specific legumes to be identified. Results showed that soybean, chickpea and lentil had distinct volatile and fatty acid profiles, with discriminating volatiles including lactone, ester and ketone, respectively. While all three *Phaseolus* cultivars shared similar volatile profiles, 3-methyl-1-butanol was found to be the only volatile differentiating them against the other eight legumes. Overall, this is the first time a multivariate data analysis has been used to characterise the volatile and fatty acid profiles across different legume seeds, while also identifying discriminating compounds specific for certain legume species. Such information can contribute to the creation of legume-based ingredients with specific volatile characteristics while reducing undesirable odours, or potentially inform relevant breeding programs.

Keywords: legumes; volatiles; fatty acids; characterisation; fingerprinting; multivariate data analysis

1. Introduction

The seeds of legume plants (usually referred to as 'legumes') are a nutritious source of proteins, carbohydrates, lipids, vitamins and minerals [1]. Unprocessed legumes have a distinct odour due to their inherent plant metabolism [2]. Regrettably, legumes are not widely utilised to their full potential due to various factors, such as low protein digestibility, their hard-to-cook-effect and undesirable odours. The undesirable odours are largely influenced by lipoxygenase-catalysed unsaturated fatty acid oxidation that occurs during several postharvest processes [3,4]. Trained sensory panels have used negative descriptors such as beany, musty, haylike, grassy and green to describe the odours of legumes such as soybean and peas. It has been recognised that these odours are majorly derived from volatile compounds such as hexanal and 1-octen-3-ol [3–5]. Previous studies [3,6–9] have reported volatile compounds and their formation mainly in soybean, and there is still a limited understanding on the volatile profile of other commercially relevant legumes such as cowpea, lentil, common bean and pea [10–13]. However, these studies used targeted analysis without providing a holistic picture of all low molecular weight compounds present in the legume's volatile fraction. Advancements in volatile analysis and instrument sensitivity are able to capture a wider range of volatile compounds; hence, they enable researchers to take an untargeted fingerprinting approach. By definition, fingerprinting is an untargeted analytical approach aiming to detect as many compounds as possible in a particular food fraction [14]. This untargeted fingerprinting approach has been used as a tool to study the effect of cooking on legumes [10,15]. To date, there is still a gap in the detailed, holistic characterisation of legume volatiles using this approach, and hence, it is essential to examine whether this analytical approach could differentiate different types of legumes. As previously mentioned, undesirable legume volatiles are primarily formed from fatty acid oxidation. In spite of this connection, no previous investigation has examined fatty acid composition of legumes in conjunction with their volatile profile. There is an opportunity here to elucidate the connection of fatty acid composition and volatile profile in the whole bean matrix.

Therefore, the objective of the present study was to comprehensively characterise the volatile and fatty acid profiles in commercially relevant legumes. The headspace solid phase-gas chromatography–mass spectrometry (HS-SPME-GC-MS) approach was implemented to detect volatile compounds, whereas the fatty acid was profiled using gas chromatography flame ionisation detection (GC-FID). The novelty of this study lies on the fact that (i) the fingerprinting approach is being applied for the first time on a wide range of commercial relevant legumes, and (ii) volatile and fatty acid components are analysed in an integrated fashion using multivariate data analysis.

2. Materials and Methods

2.1. Raw Material Handling and Storage

Eleven types of dry legume seeds, namely, soybean (*Glycine max*), pea (*Pisum sativum*), chickpea (*Cicer arretium* var. kabuli), orange lentil (*Lens culinaris*), mung bean (*Vigna radiata*), fava bean (*Vicia faba*), cowpea (*Vigna unguiculata*), adzuki bean (*Vigna angularis*), kidney bean (*Phaseolus vulgaris*), navy bean (*Phaseolus vulgaris*) and black bean (*Phaseolus vulgaris*), were purchased in a single batch from the local market in Dunedin (New Zealand). Seeds with physical damages and discolouration were discarded. The remaining seeds were vacuum packed in opaque aluminium bags and stored at 4 °C until analysis.

2.2. Sample Preparation

Legume seeds (30 g) were ground using a laboratory blender (Waring, Auckland, New Zealand) for 60 s, with a pause every 30 s, at room temperature (20 ± 2 °C). The resulting flour was sieved to pass through an 850 μm mesh. Flour retaining between 450 and 850 μm mesh size was used for lipid analysis for consistent extraction of lipids. For volatile profiling, the grounding time was increased to 180 s to maximise flour surface area.

2.3. Moisture Determination

The legume flour sample (0.2 g) was weighed and transferred into a glass petri dish (Steriplan, Kimax, Auckland, New Zealand). The dish was subsequently covered with perforated aluminium foil. The drying was carried out at 130 °C in a convection oven for 16 h (Qualtex, Andrew Thom, Sydney, Australia). The samples were removed and cooled in a desiccator lined at the bottom with silica beads. The percentage of moisture content was estimated based on the weight loss after drying and cooling. The moisture content determination was conducted in five independent replicates.

2.4. Headspace Volatile Analysis with HS-SPME-GC-MS Fingerprinting

Headspace solid phase micro-extraction gas chromatography mass spectrometry (HS-SPME-GC-MS) was conducted according to the work of Liu and others [16] with modifications, consisting of sample preparation, incubation, volatile extraction, injection and GC-MS analysis. Prior to analysis, method parameters were optimised in order to capture a wide range of volatile compounds. The optimisation included sample weight, sample dilution, sample to salt ratio and type of GC column.

Upon sample analysis, legume flour was weighed (2.5 g) into a 20 mL glass vial, and 5 mL of saturated sodium chloride solution (360 g/L) was added to increase the solution's ionic strength and drive the legume volatiles into the headspace. The vial was then tightly sealed with PTFE-coated silicon septa screw cap (Supelco, Sigma-Aldrich, St. Louis, MO, USA). The sealed vials were then vortexed for 30 s.

Using the Gerstel MPS Maestro autosampler (Gerstel, Linthicum Heights, MD, USA), each sample was incubated at 40 °C for 5 min, with agitation at 250 rpm. Thereafter, headspace volatile compounds were extracted using headspace-solid phase microextraction (HS-SPME). A preconditioned (according to the manufacturer's instructions) SPME fibber with a 30/50 μm divinylbenzene/carboxen/polydimethylsiloxane (DVB/CAR/PDMS) sorptive coating (Stableflex, Supelco, Bellefonte, PA, USA) was used to extract a wide range of volatile compounds from the headspace of the vial for 30 min at 40 °C.

For the GC-MS analysis (Agilent 6890N, Agilent Technologies, Santa Clara, CA, USA), extracted volatiles were desorbed in the injection port at 230 °C for 2 min, then injected in splitless mode onto a ZB-Wax capillary column (30 m × 0.25 mm × 0.25 μm; Agilent Technologies, Santa Clara, CA, USA) for separation with helium as the carrier gas at 1.5 mL/min. To facilitate separation and elution of the injected headspace volatile compounds, the GC oven was maintained at 50 °C for 5 min before the temperature was ramped up to 210 °C at 5 °C/min, after which it was again ramped to 240 °C at the rate of 10 °C/min, for a total GC-MS run time of 37 min. For the MS, the quadrupole was set at 70 eV, and the ion sources were 150 °C and 230 °C, respectively, with a mass-to-charge ratio scanning range of 30–300 m/z. Thereafter, the SPME fibre was regenerated according to the manufacturer's instruction. The same SPME fibre was used across all samples. The volatile profiling for each legume seed was conducted in five independent replicates.

With regard to preprocessing of GC-MS chromatograms, volatile fingerprinting chromatograms often contain co-eluting peaks, which can confound data analysis. Therefore, an automated mass deconvolution and identification system (AMDIS; version 2.72, build 140.24, Agilent Technologies, Santa Clara, CA, USA) was used to deconvolute potential co-eluting peaks. The spectra obtained were further processed by mass profiler professional (MPP; version 14.9.1, build 1316, Agilent Technologies, Santa Clara, CA, USA), a peak filtering and alignment software. This creates aligned peaks lacking nonreproducible and background peaks. Afterwards, a table of retention time and volatile amount expressed as peak area was obtained.

Tentative identification of volatile compounds was conducted manually. In the present work, three criteria were employed to increase the power of compound identification: (i) match and reverse match with the NIST library of no less than 90%; (ii) comparison of experimental retention index with RI according to literature; and (iii) matching retention time and spectra with authentic standards from chemical groups of detected volatiles (alcohol, aldehyde, terpene and acid) (See Table S1).

2.5. *Determination of Fatty Acids in Legume Seeds Using FAME-GC-FID*

For fatty acid analysis, legume lipid was extracted and converted to fatty acid methyl esters (FAME) and detected using gas chromatography flame ionisation detection (GC-FID) according to AOAC method 963.22 [17] with modifications.

2.5.1. Total Lipid Extraction Based on Soxhlet Method

Legume flour sample (2 g) was weighed and placed inside a cellulose extraction thimble (26 × 60 mm, Whatman, Buckinghamshire, UK). The filled extraction thimble was fitted onto a Soxtec distillation apparatus (Tecator, Hilleroed, Denmark). Meanwhile, aluminium cups were filled with five to ten antibumping granules and weighed before adding with 25 mL of organic solvent mix consisting of a 2:1 (*v*/*v*) ratio of chloroform (EMPARTA, Merck, Darmstadt, Germany) and methanol (Ajax Univar,

North Shore, New Zealand). The cups were then fitted underneath the thimble and above the heating plate of the Soxtec distillation apparatus. Continuous reflux distillation of the samples was then carried out for 1 h with the heating plate set at 160 °C. After that, solvent was evaporated from the sample, and then the cups were released from the apparatus, and the residual solvent was allowed to evaporate inside a convection oven (Sanyo MOV-212F, New South Wales, Australia) set at 50 °C for 15 min.

The lipid yield of each sample was estimated by weight difference between the weight of the empty cups (filled with antibumping granules) and the weight of the cup filled with lipids upon completion of the solvent extraction. After weighing, 15 mL of hexane (Ajax Finechem, Auckland, New Zealand) was used to resuspend the lipids in each aluminium cup. The hexane containing lipid solution was then stored in a refrigerator at 4 °C in a tightly sealed glass tube before proceeding to the lipid purification step. The lipid extraction of each legume seed was conducted in four independent replicates.

2.5.2. Lipid Purification

Fatty acid methyl esters (FAMEs) were obtained by purifying lipids by saponification to remove nonsaponifiable materials, followed by esterification into FAMEs. A volume of lipid containing hexane solution equivalent to 5 mg lipid was pipetted into a sealable glass tube. Thereafter, 5 mL of a solution containing potassium hydroxide (0.5 M, AnalaR, Leuven, Belgium) dissolved in methanol was added, and the tube was sealed immediately. The fatty acid saponification was carried out for 20 min at 80 °C on a heating block. The tubes were removed from the heating block and allowed to cool in ambient air for 10 min.

To the cooled solution, 3 mL of diethyl ether (LabServ, Auckland, New Zealand) and 7 mL of milliQ water were added, and the test tube was inverted to mix. The mixture was allowed to stand for 2 min to allow separation between water and organic solvent layers. The top layer of diethyl ether was then discarded to remove any nonsaponifiable material. The fatty acids were liberated by neutralisation to ~pH 7 with concentrated hydrochloric acid (37%; EMSURE, Merck, Darmstadt, Germany). Then, another 4 mL of diethyl ether was added and inverted to mix. The top diethyl ether layer formed was collected in a clean glass test tube for derivatisation.

2.5.3. Lipid Derivatisation to Fatty Acid Methyl Esters (FAMEs)

One millilitre of boron trifluoride (14%) in methanol (Sigma-Aldrich, St. Louis, MO, USA) was promptly added as derivatisation agent. Fatty acid esterification was carried out for 20 min at 80 °C on a heating block and then cooled. After cooling, 7 mL of saturated sodium chloride solution (360 g/L) was added and vortexed for 15 s. The top diethyl ether formed was then collected for FAME GC-FID analysis.

2.5.4. Fatty Acid Profiling Using GC-FID

For fatty acid analysis, gas chromatography flame ionisation detection (GC-FID) was conducted [17] with modifications. A GC-FID system (6890A G1530A; Agilent Technologies, Santa Clara, CA, USA) was used. It was equipped with an autosampler (7683 series injector, Agilent Technologies, Santa Clara, CA, USA) and fitted with a BPX70 capillary column (70% Cyanopropyl Polysilphenylene-siloxane, SGE, Victoria, Australia).

Samples (1 μL) were injected in split mode (20:1 ratio) at 240 °C for separation with hydrogen gas at 2.2 mL/min. To ensure good separation of fatty acid methyl esters, the GC oven temperature was increased from its initial temperature of 120 °C to 225 °C at the rate of 3 °C/min, then ramped to 245 °C at 10 °C/min. Once the GC oven temperature reached 245 °C, the column was held at this temperature for another 2 min. For the FID, the detector temperature was set at 250 °C.

2.5.5. Identification and Data Preprocessing of FAME

Chromatograms obtained from GC-FID were analysed with GC ChemStation (Build 4.01, Agilent Technologies, Santa Clara, CA, USA) and individual peaks manually identified by matching retention time with commercial standards (FAMQ-005, AccuStandards, New Haven, CT, USA). Following manual peak alignment and removal of interfering background compounds, the proportion of signal abundance of each fatty acid was calculated in % abundance of total signal abundance. Thereafter, a table of fatty acid profile for each legume was obtained.

2.6. Multivariate Data Analysis and Identification of Compounds Relevant to Specific Legume Type

Multivariate data analysis, marker selection and marker identification were performed on the combined data sets comprising legume fatty acid and headspace volatile data sets.

Using both volatile and fatty acid data, multivariate data analysis was conducted using principle component analysis (PCA), followed by partial least square discriminant analysis (PLS-DA), utilising Solo software (Version 8.2.1, Eigenvector Research, Manson, WA, USA). Firstly, PCA was used as an unsupervised, exploratory technique to determine grouping/separation in the data, as well as to detect outlier. Secondly, PLS-DA was used as a supervised technique to detect similarities and differences between different legume seeds, as well as correlation between volatile compounds and legumes. Thereafter, a bi-plot was generated as a visual representation of the information obtained (OriginPro, OriginLab, Northampton, MA, USA).

Volatile compounds that showed a clear discriminant correlation with each legume were selected through determination of variable identification (VID) coefficients [18]. VID values are the corresponding correlation coefficients between X-variables (volatile compounds and fatty acids) and predicted Y-variables (Legume type). An absolute threshold value of |0.800| was selected. Therefore, volatiles with an absolute VID coefficient higher than 0.800 were plotted as bar graphs, and statistical significance between the means was determined using analysis of variance, conducted through SPSS Statistics (IBM, Version 26), followed by Tukey's post-hoc test ($p < 0.05$). Those compounds were considered important discriminant compounds associated with each legume seed.

Discriminant volatile compounds were identified by comparing the deconvoluted mass spectra with an established mass spectra library using National Institute of Standards and Technology (NIST) Mass Spectral Search Program (Version 2.2, build June 10, 2014). The identities were also rechecked with a minimal 90% match and reverse match on NIST, as well as comparison of retention index with literature.

3. Results

3.1. Moisture Content of Legume Seeds

The moisture content of legume seeds was within the range of 7.4% (soybean) to 14.9% (kidney bean). Pea, navy bean, orange lentil and chickpea contained 9.9%, 9.6%, 8.7% and 8.4% moisture, respectively. In comparison, adzuki bean, black bean, fava bean, mung bean and cowpea were slightly moister, at 12.4%, 11.3%, 11.0%, 10.3% and 10.3%, respectively.

3.2. Fatty Acid Analysis of Legume Seeds

GC-FID analysis of the fatty acid methyl esters was able to detect five clearly separated peaks in eleven legume samples. The legume lipid fractions consisted of palmitic (C16:0), stearic (C18:0), oleic (C18:1), linoleic (C18:2) and α-linolenic (C18:3) acids (Table 1). A commonality of the lipid profiles is that the level of saturated fatty acids such as palmitic and stearic acids is low, with the ratio of saturated to unsaturated fatty acids at 1:2 in cowpea and mung bean, and up to 1:5.4 and 1:6 in soybean and chickpea. On the other hand, all legumes had a high level of essential polyunsaturated fatty acids, namely, linoleic and α-linolenic, ranging from 47.3% in orange lentil to 71.0% in black bean. Out of the eleven types of legumes, four of them, i.e., chickpea, orange lentil, pea and fava bean, had a high (>20%) proportion of oleic acid, a monounsaturated fatty acid. This affirms that legumes are a good source of unsaturated fatty acids [19], with their fatty acid profile favourable from a cardioprotective perspective [20].

Table 1. Relative fatty acid abundance of 11 types of legume seeds, as analysed by fatty acid methyl ester gas chromatography coupled with a flame ionisation detector (FAME-GC-FID).

Legumes	Total Lipid Extracted (g/100g Sample)	C16:0 (g/100g Lipid)	C18:0 (g/100g Lipid)	C18:1 (g/100g Lipid)	C18:2n-6 (g/100g Lipid)	C18:3n-3 (g/100g Lipid)	SFAs (g/100g Lipid)	MUFAs (g/100g Lipid)	PUFAs (g/100g Lipid)	n-6/n-3 Ratio
Soybean	19.20 ^{d} ± 1.98	11.76 ^{a} ± 0.13	3.68 ^{c,d} ± 0.01	19.20 ^{c} ± 0.28	55.15 ^{g} ± 0.15	8.88 ^{b} ± 0.02	15.44 ^{a,b} ± 0.13	19.20 ^{c} ± 0.28	64.03 ^{c,d} ± 0.14	6.21 ^{c} ± 0.02
Chickpea	7.73 ^{c} ± 0.73	10.94 ^{a} ± 0.20	1.80 ^{a} ± 0.0.16	37.87 ^{e} ± 0.16	45.78 ^{f} ± 0.42	2.33 ^{a} ± 0.05	12.74 ^{a} ± 0.35	37.87 ^{e} ± 0.16	48.11 ^{a} ± 0.46	19.67 ^{e} ± 0.36
Lentil	3.90 ^{a,b} ± 0.21	21.40 ^{e} ± 0.41	2.77 ^{a,b,c} ± 0.10	28.06 ^{d} ± 0.35	38.21 ^{c,d} ± 0.55	9.07 ^{b} ± 0.09	24.29 ^{e} ± 0.40	28.06 ^{d} ± 0.35	47.27 ^{a} ± 0.64	4.21 ^{b,c} ± 0.03
Cowpea	3.46 ^{a,b} ± 0.10	27.68 ^{f} ± 0.26	4.76 ^{e} ± 0.53	7.35 ^{a,b} ± 1.64	35.97 ^{b,c} ± 0.85	23.34 ^{e} ± 0.48	33.34 ^{f} ± 0.97	7.35 ^{a,b} ± 1.64	59.31 ^{b,c} ± 1.05	1.54 ^{a} ± 0.04
Pea	3.41 ^{a,b} ± 0.12	13.48 ^{a,b} ± 0.35	4.50d ^{e} ± 0.11	34.40 ^{e} ± 2.30	38.66 ^{c,d} ± 2.38	8.78 ^{b} ± 0.67	17.98 ^{b,c} ± 0.44	34.40 ^{e} ± 2.30	47.44 ^{a} ± 3.04	4.41 ^{b,c} ± 0.07
Mung bean	3.20 ^{a,b} ± 0.15	27.04 ^{f} ± 1.56	5.73 ^{f} ± 0.12	6.54 ^{a} ± 37.78	43.71 ^{e,f} ± 2.23	15.82 ^{c} ± 1.17	33.75 ^{f} ± 2.22	6.54 ^{a} ± 3.78	59.53 ^{b,c,d} ± 3.18	2.76 ^{a,b} ± 0.14
Fava bean	2.75 ^{a,b} ± 0.19	15.25b ^{c} ± 0.39	3.64 ^{c,d} ± 0.93	24.57 ^{c,d} ± 0.34	52.68 ^{g} ± 0.88	3.61 ^{a} ± 0.57	19.14 ^{c,d} ± 1.02	24.57 ^{c,d} ± 0.34	56.29 ^{b} ± 1.17	14.59 ^{d} ± 2.06
Navy bean	3.87 ^{b} ± 0.17	18.00 ^{d} ± 0.22	3.12 ^{bc} ± 0.13	19.49 ^{c} ± 0.81	28.30 ^{a} ± 0.45	31.09 ^{f} ± 0.33	21.12 ^{d} ± 0.15	19.49 ^{c} ± 0.81	59.39 ^{b,c,d} ± 0.73	0.91 ^{a} ± 0.01
Kidney bean	3.59 ^{a,b} ± 0.14	17.97 ^{d} ± 0.49	2.40 ^{a,b} ± 0.27	12.71 ^{b} ± 2.00	29.02 ^{a} ± 1.16	35.71 ^{g} ± 1.39	20.49 ^{c,d} ± 0.65	12.71 ^{b} ± 2.00	64.74 ^{d} ± 2.55	0.81 ^{a} ± 0.00
Black bean	3.20 ^{a,b} ± 0.16	17.17c ^{d} ± 0.48	2.39 ^{a,b} ± 0.29	8.29 ^{a,b} ± 0.16	33.65 ^{b} ± 0.34	37.30 ^{g} ± 1.22	19.56 ^{c,d} ± 0.71	8.29 ^{a,b} ± 0.16	70.95 ^{e} ± 1.50	0.90 ^{a} ± 0.02
Adzuki bean	1.96 ^{a} ± 0.10	27.96 ^{f} ± 1.02	3.36 ^{c} ± 0.40	3.92 ^{a} ± 0.52	41.88 ^{d,e} ± 1.12	20.31 ^{d} ± 0.73	31.35 ^{f} ± 1.45	3.92 ^{a} ± 0.52	62.19 ^{c,d} ± 1.96	2.06 ^{a} ± 0.01
F-value	170.887	229.2	33.7	93.3	114.7	1047.5	148.5	93.3	46.0	218.0
Significant	0.00	0.00	0.00	0.00	0.00	0.00	0.00	0.00	0.00	0.00

Values expressed as mean ± standard deviation ($n = 4$). C16:0 = palmitic acid; C18:0 = stearic acid; C18:1 = oleic acid; C18:2 = linoleic acid; C18:3 = α-linolenic acid; SFAs = saturated fatty acids; MUFAs = monounsaturated fatty acids; PUFAs = polyunsaturated fatty acid; n-6/n-3 = ratio of omega-6 to omega-3 fatty acids. Means with different superscripts in the same column indicate significant difference ($p < 0.05$).

3.3. Volatile Analysis of Legume Seeds

The HS-SPME-GC-MS fingerprinting method was able to detect an increased number of volatile compounds, totalling 97 different volatiles across all 11 legumes. Visually, the total ion chromatograms appear to be different based on the number and intensity of the peaks present amongst the samples. Some representative total ion chromatograms of the samples are shown in Figure 1.

The chemical classes of detected volatile compounds consisted of alcohols, aldehydes, ketones, esters, lactones, terpenes, hydrocarbons, furans, pyrroles and sulphur-containing compounds. Note that the percentage of specific compounds mentioned in this section refers to their relative abundance, not absolute concentration. It is also important to note that soybean and chickpea had volatiles with the highest total peak area compared to other legumes.

In soybean, a total of 63 volatile compounds were detected with the headspace fingerprinting method, consisting mainly of aldehydes, ketones, alcohols, esters, furans and sulphur and hydrocarbons. Hexanal (40.9%), 1-octen-3-ol (21.1%) and 1-hexanol (5.9%) were volatiles with the highest abundance. A total of 76 compounds were detected in chickpea with aldehydes and alcohols as two dominant chemical classes, consisting of hexanal (56.3%), nonanal (8.7%) and 1-hexanol (4.4%). Acid, ketones, furans, esters and terpenes were also present, with hexanoic acid (1.7%) being the most abundant. In cowpea, 80 headspace volatiles were detected in this study. Hexanal (22.6%), 4,1 methylethyl benzaldehyde (22.0%) and 1-hexanol (10.4%) were prominent. Other aldehydes, terpene and benzene compounds were also detected in cowpea. A total of 65 compounds were detected in pea. Alcohols and aldehydes are the majority of volatile detected in pea, such as hexanal (42.5%), 1-penten-3-ol (9.0%), 1-hexanol (7.7%) and nonanal (6.0%). In orange lentil, 82 volatile compounds were detected. Hexanal (32.7%), 1-hexanol (16.4%), 2-hexenal (5.6%) and o-cymene (4.6%) make up the top four most abundant volatile compounds, accounting for 59.3% of total volatile detected. Uniquely, lentil had the largest number (13) of terpenes of all the samples. Headspace volatiles detected in mung bean (63) comprised aldehydes, alcohols, terpenes, ketones and sulphur compounds. The top three volatiles consisted of hexanal (38.8%), 1-hexanol (13.9%) and 1, 3-dimethyl-benzene (6.3%). Interestingly, mung bean had the highest abundance of xylene (5.7%) detected out of all samples. Fava bean had the fewest (55) volatiles detected in the headspace. Aldehydes and alcohols are the main volatile detected, with terpenes, terpene derivatives, furans and esters comprising a minor fraction. Hexanal (40.4%), 3-methylbutanol (19.1%) and 3-methylbutanoic acid (10.3%) were the top three volatile compounds detected in the fava bean headspace fraction. Major volatiles detected in adzuki bean (67) headspace fraction consisted of aldehyde, alcohol and, interestingly, furans, including hexanal (21.1%), 3-furaldehyde (19.5%), 3-furanmethanol (7.5%) and 1-hexanol (6.0%). Black, navy and kidney beans had 68, 76 and 72 volatile compounds detected in the headspace, respectively, with the most abundant fraction being aldehyde and alcohol, followed by terpenes, acids and ketones. Hexanal, 1-hexanol, 1-penten-3-ol and 3-methylbutanol are similarly the major volatile compounds detected in the three *Phaseolus* samples.

3.4. Comparison of the Volatile and Fatty Compositions Among the Eleven Legumes and Identifying Discriminant Compounds

Multivariate data analysis (MVDA), which is an advanced chemometrics technique, was used to compare the volatile and fatty acid profiles among the 11 legume samples and identify discriminating compounds. In order to investigate the interdependence and relation among the measured attributes, the volatile and fatty acid data were merged into a single data matrix and analysed with MVDA. A principle component analysis (PCA) was first used as an unsupervised exploratory technique to detect groupings, separations or outliers within the volatile and sample data. From the PCA modelling (results not shown), it was able to be determined that there is indeed some distinct grouping and separation within the samples and that there were no outliers.

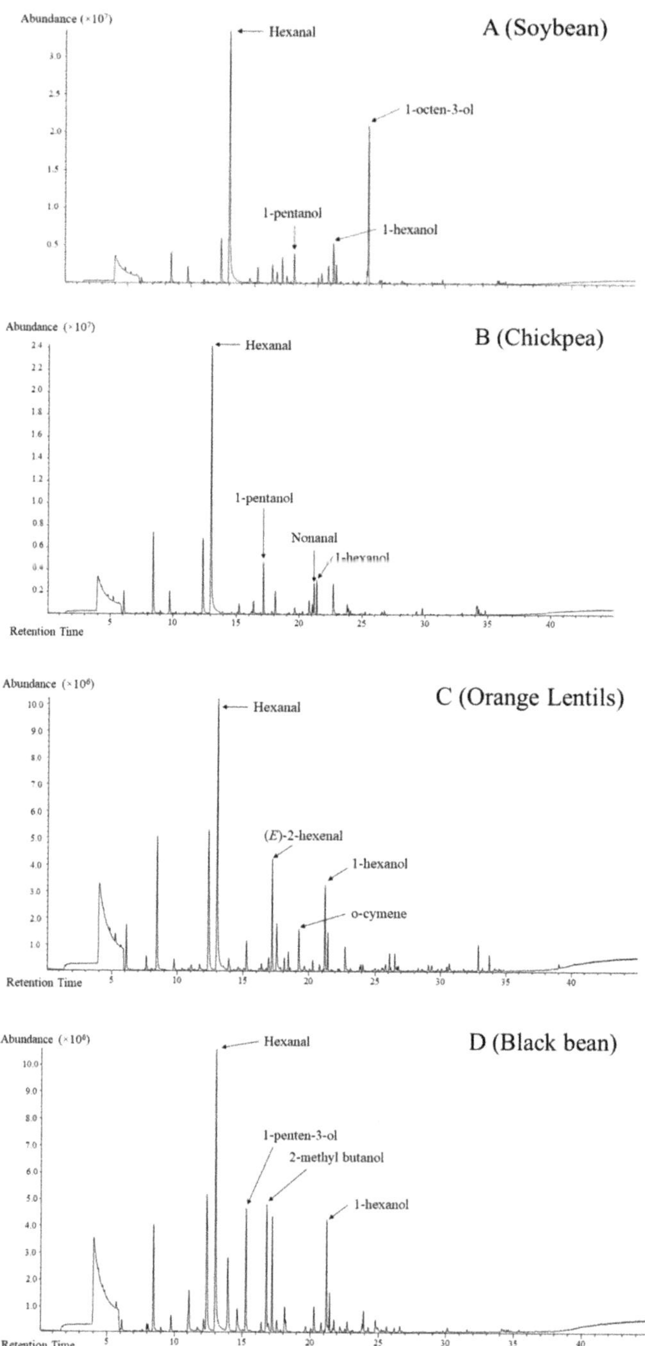

Figure 1. Representative total ion chromatograms of soybean (**A**), chickpea (**B**), orange lentil (**C**) and black bean (**D**) obtained with the headspace solid-phase microextraction gas chromatography–mass spectrometry (HS-SPME-GC-MS) fingerprinting method.

Thereafter, a partial least squares discriminant analysis (PLS-DA) model was constructed using the volatile and fatty acid profiles as X-variables and the 11 types of legumes as categorical Y-variables. A bi-plot constructed using the first two latent variables (LVs) is shown in Figure 2. On the bi-plot, samples that are close to each other are considered similar, whereas samples that are further apart are considered different [18]. Figure 2 clearly shows that soybean and chickpea are projected further away from other samples in their own quadrant, indicating a large difference compared to the other legumes. The third quadrant is shared by cowpea and lentil, again indicating differences from other legumes. The fourth quadrant is occupied by pea, mung bean, fava bean, adzuki bean and all three *Phaseolus* beans, indicating similarity between the samples, especially between kidney, navy and black beans. This similarity may be attributed to them belonging to the same species.

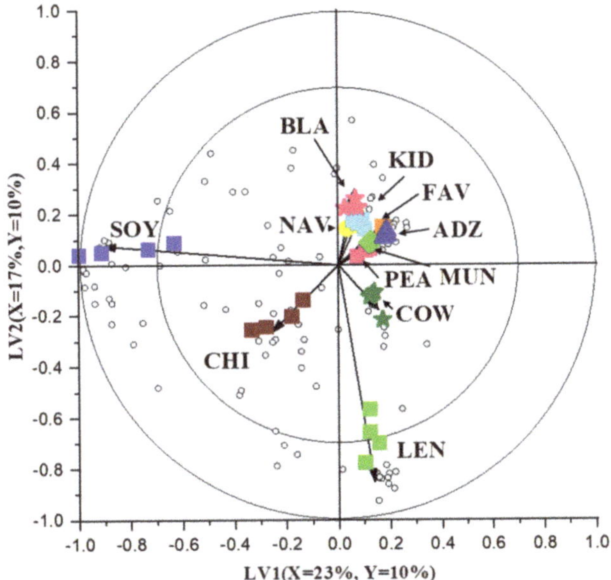

Figure 2. A bi-plot based on partial least square discriminant analysis (PLS-DA) comparing the volatile and fatty acid profiles among the 11 types of legumes. The variance explained is (X = 22%, Y = 10%) and (X = 17%, Y = 10%) for the first and second latent variable, respectively. ▲ = Adzuki bean (ADZ) A. ■ = Chickpea (CHI) ★ = Black bean (BLA) ■ = Navy bean (NAV) ■ = Kidney bean (KID) ■ = Fava bean (FAV) ♦ = Mung bean (MUN) ■ = Pea (PEA) ■ = Soybean (SOY) ★ = Cowpea (COW) ■ = Orange lentil (LEN).

In addition to the legume samples, unfilled circles on the bi-plot represent volatile and fatty acid compounds (X-variables). The location of each circle represents its relation to other measured attributes (X-variables) or samples (Y-variables). Hence, a PLS-DA bi-plot provides a graphical representation of the relation between measured attributes and legume types. To gain further understanding into the specific volatiles or fatty acids which are clearly different between legume samples, variable selection was performed using a VID technique. The selected discriminant compounds are listed in Table 2. To illustrate the differences amongst the seeds, some representative discriminant volatiles and fatty acids are also visually presented in Figure 3, with significant difference ($p < 0.05$) determined using analysis of variance and Tukey's post-hoc test. Key points are discussed in Section 4.

Table 2. List of discriminant volatile compounds/fatty acids for individual legume samples.

VID	Identity	RI	Chemical Group
	Soybean (18)		
0.989	α-methyl-γ-butyrolactone	1621	Ester & Lactone
0.977	1-octen-3-one	1321	Ketone
0.971	β-methyl-γ-butyro-lactone	1644	Ester & Lactone
0.967	Heptanal	1189	Aldehyde
0.964	Linoleic acid	*	Fatty Acid
0.959	2(*Z*)-heptenal	1349	Aldehyde
0.958	Stearic acid	*	Fatty Acid
0.941	1-octen-3-ol	1461	Alcohol
0.934	1-pentanol	1251	Alcohol
0.934	2(*E*)-octenal	1458	Aldehyde
0.920	Palmitic acid	*	Fatty Acid
0.902	3,5-octadien-2-ol	1433	Alcohol
0.889	2(*Z*)-penten-1-ol	1330	Alcohol
0.878	2,4-nonadienal	1710	Aldehyde
0.874	5-ethylcyclopent-1-enecarboxaldehyde	1451	Aldehyde
0.867	3-octanone	1268	Ketone
0.823	Pentanal	949	Aldehyde
0.804	Oleic acid	*	Fatty Acid
	Lentil (15)		
0.984	2-butanone	873	Ketone
0.982	Pyrrole	1540	Pyrrole
0.976	Menthol	1647	Alcohol
0.974	2-methoxyethylbenzene	1519	Hydrocarbon
0.970	Anethole	1815	Hydrocarbon
0.966	Caryophyllene	1630	Terpene
0.958	o-cymene	1291	Terpene
0.953	α-copaene	1529	Terpene
0.944	Linalool	1554	Terpene
0.941	Terpinen-4-ol	1620	Terpene
0.898	α-terpinyl acetate	1705	Terpene

Table 2. *Cont.*

VID	Identity	RI	Chemical Group
0.892	γ-Terpinene	1262	Terpene
0.820	p-Cymen-7-ol	2009	Terpene
0.819	D-limonene	1208	Terpene
0.806	2(*E*)-hexenal	1230	Aldehyde
	Chickpea (8)		
0.967	Allyl nonanoate	1632	Ester & Lactone
0.943	1,4-dichlorobenzene	1478	Hydrocarbon
0.884	6-methyl-5-hepten-2-one	1358	Ketone
0.883	1-octanol	1564	Alcohol
0.883	Nonanal	1418	Alcohol
0.859	Hexanoic acid	1831	Acid
0.833	2(*E*)-decenal	1657	Aldehyde
0.800	6-methyl-3,5-heptadiene-2-one	1613	Ketone
	Cowpea (6)		
0.975	α-muurolene	1734	Hydrocarbon
0.971	4,1 methylethyl benzaldehyde	1785	Aldehyde
0.967	α-terpinen-7-al	1796	Aldehyde
0.953	γ-ethyl-γ-butyrolactone	1719	Ester & Lactone
0.925	3-p-menthen-7-al	1598	Aldehyde
0.803	γ-methyl-γ-butyrolactone	1639	Ester & Lactone
	Mung bean (5)		
0.967	p-xylene	1194	Hydrocarbon
0.954	Isophorone	1624	Ketone
0.900	o-xylene	1133	Hydrocarbon
0.877	1,3-dimethylbenzene	1141	Hydrocarbon
0.868	Toluene	1019	Hydrocarbon

Table 2. *Cont.*

VID	Identity	RI	Chemical Group
	Fava bean (2)		
0.904	7-epi-silphiperfol-5-ene	1490	Ester & Lactone
0.863	3-methylbutanoic acid	1684	Acid
	Pea (1)		
0.875	2-pentanone	947	Ketone
	Adzuki bean (6)		
0.934	3-furaldehyde	1455	Aldehyde
0.925	3-furanmethanol	1680	Furan
0.846	2-methylfuran	869	Furan
0.810	Methyl salicylate	1782	Ester & Lactone
	Kidney bean (1)		
0.838	3-methyl-1-butanol	1202	Alcohol
	Black bean (5)		
0.900	β-pinene	1099	Terpene
0.873	α-pinene	997	Terpene
0.869	3-methyl-1-butanol	1202	Alcohol
0.859	4(Z)-heptenal	1254	Aldehyde
0.847	1-penten-3-ol	1148	Alcohol
	Navy bean (1)		
0.871	3-methyl-1-butanol	1202	Alcohol

* Experimental retention index was calculated for each volatile compound, while individual fatty acids were identified by matching retention time with commercial standards (FAMQ-005, AccuStandards, New Haven, CT, USA).

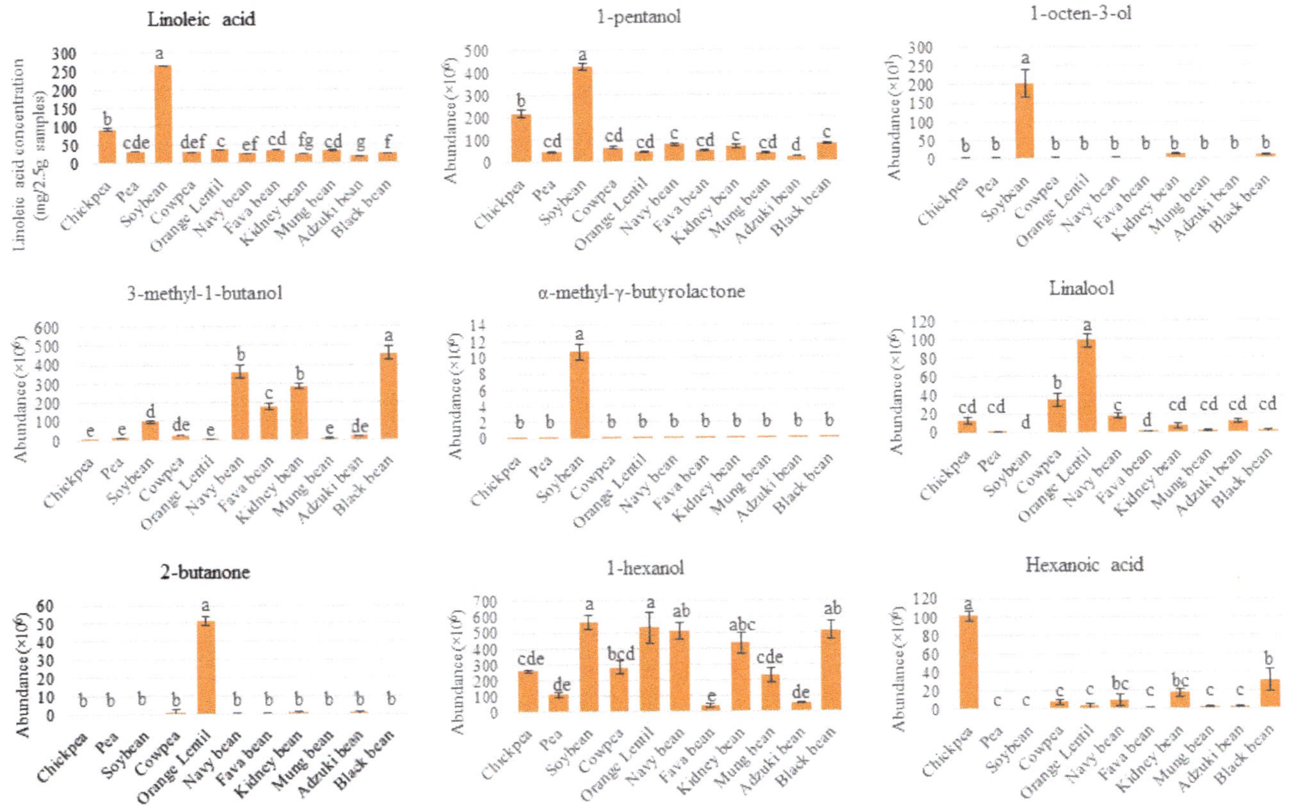

Figure 3. Individual plots of some representative discriminant compounds driving the distinction between legume samples observed in the PLS-DA biplot in Figure 2. Result expressed as mean ± standard deviation of four independent measurements per legume type ($n = 4$). Samples with different letters indicate significant difference ($p < 0.05$).

4. Discussion

The information gathered in this study has shown some key trends emerging from data analysis. These subsections discuss them in relation to the existing literature and known reaction pathways.

4.1. Ratio of Unsaturated Fatty Acids in Studied Legumes

The *Phaseolus* group (navy, kidney and black bean) has a low ratio (<1) of omega-6 (n-6; linoleic acid) to omega-3 (n-3; α-linoleic acid) fatty acids. Adzuki bean, mung bean and cowpea also have relatively low n-6/n-3 ratio. This is in contrast to chickpea and fava bean, which have a very high n-6/n-3 ratio of more than 18 and 14, respectively (Table 1). Diets where the ratio of fat consumed has a high n-6/n-3 ratio (≥10:1) have been linked to risk of developing chronic noncommunicable diseases, such as autoimmune and inflammatory disease, cardiovascular disease and cancer, whereas a ratio of less than 4:1 generally indicates better health outcomes [21]. Therefore, legumes, excluding chickpea and fava bean, are a good source of essential fatty acid with a desirable n-6/n-3 ratio.

However, despite their health benefits, unsaturated fatty acids are known to be more susceptible to oxidation than saturated fatty acids, due to the presence of one or more double bonds. The extent of lipid oxidation in foods can be reliably measured by determining the peroxide value so that preventive measures can be taken to delay oxidation in legume seeds after harvest while preventing the production of undesirable rancid flavour in legumes [22]. Endogenous enzymes present in legumes such as lipoxygenase utilise unsaturated fatty acids as substrate, producing volatile compounds, some of which possess undesirable odours. This is especially a problem for legumes with a high proportion of unsaturated fatty acids (>80%), such as soybean and chickpea (Table 1). The evolution and presence of volatile compounds are discussed further in the next section.

4.2. Aldehydes, Alcohols, Ketones and Terpenes in Studied Legumes

Aldehydes were the most abundant chemical class detected in all samples, except for navy, kidney and black beans; these beans seem to be rich in alcohols. Hexanal was the most abundant compound detected in all samples, except for navy bean. Nonanal and 2-hexenal were the second and third most abundant aldehyde in the legumes studied in this work. This is consistent with previous studies in soybean, winged bean [23] and three cultivars of common beans (black, pinto and dark red kidney) [10]. The difference in hexanal abundance may be attributed to the difference in the amount of linoleic and linolenic acid (Table 1), since they are the precursor for lipoxygenase-catalysed evolution of hexanal [3].

Lipoxygenases have been well characterised for soybean and chickpea [24]. Soybean and chickpea lipoxygenase isozymes form C_9 and C_{13} hydroperoxides of PUFAs, acting on linoleic and α-linolenic acid, while chickpea lipoxygenase also exhibited high co-oxidation with carotene and retinol compounds [25]. Hydroperoxide lyase isozymes then cleave the aforementioned C9 and C13 hydroperoxides into isomeric nonenals, 4-hydroxynonenal and 9-oxo-nonanoic acid [26] and hexanal, cis-3-hexenal and 12-oxo-cis-9-dodecenoic acid [27,28], respectively. These products can degrade and/or be further acted on by isomerase and other enzymes, generating additional volatile aldehydes [29], which may help to explain the presence of other analogous aldehydes such as pentanal, heptanal, octanal, 2-hexenal, 2-heptenal and 2-octenal in soybean and chickpea.

Alcohols comprised the second-most abundant chemical class of volatile compounds in the headspace of legume seeds, except for navy bean, where hexanol is the most abundant volatile compound. This is not surprising, as legumes contain alcohol dehydrogenases that act on products of the lipoxygenase pathway described above. For example, three isozymes of alcohol dehydrogenase have been described in chickpea, which catalyse the interconversion of aldehydes, alcohol and acid [30], possibly explaining the abundance of 1-hexanol and 1-pentanol in chickpea. Other alcohols prominent in legumes possibly arising from enzymatic actions include 1-penten-3-ol and 1-octen-3-ol, two compounds which had been described as having undesirable odour.

Ketones and hydrocarbons are also detected in the headspace volatile fractions of legumes, though less so than alcohol and aldehydes. Ketones and hydrocarbons are also derived from lipid oxidation, from both (non) enzymatic oxidative degradations [2]. For example, 2,3-pentanedione is found in oxidised soybean oil and has a buttery flavour (Seals and Hammond 1970). Acetone, 6-methylhept-5-en-2-one (methyl heptenone) and (E,E)-3,5-octadien-2-one detected in this study have been found in large quantities in dry beans [10].

Terpenes were also detected in the headspace volatile of legume seeds; the largest number is detected in lentils. Unlike aldehydes, alcohols, ketones and hydrocarbons, which are products of fatty acid oxidation, terpenes are naturally present/synthesised by the plant [2]. α-Pinene and β-pinene are the two most common terpenes in legume samples investigated.

4.3. Lipoxygenase is the Most Substantial Contributor for Volatile Evolution

Out of the 11 samples, soybean contains the highest number (18) of discriminating compounds (including volatiles and fatty acids), and this can also be seen on the bi-plot (Figure 2) as soybean samples are projected the furthest away from other legumes. Volatiles with high VID coefficients in soybean comprise aldehydes (e.g., heptanal), ketones (e.g., 1-octen-3-one) and alcohols (e.g., 1-octen-3-ol and 1-pentanol). In literature, these compounds are associated with undesirable odours in soybean. For example, 1-pentanol has a pungent fusel or solvent-like odour, while 1-octen-3-ol has a mushroom and earthy odour with a low detection threshold at 1 ppb in water and 1 ppm in soymilk [31]. Individual plots presented in Figure 3 illustrate their intensity/abundance in comparison to other legume samples. These compounds are products of lipoxygenase-catalysed fatty acid oxidation. Linoleic acid is a precursor of many aldehydes, ketones and alcohols, as it is susceptible to autoxidation and lipoxygenase-catalysed oxidation [3]. Accordingly, linoleic, stearic, palmitic and oleic acids are also detected in a higher amount in soybean, in descending order (Table 2 and Figure 3). This shows the potential of the multiplatform approach followed by chemometrics, such as multivariate data analysis, to demonstrate relationships amongst different measured attributes—in this case, fatty acids and volatile profile.

However, other legume seeds also contain fatty acids and their own isozymes of lipoxygenases [24]. Even a low amount of fatty acid may generate volatiles contributing to off-odours [4,32,33]. Therefore, while it holds that lipoxygenase-catalysed fatty acid oxidation products are detected in much higher abundance in soybean because of its high linoleic acid (and other fatty acid) content, these volatiles can also be present in other legume seeds.

4.4. Soybean and Cowpea Contain Distinctive Butyrolactones

Three butyrolactone compounds were selected as discriminant compounds in soybean (α-methyl-γ-butyrolactone and β-methyl-γ-butyro-lactone) and cowpea (γ-ethyl-γ-butyrolactone). β-scission of fatty acid hydroperoxides can yield carbonyl compounds, which can participate in Maillard and Strecker degradation reactions to yield the aforementioned cyclic volatile compound [34]. Perhaps it is due to the combination of lipoxygenase isozyme specific to soybean, combined with plentiful precursor (Table 1), that α-methyl-γ-butyrolactone was able to be detected exclusively in soybean (Figure 3). As for the other three lactones, their presence in *Phaseolus* (β-methyl-γ-butyro-lactone) and all samples except for fava bean (γ-ethyl-γ-butyrolactone) indicates the presence of lipoxygenase with similar specificity. This is supported by Chigwedere, Tadele, Yi, Wibowo, Kebede, Van Loey, Grauwet and Hendrickx [15], with β-methyl-γ-butyro-lactone also detected in *Phaseolus* samples, whereby it was identified as a marker distinguishing aged and fresh beans cooked for 270 min.

4.5. Orange Lentil Contains Discriminant Terpene and Carotenoid Degradation Products

Based on Figure 2, orange lentil has a distinct volatile and fatty acid profile compared to other legumes, with 15 discriminant volatile compounds. Specifically, orange lentil is distinct for the presence of a high number of discriminating terpenes, terpene derivatives and cyclic hydrocarbons. This result is not surprising, as the orange colour of the lentil suggests the presence of carotenoids. A previous study conducted by Zhang, et al., [35] showed that the total carotenoid content of 20 cultivars of red lentil grown in Canada by dry weight ranged between 5.32 and 28.1 µg/g. In addition to being secondary metabolites, volatile terpenes may arise from degradation of carotenes by either legume lipoxygenases or hydroperoxides generated from autolytic and enzyme-catalysed lipid oxidation [2]. Linalool (Figure 3) and D-limonene can also be considered as potential discriminant compounds for orange lentil, though they were also present in other samples. These volatiles are associated with a citrus and fresh odour [36].

4.6. Presence of 2-Butanone and Methylated Compounds in Orange Lentil

In orange lentils, 2-butanone and pyrrole were selected as discriminant compounds with high VID coefficients (Table 2). The ketone 2-butanone has a "moderately sharp, sweet, pungent, and acetone-like" odour and has been reported as a product of lipoxygenase-catalysed oxidation of unsaturated fatty acid hydroperoxides [37], as well as temperature-accelerated oxidation of saturated fatty acids [3]. According to individual plot (Figure 3), a high relative amount of 2-butanone was detected in orange lentil in the current study and has previously been reported in lentils, navy bean, red kidney bean and peas [2,12,37]. With orange lentil only having 3.9% lipid content (Table 1), the high relative abundance of 2-butanone suggests that the sample may have undergone heat treatment in its distribution chain, or that some components of orange lentil are especially vulnerable to heat treatment. This hypothesis is supported by the presence of pyrrole. Ma, Boye, Azarnia and Simpson [37] similarly reported pyrrole in navy bean, red kidney bean, green lentil and yellow pea that had undergone heat treatment.

4.7. Presence of Acid in Chickpea and Fava Bean Suggests Alcohol Dehydrogenase Activity

Hexanoic acid is selected as a discriminant compound in chickpea (Table 2). This compound has been linked with the alcohol dehydrogenase pathway, which converts alcohol (in this case, 1-hexanol) into the corresponding acid [38]. From a visual inspection of the individual plots of 1-hexanol and hexanoic acid (Figure 3), chickpea only had a moderate amount of 1-hexanol, but it had the highest amount of hexanoic acid detected. This seems to suggest that alcohol dehydrogenase was present in a high concentration or had relatively high activity compared to other isozymes. A similar observation can be seen with fava bean, where it had a distinctly higher relative amount of 3-methyl-butanoic acid (VID coefficient 0.863), corresponding to 3-methyl-1-butanol contributing to 19.1% of its headspace volatile. Similar to lipoxygenase, alcohol dehydrogenase also has isozymes. Gomes, Jadrić, Winterhalter and Brkić [30] have reported three isozymes in chickpea cotyledon. This could suggest that chickpea and fava bean contain alcohol dehydrogenase which converts aldehydes to alcohols.

4.8. Members of the Phaseolus Group Appears to Contain Similar Dominant Volatile Compounds

In this study, 3-methyl-1-butanol was identified as a discriminant compound common to kidney, black and navy beans. Figure 3 visually illustrates its abundance. As previously stated in Section 3.3, hexanal and 1-penten-3-ol are volatile common to the *Phaseolus* legumes and have previously been identified as marker compounds in three cultivars of common beans, having high scores in the principle components [10]. This indicates that despite being of different cultivars and colours, kidney, navy and black beans have similar characterising headspace volatiles.

5. Conclusions

The approach used in this paper, integrating fingerprinting and profiling, is effective to characterise the volatile and fatty acid profiles of the eleven legumes seeds selected. The detected volatiles can be grouped into aldehyde, alcohol, ketone, terpene, ester and lactone and hydrocarbon chemical classes. The lipid profiles comprised palmitic, stearic, oleic, linoleic and α-linolenic acids. Advanced chemometrics utilising multivariate data analysis were used to determine distinctive volatile compounds for different legume species. The occurrence of specific discriminant compounds is hypothesised to be majorly derived from the action of species-specific isozymes, especially lipoxygenases. While findings from this result majorly emphasised how certain volatile compounds discriminate each legume and how they are linked to fatty acids, this information may aid in choosing legume-based ingredients with the desired volatile profile. This insight can also be valuable to legume breeders in selecting legumes with certain fatty acid profile, aiming for higher n-6/n-3 ratios and/or oxidative stability. Lastly, since legumes are typically processed (such as soaking and cooking) prior to consumption, the effects of processing on these legume-specific volatiles is worth further investigation.

Supplementary Materials: The following are available online at http://www.mdpi.com/2304-8158/8/12/651/s1, Table S1: Pure standards injected to confirm identity of selected compounds of interest (10 ppm).

Author Contributions: Conceptualisation, B.K., I.O. and P.K.; Methodology, B.K. and P.K.; Software, B.K.; Validation, B.K. and I.O.; Formal Analysis, P.K.; Investigation, P.K.; Resources, B.K., I.O. and S.Y.L.; Data Curation, B.K. and P.K.; Writing—Original Draft Preparation, P.K.; Writing—Review and Editing, B.K., I.O. and S.Y.L.; Visualisation, P.K.; Supervision, B.K., I.O. and S.Y.L.; Project Administration, I.O.; Funding Acquisition, I.O.

Funding: This research was supported by the Riddet Institute, a New Zealand Centre of Research Excellence, funded by the Tertiary Education Commission. The author also gratefully thanks the New Zealand Aid Scholarship funded through the Ministry of Foreign Affairs and Trade.

Acknowledgments: Khrisanapant would like to thank the excellent team at the Department of Food Science at the University of Otago: John Birch, who taught him all about fatty acid analysis and Michelle Leus, who helped him to gain competency in gas chromatography, as well as Ivy Salih from the Department of Human Nutrition for teaching him the operation of the Soxtec apparatus. The authors also acknowledge Agilent Technologies, for giving a discount on the Mass Profiler Professional (MPP) software.

Conflicts of Interest: The authors declare no conflict of interest.

References

1. Tharanathan, R.N.; Mahadevamma, S. Grain legumes—A boon to human nutrition. *Trends Food Sci. Technol.* **2003**, *14*, 507–518. [CrossRef]
2. Azarnia, S.; Boye, J.I. Flavour Compounds in Legumes: Chemical and Sensory Aspects. In *Progress in Food Science and Technology*; Azarnia, S., Boye, J.I., Eds.; Nova Science Publishers, Inc.: New York, NJ, USA, 2011.
3. MacLeod, G.; Ames, J.; Betz, N.L. Soy flavor and its improvement. *Crit. Rev. Food Sci. Nutr.* **1988**, *27*, 219–400. [CrossRef] [PubMed]
4. Rackis, J.J.; Sessa, D.J.; Honig, D.H. Flavor problems of vegetable food proteins. *J. Am. Oil Chem. Soc.* **1979**, *56*, 262–271. [CrossRef]
5. Roland, W.S.U.; Pouvreau, L.; Curran, J.; van de Velde, F.; de Kok, P.M.T. Flavor aspects of pulse ingredients. *Cereal Chem.* **2017**, *94*, 58–65. [CrossRef]
6. Zhang, Y.; Guo, S.; Liu, Z.; Chang, S.K.C. Off-flavor related volatiles in soymilk as affected by soybean variety, grinding, and heat-processing methods. *J. Agric. Food Chem.* **2012**, *60*, 7457–7462. [CrossRef] [PubMed]
7. Li, Y.-Q.; Zhang, W. Effects of pulsed electric fields on volatile constituents of soymilk. *Agro Food Ind. Hi Tech* **2012**, *23*, 6.
8. Achouri, A.; Boye, J.I.; Zamani, Y. Soybean variety and storage effects on soymilk flavour and quality. *Int. J. Food Sci. Technol.* **2008**, *43*, 82–90. [CrossRef]
9. Min, S.; Yu, Y.; Yoo, S.; Martin, S.S. Effect of soybean varieties and growing locations on the flavor of soymilk. *J. Food Sci.* **2005**, *70*, C1–C7. [CrossRef]
10. Oomah, B.D.; Liang, L.S.Y.; Balasubramanian, P. Volatile Compounds of dry beans (*Phaseolus vulgaris* L.). *Plant Foods Hum. Nutr.* **2007**, *62*, 177. [CrossRef]
11. Rembold, H.; Wallner, P.; Nitz, S.; Kollmannsberger, H.; Drawert, F. Volatile components of chickpea (*Cicer arietinum* L.) seed. *J. Agric. Food Chem.* **1989**, *37*, 659–662. [CrossRef]
12. Lovegren, N.V.; Fisher, G.S.; Legendre, M.G.; Schuller, W.H. Volatile constituents of dried legumes. *J. Agric. Food Chem.* **1979**, *27*, 851–853. [CrossRef]
13. Fisher, G.S.; Legendre, M.G.; Lovgren, N.V.; Schuller, W.H.; Wells, J.A. Volatile constituents of southernpea seed (*Vigna unguiculata* (L.) Walp.). *J. Agric. Food Chem.* **1979**, *27*, 7–11. [CrossRef]
14. Vervoort, L.; Grauwet, T.; Kebede, B.T.; Van der Plancken, I.; Timmermans, R.; Hendrickx, M.; Van Loey, A. Headspace fingerprinting as an untargeted approach to compare novel and traditional processing

technologies: A case-study on orange juice pasteurisation. *Food Chem.* **2012**, *134*, 2303–2312. [CrossRef] [PubMed]
15. Chigwedere, C.M.; Tadele, W.W.; Yi, J.; Wibowo, S.; Kebede, B.T.; Van Loey, A.M.; Grauwet, T.; Hendrickx, M.E. Insight into the evolution of flavor compounds during cooking of common beans utilizing a headspace untargeted fingerprinting approach. *Food Chem.* **2019**, *275*, 224–238. [CrossRef] [PubMed]
16. Liu, F.; Grauwet, T.; Kebede, B.T.; Van Loey, A.; Liao, X.; Hendrickx, M. Comparing the Effects of High Hydrostatic Pressure and Thermal Processing on Blanched and Unblanched Mango (*Mangifera indica* L.) Nectar: Using Headspace Fingerprinting as an Untargeted Approach. *Food Bioprocess Technol.* **2014**, *7*, 3000–3011. [CrossRef]
17. Association of Official Analytical Chemists. *AOAC Official Method 963.22 Methyl Esters of Fatty Acids in Oils and Fats*; AOAC International: Rockville, MD, USA, 2005.
18. Kebede, B.T.; Grauwet, T.; Tabilo-Munizaga, G.; Palmers, S.; Vervoort, L.; Hendrickx, M.; Van Loey, A. Headspace components that discriminate between thermal and high pressure high temperature treated green vegetables: Identification and linkage to possible process-induced chemical changes. *Food Chem.* **2013**, *141*, 1603–1613. [CrossRef] [PubMed]
19. Ryan, E.; Galvin, K.; O'Connor, T.P.; Maguire, A.R.; O'Brien, N.M. Phytosterol, Squalene, Tocopherol Content and Fatty Acid Profile of Selected Seeds, Grains, and Legumes. *Plant Foods Hum. Nutr.* **2007**, *62*, 85–91. [CrossRef]
20. Mensink, R.P.; Zock, P.L.; Kester, A.D.; Katan, M.B. Effects of dietary fatty acids and carbohydrates on the ratio of serum total to HDL cholesterol and on serum lipids and apolipoproteins: A meta-analysis of 60 controlled trials. *Am. J. Clin. Nutr.* **2003**, *77*, 1146–1155. [CrossRef]
21. Simopoulos, A.P. The importance of the ratio of omega-6/omega-3 essential fatty acids. *Biomed. Pharmacother.* **2002**, *56*, 365–379. [CrossRef]
22. Damodaran, S.; Parkin, K.L. *Fennema's Food Chemistry*, 4th ed.; CRC Press: Boca Raton, FL, USA, 2017.
23. Del Rosario, R.; de Lumen, B.O.; Habu, T.; Flath, R.A.; Richard Mon, T.; Teranishi, R. Comparison of headspace volatiles from winged beans and soybeans. *J. Agric. Food Chem.* **1984**, *32*, 1011–1015. [CrossRef]
24. Baysal, T.; Demirdöven, A. Lipoxygenase in fruits and vegetables: A review. *Enzym. Microb. Technol.* **2007**, *40*, 491–496. [CrossRef]
25. Sanz, L.C.; Perez, A.G.; Olias, J.M. Purification and catalytic properties of chickpea lipoxygenases. *Phytochemistry* **1992**, *31*, 2967–2972. [CrossRef]
26. Gardner, H.W.; Weisleder, D.; Plattner, R.D. Hydroperoxide lyase and other hydroperoxide-metabolizing activity in tissues of soybean. *Glycine Max Plant Physiol.* **1991**, *97*, 1059. [CrossRef] [PubMed]
27. Olias, J.M.; Rios, J.J.; Valle, M.; Zamora, R.; Sanz, L.C.; Axelrod, B. Fatty acid hydroperoxide lyase in germinating soybean seedlings. *J. Agric. Food Chem.* **1990**, *38*, 624–630. [CrossRef]
28. Matoba, T.; Hidaka, H.; Kitamura, K.; Kaizuma, N.; Kito, M. Contribution of hydroperoxide lyase activity to n-hexanal formation in soybean. *J. Agric. Food Chem.* **1985**, *33*, 856–858. [CrossRef]
29. Liu, K. *Soybeans: Chemistry, Technology, and Utilization*; Springer: Boston, MA, USA, 1997.
30. Gomes, J.; Jadrić, S.; Winterhalter, M.; Brkić, S. Alcohol dehydrogenase isoenzymes in chickpea cotyledons. *Phytochemistry* **1982**, *21*, 1219–1224. [CrossRef]
31. Stark, W.; Forss, D. A compound responsible for mushroom flavour in dairy products. *J. Dairy Res.* **1964**, *31*, 253–259. [CrossRef]
32. Sessa, D.; Honig, D.; Rackis, J. Lipid oxidation in full-fat and defatted soybean flakes as related to soybean flavor. *Cereal Chem.* **1969**, *46*, 675–686.
33. Sessa, D.J.; Rackis, J.J. Lipid-Derived flavors of legume protein products. *J. Am. Oil Chem. Soc.* **1977**, *54*, 468–473. [CrossRef]
34. Blank, I.; Fay, L.B. Formation of 4-hydroxy-2,5-dimethyl-3(2H)-furanone and 4-hydroxy-2(or 5)-ethyl-5(or 2)-methyl-3(2H)-furanone through Maillard reaction based on pentose sugars. *J. Agric. food Chem.* **1996**, *44*, 531–536. [CrossRef]
35. Zhang, B.; Deng, Z.; Tang, Y.; Chen, P.; Liu, R.; Ramdath, D.D.; Liu, Q.; Hernandez, M.; Tsao, R. Fatty acid, carotenoid and tocopherol compositions of 20 Canadian lentil cultivars and synergistic contribution to antioxidant activities. *Food Chem.* **2014**, *161*, 296–304. [CrossRef] [PubMed]
36. Mosciano, G. Organoleptic characteristics of flavor materials. *Perfum. Flavorist* **2000**, *25*, 26.

37. Ma, Z.; Boye, J.I.; Azarnia, S.; Simpson, B.K. Volatile flavor profile of saskatchewan grown pulses as affected by different thermal processing treatments. *Int. J. Food Prop.* **2016**, *19*, 2251–2271. [CrossRef]
38. Grosch, W. Lipid oxidation product and flavour. In *Food Flavors: Part A, Introduction*; Morton, I.D., MacLeod, A.J., Eds.; Elsevier: New York, NY, USA, 1982; p. 325.

Article

Tailoring Physical and Sensory Properties of Tofu by the Addition of Jet-Milled, Superfine, Defatted Soybean Flour

Ye-Na Kim [1], Syahrizal Muttakin [1,2], Young-Min Jung [1], Tae-Yeong Heo [1] and Dong-Un Lee [1,*]

1 Department of Food Science and Technology, Chung-Ang University, 17546 Anseong, Korea; kyn6874@naver.com (Y.-N.K.); smuttakin@gmail.com (S.M.); jym20023@hanmail.net (Y.-M.J.); xodud1485@naver.com (T.-Y.H.)

2 Indonesian Agency for Agricultural Research and Development (IAARD), 12540 Jakarta, Indonesia

* Correspondence: dong-un.lee@cau.ac.kr; Tel.: +82-(0)31-670-3034

Received: 1 November 2019; Accepted: 22 November 2019; Published: 25 November 2019

Abstract: The use of defatted soybean flour (DSF) in food as a source of dietary fiber has been limited due to its rough texture and bitter taste. Our previous work indicates that superfine DSF prepared by jet milling could overcome these problems, as it positively affected physical and sensory properties. Therefore, differently sized DSFs were incorporated in tofu, and their impacts on physical and sensory properties were investigated in this study. Coarse DSF (Dv_{50} = 341.0 μm), fine DSF (Dv_{50} = 105.3 μm), and superfine DSF (Dv_{50} = 5.1 μm) were prepared by conventional sifting and jet milling. Tofu was made with a 5% addition of differently sized DSFs and without DSF (control tofu). The quality of tofu was evaluated by scanning electron microscopy, color measurement, texture profile analysis, and quantitative descriptive analysis. The tofu made with coarse and fine DSF showed negative changes in its physical and organoleptic qualities, such as reduced yields, a less pure color, a harder texture, and a rougher mouthfeel. However, the tofu made with superfine DSF showed only minimal changes in its qualities compared to the control. Therefore, superfine DSF is a promising fiber supplement that does not change the physical and sensory properties in the making of high-quality tofu.

Keywords: defatted soybean flour; jet mill; super-fine powder; tofu; quantitative descriptive analysis; texture profile analysis

---

1. Introduction

Tofu is a well-known Asian food made by coagulating soymilk and then pressing the resulting curds into soft white blocks. The coagulation of soymilk is done with coagulants such as $CaSO_4$, $CaCl_2$, or glucono-δ-lactone (GDL). Transglutaminase, which can catalyze the cross-linking reactions between soy proteins, is also being investigated as a coagulant for tofu [1]. During the process of making tofu, soymilk is heated to 90 °C. Thermal treatment is routinely used in the production of tofu to dissociate, denature, and aggregate the soy protein, inhibit microbial growth, reduce the beany flavor, and inactivate undesirable biological compounds such as trypsin inhibitors and lipoxygenase [2,3].

Tofu is a good source of protein. Its general composition is 84.6% water, 8.1% protein, 4.8% lipids, and 1.9% carbohydrates [4]. As it is derived from soybeans, tofu is also rich in phenolic compounds, such as isoflavones, which have antioxidant activities and health benefits [5–7]. Although tofu contains valuable nutritional compounds, it has little dietary fiber because the soybean pulp is removed during production. Aside from the health benefits of dietary fiber, consumer demand for unrefined or minimally processed whole foods is increasing. The small amount of dietary fiber in tofu and increasing consumer demand for whole foods could be resolved if a fiber-enriched tofu could be produced with dietary fiber from unpolished beans.

Significant effort has been made to obtain dietary fiber from vegetable products as a functional ingredient. Examples include baked foods with added dietary fiber [8,9], fiber-enriched cocoa powder [10], dietary fiber-enriched biscuits using okra flour [11], whole wheat/soy flour bread [12], and breads containing corn bran [13]. However, there has been little research to verify the effects of the addition of dietary fiber to tofu.

One of the dietary ingredients derived from soybeans is defatted soybean flour (DSF). DSF contains high levels of dietary fiber, along with proteins, carbohydrates, and fats [14]. Although DSF is produced in large quantities as a by-product of soybean oil, the addition of DSF to food as a source of fiber has been restricted due to its rough texture and bitter taste.

We have previously shown that superfine DSF powder could be prepared by conventional milling and sifting followed by jet milling [15]. Jet-milled DSF showed significant reductions in bitterness and roughness by descriptive analysis with trained panelists, and its water-holding capacity, water-solubility index, and swelling capacity were significantly increased. These results indicate the possibility of using superfine DSF as a functional ingredient to modify the physical properties of food without any negative changes to its sensory properties. The objective of this study was to apply jet-milled, superfine DSF powder to the preparation of fiber-enriched tofu and to verify its effects on the physical and sensory properties of tofu.

2. Materials and Methods

2.1. Materials

Soybeans (*Glycine max* Merr.) were obtained from a local market (NH market, Anseong, Korea). Food-grade DSF was obtained from Sam Chang Industry Co. (Anseong, Korea). Magnesium chloride, glucono-δ-lactone, and sodium chloride were obtained from Samchun Chemical Co. (Seoul, Korea). All of those chemicals were food grade.

2.2. Preparation of Coarse, Fine, and Superfine Fractions of DSF

The DSF powder was sieved twice (150 and 63 μm testing sieves; Nonaka Rikaki, Tokyo, Japan) to obtain coarse and fine fractions (Figure 1). The powders that could not pass through the 150 μm sieve were collected as the coarse fraction. The powders that passed through the 150 μm sieve but were retained by the 63 μm sieve were collected as the fine fraction. The coarse fraction of DSF was then further pulverized by a fluidized-bed jet mill (CGS-10, Netzsch GmbH, Selb, Germany), yielding a superfine fraction. Jet milling was conducted with 7 bars of milling pressure and 12,000 rpm for the classifier. The particle size distributions of the DSFs were determined using a laser diffraction particle sizer (Mastersizer 3000, Malvern Instruments, Malvern, UK).

2.3. Tofu Preparation

Tofu was prepared as described with some modifications [16,17]. Soybeans (9 kg) were soaked in tap water with a soybean-to-water ratio of 1:3 (w/w) for 12 h at room temperature. The swollen soybeans were drained and poured into a soymilk grinder equipped with a soy pulp separator (JH3211A, Junghoon Co., Seoul, Korea) using distilled water (1:8, w/w). The resulting soymilk was heated to 95 °C for 5 min. The hot soymilk was then poured into a stainless steel container, and the respective DSFs (coarse, fine, and superfine DSF) were added gradually to the soymilk to final concentrations of 5%. The final volume of soymilk was 28.8 L. The soymilk with DSF was cooled to 80 °C and combined with coagulant solution. The coagulant solution was composed of 45 g of magnesium chloride, 74 g of GDL, and 72 g of sodium chloride in 1 L of distilled water. The curd was left at ambient temperature to coagulate for 10 min before being transferred to a perforated stainless steel container lined with cheesecloth. The whey in the curd was removed by pressing at 0.2 kg/cm^2 for 25 min. The tofu yield was expressed as the kg of tofu per 28.8 L of soymilk. Ordinary tofu without any addition of DSF was used as a control. The fibrous tofu varieties containing 5% coarse, fine, and superfine DSFs were labeled as tofu with coarse DSF, tofu with fine DSF, and tofu with superfine DSF, respectively.

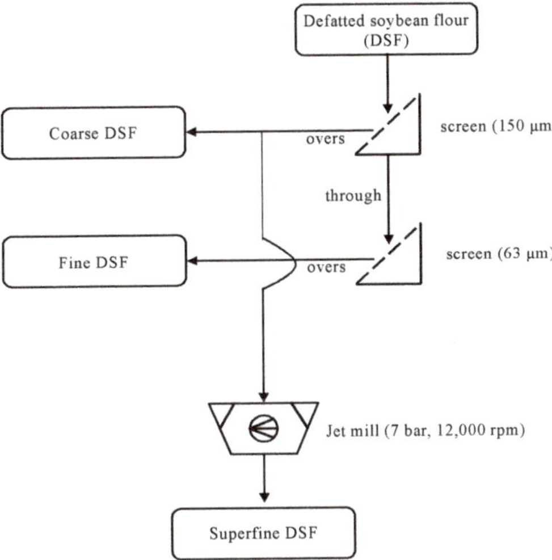

Figure 1. The process flow for the preparations of coarse, fine, and superfine defatted soybean flour (DSF) using serial screening and jet milling.

2.4. Moisture Determination

The amount of moisture in each tofu sample was determined according to the AACC method [18] with some modifications. A sliced, 1.0 g amount of freshly prepared tofu was placed on an aluminum dish. The sample was then dried at 105 °C in a drying oven for 24 h. The moisture content was calculated as the percentage of water content per 1 g of fresh tofu.

2.5. Microstructure Observation

The tofu sample for scanning electron microscopy (SEM) was cut into cubes (5 × 5 × 5 mm), and freeze-dried. Lyophilized samples were coated with platinum–lead (Pt–Pb) by an ion coater (E-1010, Eiko Co., Hyogo, Japan) and observed under a SEM system (S-3400N, Hitachi, Tokyo, Japan).

2.6. Color Measurement

The color of the tofu was determined based on the CIE L^* (lightness), a^* (redness/greenness), and b^* (yellowness/blueness) values using a colorimeter (UltraScan Pro, HunterLab, Reston, VA, USA). A standard white plate with $L^* = 97.49$, $a^* = 0.13$, and $b^* = 0.04$ was used for calibration. The tofu samples were cut into 5.0 × 5.0 × 1.0 cm cubes, and their colors were measured. Each sample was measured three times. The color differentiation (ΔE) between the control tofu and tofu containing 5% DSF was calculated as follows:

$$\Delta E = \sqrt{(\Delta L^*)^2 + (\Delta a^*)^2 + (\Delta b^*)^2}$$

2.7. Texture Profile Analysis (TPA)

The texture of the tofu was analyzed according to a texture profile analysis (TPA) using a texture analyzer (TA-XT2i, Stable Micro Systems, Surrey, UK) with a 35 mm diameter compression plunger. Each tofu sample was cut into a cylindrical shape with a diameter of 20 mm and a height of 15 mm. Each cylindrical sample was placed on the center of the TPA plate and compressed twice to 50% of its original height by a cylinder probe (20 mm in diameter, P/20) at a constant speed of 2 mm/s. The texture profile analysis curve was recorded, and the hardness, adhesiveness, springiness, cohesiveness, gumminess, and chewiness were calculated automatically [19].

2.8. Sensory Evaluation Using Descriptive Analysis

The sensory properties of the tofu samples were evaluated by the descriptive analysis method [20] with some modifications made to use standard reference products available in local markets. For the sensory evaluation, freshly made tofu was cut into 3.0 cm cubes and placed in white dishes with three-digit random numbers. The evaluation was repeated three times for each panelist ($n = 3 \times 15$). The samples were presented monadically to the judges at room temperature under white fluorescent lighting. Fifteen panelists, consisting of Chung-Ang University graduate students who were experienced in descriptive analysis, were first trained with previously identified standard references (Table 1). Four sensory attributes (hardness, springiness, mouthfeel, and beany flavor) were adapted and slightly modified from previous studies [21]. The trained panelists were asked to score the sensory attributes of the tofu samples. Before and between the tests, the panelists were instructed to drink water to clean their mouths. A nine point intensity scale was used to express the intensity of each sensory attribute.

Table 1. Descriptors used for the sensory evaluation of tofu by descriptive analysis.

Descriptors	Definition		Position	
			1	9
Hardness	Force required to compress the sample with molars	Intensity Standard	Soft Cream cheese	Hard Raisin
Springiness	Degree to which sample returns to the original shape after compression with molars	Intensity Standard	Not springy Cream cheese	Very springy Chewy cheese
Mouthfeel	Perception of the particles against the roof of the mouth	Intensity Standard	Smooth Soft curd	Rough Defatted rice bran
Beany flavor	Flavor associated with soy foods	Intensity Standard	No bean flavor Soymilk, 50%	Strong bean flavor Soymilk, 99%

2.9. Statistical Analysis of the Data

All experiments were performed in triplicate. The results are expressed as the means ± standard deviations. ANOVA and Duncan's multiple range comparison tests were performed using the SPSS version 20.0 software (IBM Corp., Armonk, NY, USA) for statistical analyses; p-values < 0.05 were considered significant.

3. Results and Discussion

3.1. Particle Size Distribution of Defatted Soy Flour

The particle size distributions in the DSF samples are shown in Table 2. The results for each are presented as the mean of the volume-weighted diameter ($D_{[4,3]}$), the equivalent diameter at a cumulative volume of 10% (Dv_{10}), the equivalent diameter at a cumulative volume of 50% (Dv_{50}), the equivalent diameter at a cumulative volume of 90% (Dv_{90}), and the homogeneity (span value). The span value shows the uniformity of the particle size. A smaller span value indicates a more homogenous size and a narrower particle-size distribution. All DSF samples had low span values below 1.50.

Table 2. Particle size distributions of the defatted soybean flours (DSFs).

Sample	$D_{[4,3]}$	Dv_{10}	Dv_{50}	Dv_{90}	Span
Coarse DSF	344.0 ± 3.0 ^{a}	154.3 ± 1.5 ^{a}	341.0 ± 2.0 ^{a}	546.3 ± 6.0 ^{a}	1.15 ± 0.01
Fine DSF	106.3 ± 0.6 ^{b}	64.03 ± 0.4 ^{b}	105.3 ± 0.6 ^{b}	158.0 ± 1.0 ^{b}	0.89 ± 0.01
Superfine DSF	5.7 ± 0.1 ^{c}	2.4 ± 0.1 ^{c}	5.1 ± 0.1 ^{c}	9.7 ± 0.1 ^{c}	1.44 ± 0.00

Values are the means of triplicates and expressed as the means ± standard deviations. No significant difference was observed between the means designated by the same letter (Duncan's $p < 0.05$).

3.2. Yield and Moisture Content of Tofu

The yields and moisture contents of the tofu samples made with the different DSFs are presented in Table 3. The yield of tofu increased slightly with a reduction in DSF particle size. There were no significant changes in the tofu yield between the control tofu (without DSF) and the tofu samples with coarse, fine, and superfine DSF. The moisture content of the tofu was increased significantly by the application of jet milling to the DSF. The moisture contents of the tofu with coarse and fine DSF were almost identical (69.9% and 69.8%, respectively). However, the moisture content of the tofu made with superfine DSF was 75.6%, which is similar to that of the control tofu.

Table 3. The effect of the DSF particle size on the yield, moisture content, and color of the tofu.

Sample	Yield (%)	Moisture (%)	Tofu Color			
			L^*	a^*	b^*	ΔE
Tofu without DSF	50.1 ± 4.4 ^{ab}	75.8 ± 1.1 ^{a}	87.55 ± 0.24 ^{a}	0.32 ± 0.07 ^{b}	14.22 ± 0.36 ^{b}	-
Tofu with coarse DSF	47.5 ± 2.5 ^{b}	69.9 ± 2.0 ^{b}	86.16 ± 0.57 ^{b}	0.87 ± 0.45 ^{a}	14.67 ± 0.28 ^{a}	1.67 ± 0.59 ^{a}
Tofu with fine DSF	48.7 ± 3.5 ^{ab}	69.8 ± 3.9 ^{b}	86.11 ± 0.31 ^{b}	0.75 ± 0.15 ^{a}	14.94 ± 0.13 ^{a}	1.69 ± 0.29 ^{a}
Tofu with superfine DSF	51.0 ± 2.0 ^{a}	75.6 ± 1.5 ^{a}	87.10 ± 0.40 ^{a}	0.50 ± 0.08 ^{ab}	14.93 ± 0.36 ^{ab}	0.88 ± 0.18 ^{b}

Values are the means of triplicates and expressed as the means ± standard deviations. No significant difference was observed between the means designated by the same letter (Duncan's $p < 0.05$).

A previous report indicated that the yield of tofu is strongly related to the aggregation of soy protein, which traps water in the tofu structure [22]. However, an increased solid content seems to have a negative effect on the moisture content of tofu. Cai et al. [23] showed that tofu containing a high proportion of solids had a low moisture content. The same result, that the interaction of the components of soybeans and additional solid particles decreased the water holding capacity of tofu, was also reported by Lim et al. [24]. These results suggest that the addition of DSF to tofu could reduce its ability to retain water and form a gel and might result in tofu with a reduced moisture content. In this study, the addition of coarse and fine DSF to tofu resulted in a moisture content of 70%, whereas the moisture content of tofu with superfine DSF was 75.6%. The increased moisture content of tofu with superfine DSF can be explained by the increased hydration properties of the superfine DSF. Our previous studies have shown that superfine DSF has increased hydration properties, such as a water-holding capacity and swelling capacity [15]. The water-holding capacity and swelling capacity explain the ability of a moist material to retain water. The higher water-holding capacity of the jet-milled, superfine DSF may explain the increased moisture content and yield of the tofu samples.

3.3. Microstructure and Color of Tofu

Figure 2 shows scanning electron micrographs of the tofu with different DSFs. As shown in Figure 2a, tofu made without DSF has an obvious gel network and forms a large cell-like structure with a flat wall. This gel network seems to be affected by the addition of DSFs in the other samples. Tofu made with the coarse DSF (Figure 2b) had a non-uniform structure, and a flat wall was not found. Tofu made with fine DSF (Figure 2c) showed a denser network than the control. The microstructure of the tofu made with the superfine DSF (Figure 2d) was more similar to the control than the tofu made with coarse and fine DSF. The DSFs in the tofu made with superfine DSF appear to be entrapped inside the wall structure. Ullah et al. [25] also found a change in the tofu matrix after the addition of Okara dietary fiber due to the interaction of fiber with proteins.

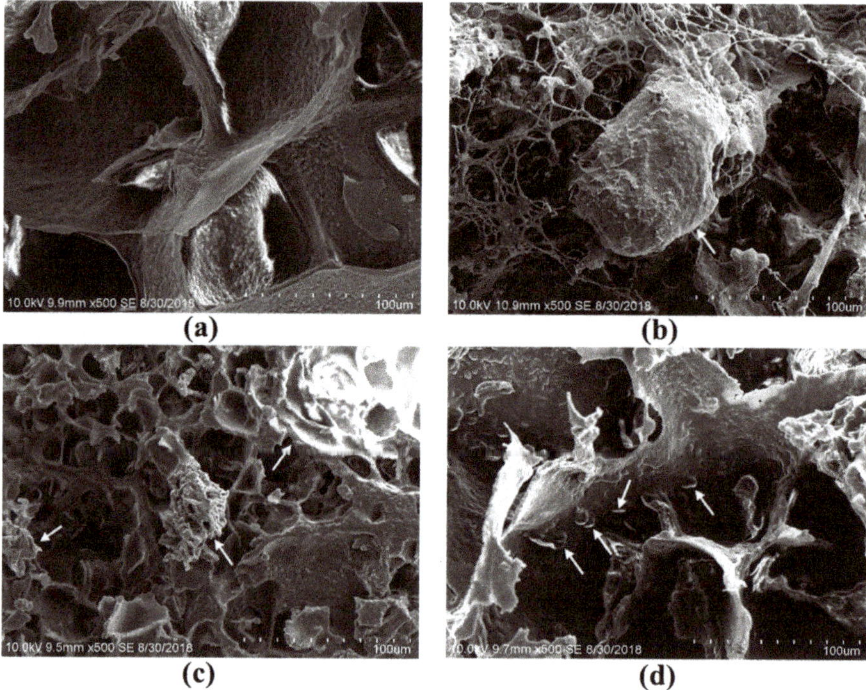

Figure 2. Scanning electron micrographs of tofu made with different fibers at 500× magnification: (**a**) tofu made without DSF, (**b**) tofu made with coarse DSF, (**c**) tofu made with fine DSF, and (**d**) tofu made with superfine DSF. The arrows represent DSFs.

Color is an important property of tofu. Good-quality tofu should have a light-yellow or white color. Table 3 shows the effect of DSF particle size on tofu's color. The lightness (L^* value) of the control tofu was 87.55 ± 0.24 and decreased with the addition of coarse and fine DSFs. The darkening of tofu by the addition of fibrous material was reported by Murugkar [26], who reported that an increased particle size of ground sprout seed in soymilk resulted in a decreased whiteness index of tofu products. However, there was no significant difference in the lightness between the control and the tofu with superfine DSF.

The same trends were observed for the a^* and b^* values of the tofu. The addition of coarse and fine DSFs resulted in increased a^* and b^* values, but the addition of superfine DSF did not change the a^* and b^* values. The tofu with coarse and fine DSFs had a greater ΔE value (meaning that the color of these samples differed significantly from the control tofu; $p < 0.05$). These results are in accordance with previous reports, showing that, compared to raw materials, jet-milled, superfine powders produce brighter and more neutral colors [15,17].

3.4. Textural Properties of Tofu

The textural properties of tofu are associated with its physical properties and play an important role in consumer acceptability [27]. Table 4 shows the results of a texture profile analysis of the control and tofu with DSFs. The addition of larger particles of DSF (coarse and fine DSFs) increased the hardness of the tofu significantly to 0.61 ± 0.05 and 0.60 ± 0.06, respectively, whereas the addition of superfine DSF resulted in no change in hardness compared to the control tofu. The same trend was observed in the springiness and cohesiveness of the tofu. The addition of coarse and fine DSFs reduced the springiness and cohesiveness of the tofu, but the addition of superfine DSF did not change the springiness and cohesiveness of the control tofu. However, DSF's addition did not change the chewiness and gumminess for any of the tofu samples.

Table 4. The effect of DSF particle size on the textural properties of tofu.

Sample	Textural Property				
	Hardness (N)	Springiness	Cohesiveness	Chewiness (J)	Gumminess (N)
Tofu without DSF	0.52 ± 0.05 [b]	0.94 ± 0.02 [ab]	0.60 ± 0.02 [ab]	0.29 ± 0.04 [a]	0.31 ± 0.04 [a]
Tofu with coarse DSF	0.61 ± 0.05 [a]	0.90 ± 0.02 [c]	0.58 ± 0.03 [b]	0.32 ± 0.03 [a]	0.35 ± 0.03 [a]
Tofu with fine DSF	0.60 ± 0.06 [a]	0.92 ± 0.02 [bc]	0.59 ± 0.02 [b]	0.32 ± 0.03 [a]	0.33 ± 0.05 [a]
Tofu with superfine DSF	0.53 ± 0.07 [b]	0.95 ± 0.02 [a]	0.62 ± 0.03 [a]	0.32 ± 0.05 [a]	0.32 ± 0.04 [a]

Values are the means of triplicates and are expressed as the means ± standard deviations. No significant difference was observed between the means designated by the same letter (Duncan's $p < 0.05$).

Hardness was measured by the absolute peak force on the first down stroke during TPA analysis. The addition of conventional DSF to soymilk resulted in an increase in the solid content of the tofu product. Moreover, the force required to break down the fibrous tofu structure was higher than for normal tofu. In contrast, the jet-milled DSF did not behave as a solid in the tofu product due to its superfine particle size. The hardness value of tofu is also related to its moisture content [28]. As described previously, the tofu with superfine DSF had as high a moisture content as the control tofu. These properties led to a soft-textured tofu with superfine DSF. Springiness is a measure of how well a product physically springs back after it has been deformed during the first compression, and products with higher springiness are thought to have higher elasticity. The control tofu and tofu made with superfine DSF had almost identical springiness values of 0.94 ± 0.02 and 0.95 ± 0.02, respectively, whereas the tofu made with coarse and fine DSFs showed slightly reduced springiness values. There were no significant differences in chewiness and gumminess between the control tofu and the tofu with different DSFs.

3.5. Sensory Analysis of Tofu by Descriptive Analysis

The sensory properties of the control and DSF tofu samples were analyzed by the descriptive analysis method (Table 5). Four sensory attributes (hardness, springiness, mouthfeel, and beany flavor) with the previously identified standards are shown in Table 1. The addition of 5% coarse DSF resulted in significant increases in hardness, mouthfeel, and beany flavor compared to the control tofu. The increases of hardness and mouthfeel are negative results, since consumers expect tofu with a soft and smooth structure. The increase of beany flavor can be good or bad depending on individual preference. There was no significant change in springiness.

Table 5. The effect of DSF particle size on the sensory properties of tofu.

Sample	Hardness	Springiness	Mouthfeel	Beany Flavor
Tofu without DSF	3.7 ± 1.6 [b]	5.0 ± 1.4 [a]	2.9 ± 0.7 [b]	3.9 ± 1.1 [c]
Tofu with coarse DSF	6.0 ± 1.2 [a]	4.6 ± 1.4 [a]	5.9 ± 1.1 [a]	6.4 ± 1.2 [a]
Tofu with fine DSF	5.1 ± 1.3 [a]	4.7 ± 1.2 [a]	5.6 ± 1.0 [a]	5.3 ± 1.5 [ab]
Tofu with superfine DSF	3.9 ± 0.9 [b]	5.2 ± 1.2 [a]	3.5 ± 0.7 [b]	4.7 ± 1.1 [bc]

Values are means of three replicates and 15 judges ($n = 3 \times 15$) and are expressed as the means ± standard deviations. No significant difference was observed between the means designated by the same letter (Duncan's $p < 0.05$).

The increased hardness in tofu by the addition of 5% DSF was relieved by the reduction of DSF's particle size. The hardness of the tofu with coarse DSF was 6.0 ± 1.2, while the hardnesses of the tofu with fine and superfine DSFs were 5.1 ± 1.3 and 3.9 ± 0.9, respectively. The mouthfeel showed a similar tendency. The addition of larger particles of DSF (coarse and fine DSFs) resulted in tofu with a rougher mouthfeel. However, the addition of superfine DSF had a reduced mouthfeel, which indicates that soft structured tofu was made due to the smaller particle size of the DSF. A smaller particle size resulted in a less beany flavor compared to the larger DSF. The control tofu had a beany flavor score of 3.9 ± 1.1. The addition of coarse DSF increased the beany flavor score to 6.4 ± 1.2. The addition of superfine DSF resulted in a beany flavor score of 4.7 ± 1.1, which is an increase in value but is not significantly different from the score of the control tofu ($p < 0.05$).

The descriptive analysis results showed the same trend as the textural property values obtained using the texture analyzer, especially for hardness and springiness. The mouthfeel attribute was defined as a rough particle size texture when panelists tasted the tofu. As expected, a larger particle size increased the mouthfeel value of the tofu. The same result was found by Kim et al. [29], who reported that the addition of shell powder during tofu processing increased the mouthfeel score. The limited increase in beany flavor of the tofu with superfine DSF was unexpected. One possible explanation is that compounds inducing a beany flavor are removed during sieving or jet milling. We previously showed that superfine DSF can be a food ingredient because it decreases bitter taste and increases the sweetness of DSF [15]. Our descriptive analysis results indicate that the addition of superfine DSF powder up to a value up to 5% does not change the sensory properties of tofu. Thus, superfine DSF could be a good raw material for enhancing the dietary fiber content of tofu products.

4. Conclusions

The results of this study show that fibrous tofu consist of up to 5% DSF. However, conventionally prepared DSFs with an average particle size larger than 100 μm negatively affected the sensory properties of tofu. The addition of such DSFs resulted in a lower yield, lower moisture content, harder texture, rougher mouthfeel, and increased beany flavor compared to the control tofu. On the other hand, the tofu made with jet-milled DSF overcame all of these disadvantages. The color, textural properties, and sensory qualities were statistically identical to those of the tofu without DSF. Our descriptive analysis results show that consumers will accept tofu with 5% superfine DSF because even trained panelists could not distinguish it from control tofu due to the ultrafine characteristics of the superfine DSF. Therefore, it can be concluded that superfine DSF is a promising additive, for enriching fiber, that does not change the physical or sensory properties during the production of high-quality tofu.

Author Contributions: Conceptualization, D.-U.L.; methodology, Y.-N.K. and S.M., Y.-M.J.; formal analysis, Y.-N.K., S.M. and T.-Y.H.; writing—original draft preparation, S.M. and Y.-M.J.; writing—review and editing, Y.-N.K. and D.-U.L.

Funding: This research was funded by IPET (Korea Institute of Planning and Evaluation for Technology in Food, Agriculture, Forestry and Fisheries), Ministry of Agriculture, Food and Rural Affairs, Republic of Korea.

Conflicts of Interest: The authors declare no conflict of interest.

References

1. Tang, C.-H. Effect of thermal pretreatment of raw soymilk on the gel strength and microstructure of tofu induced by microbial transglutaminase. *LWT-Food Sci. Technol.* **2007**, *40*, 1403–1409. [CrossRef]
2. Kumar, V.; Rani, A.; Tindwani, C.; Jain, M. Lipoxygenase isozymes and trypsin inhibitor activities in soybean as influenced by growing location. *Food Chem.* **2003**, *83*, 79–83. [CrossRef]
3. Liu, H.-H.; Chien, J.-T.; Kuo, M.-I. Ultra high pressure homogenized soy flour for tofu making. *Food Hydrocoll.* **2013**, *32*, 278–285. [CrossRef]
4. USDA National Nutrient Database for Standard Reference. Available online: https://ndb.nal.usda.gov/ndb/foods/show/16427 (accessed on 12 April 2019).
5. Rekha, C.R.; Vijayalakshmi, G. Influence of processing parameters on the quality of soycurd (tofu). *J. Food Sci. Technol.* **2013**, *50*, 176–180. [CrossRef]
6. Ishihara, M.; Singh, H.; Chung, G.; Tam, C. Content composition and antioxidant activity of isoflavones in commercial and homemade soymilk and tofu. *J. Sci. Food Agric.* **2007**, *87*, 2844–2852. [CrossRef]

7. Hui, E.; Henning, S.M.; Park, N.; Heber, D.; Liang, V.; Go, W. Genistein and Daidzein/Glycitein Content in Tofu. *J. Food. Compos. Anal.* **2001**, *14*, 199–206. [CrossRef]
8. Protonotariou, S.; Batzaki, C.; Yanniotis, S.; Mandala, I. Effect of jet milled whole wheat flour in biscuits properties. *LWT-Food Sci. Technol.* **2016**, *74*, 106–113. [CrossRef]
9. Ktenioudaki, A.; Gallagher, E. Recent advances in the development of high-fibre baked products. *Trends Food Sci. Technol.* **2012**, *28*, 4–14. [CrossRef]
10. Lecumberri, E.; Mateos, R.; Izquierdo-Pulido, M.; Rupérez, P.; Goya, L.; Bravo, L. Dietary fibre composition, antioxidant capacity and physico-chemical properties of a fibre-rich product from cocoa (Theobroma cacao L.). *Food Chem.* **2007**, *104*, 948–954. [CrossRef]
11. Grizotto, R.K.; Rufi, C.R.G.; Yamada, E.A.; Vicente, E. Evaluation of the quality of a molded sweet biscuit enriched with okara flour. *Ciência Tecnol. Aliment.* **2010**, *30*, 270–275. [CrossRef]
12. Shogren, R.I.; Mohamed, A.A.; Carriere, C.J. Sensory analysis of whole wheat/soy flour breads. *J. Food Sci.* **2003**, *68*, 2141–2145. [CrossRef]
13. Singh, M.; Liu, S.X.; Vaughn, S.F. Effect of corn bran as dietary fiber addition on baking and sensory quality. *Biocatal. Agric. Biotechnol.* **2012**, *1*, 348–352. [CrossRef]
14. Berk, Z. *Technology of Production of Edible Flours and Protein Products from Soybeans*; Food and Agriculture Organization of the United Nations: Rome, Italy, 1992.
15. Muttakin, S.; Kim, M.S.; Lee, D.-U. Tailoring physicochemical and sensorial properties of defatted soybean flour using jet-milling technology. *Food Chem.* **2015**, *187*, 106–111. [CrossRef] [PubMed]
16. Lee, C.-Y.; Kuo, M.-I. Effect of γ-polyglutamate on the rheological properties and microstructure of tofu. *Food Hydrocoll.* **2011**, *25*, 1034–1040. [CrossRef]
17. Phat, C.; Li, H.; Lee, D.U.; Moon, B.; Yoo, Y.B.; Lee, C. Characterization of Hericium erinaceum powders prepared by conventional roll milling and jet milling. *J. Food Eng.* **2015**, *145*, 19–24. [CrossRef]
18. AACC. *Approved Methods of the American Association of Cereal Chemists*, 8th ed.; Eagen, M.N., Ed.; American Association of Cereal Chemists Inc.: St. Paul, MN, USA, 1983.
19. Bourne, M.C. *Food Texture and Viscosity: Concept and Measurement*, 2nd ed.; Academic Press: Cambridge, MA, USA, 2002; pp. 400–405.
20. Kamizake, N.K.K.; Silva, L.C.P.; Prudencio, S.H. Impact of soybean aging conditions on tofu sensory characteristics and acceptance. *J. Sci. Food Agric.* **2018**, *98*, 1132–1139. [CrossRef]
21. Ziegler, G.R.; Mongia, G.; Hollender, R. The role of particle size distribution of suspended solids in defining the sensory properties of milk chocolate. *Int. J. Food Prop.* **2001**, *4*, 353–370. [CrossRef]
22. Prabhakaran, M.P.; Perera, C.O.; Valiyaveettil, S. Effect of differen coagulants on the isoflavone levels and physical properties of prepared firm tofu from South East Asia. *Int. J. Food Prop.* **2006**, *8*, 113–123. [CrossRef]
23. Cai, T.D.; Chang, K.C.; Shih, M.C.; Hou, H.J.; Ji, M. Comparison of bench and production scale methods for making soymilk and tofu from 13 soybean varieties. *Food Res. Int.* **1997**, *30*, 659–668. [CrossRef]
24. Lim, B.T.; deMan, J.M.; deMan, L. Yield and quality of tofu made from soybeans and soypeanut blends. *J. Am. Oil Chem. Soc.* **1990**, *67*, 381–387. [CrossRef]
25. Ullah, I.; Hu, Y.; You, J.; Yin, T.; Xiong, S.; Din, Z.-U.; Huang, Q.; Liu, R. Influence of okara dietary fiber with varying particle sizes on gelling properties, water state and microstructure of tofu gel. *Food Hydrocoll.* **2019**, *89*, 512–522. [CrossRef]
26. Murugkar, D.A. Effect of different process parameters on the quality of soymilk and tofu from sprouted soybean. *J. Food Sci. Technol.* **2014**, *52*, 2886–2893. [CrossRef] [PubMed]
27. Lee, Y.C.; Rosenau, J.R.; Peleg, M. Rheological characteristic of tofu. *J. Texture Stud.* **1983**, *14*, 143–154. [CrossRef]
28. Prabhakaran, M.P.; Perera, C.O.; Valiyaveettil, S. Quantification of Isoflavones in Soymilk and Tofu from South East Asia. *Int. J. Food Prop.* **2005**, *8*, 113–123. [CrossRef]
29. Kim, Y.S.; Choi, Y.M.; Noh, D.O.; Cho, S.Y.; Suh, H.J. The effect of oyster shell powder on the extension of the shelf life of tofu. *Food Chem.* **2007**, *103*, 155–160. [CrossRef]

Article

Lactic Acid Fermentation as a Pre-Treatment Process for Faba Bean Flour and Its Effect on Textural, Structural and Nutritional Properties of Protein-Enriched Gluten-Free Faba Bean Breads

Nesli Sozer [1,*], Leena Melama [1], Selim Silbir [1,2], Carlo G. Rizzello [3], Laura Flander [1,4] and Kaisa Poutanen [1]

1 VTT Technical Research Centre of Finland Ltd., Tietotie 2, FI02044 Espoo, Finland; leena.melama@unilever.com (L.M.); selim.silbir@ege.edu.tr (S.S.); laura.flander@perheleipurit.fi (L.F.); kaisa.poutanen@vtt.fi (K.P.)

2 Food Engineering Department, Ege University, Izmir 35040, Turkey

3 Department of Soil, Plant, and Food Science, University of Bari, 70121 Bari, Italy; carlogiuseppe.rizzello@uniba.it

4 Bakery Primula Ltd., 04440 Järvenpää, Finland

* Correspondence: nesli.sozer@vtt.fi; Tel.: +358-401523875

Received: 1 September 2019; Accepted: 19 September 2019; Published: 21 September 2019

Abstract: Lactic acid fermentation could be used as a potential modification tool for faba bean flour to enable its incorporation in boosting the nutritional profile of gluten-free breads. Gluten-free breads made with fermented or unfermented faba bean flours were compared with commercial soy flour. The amounts of faba- and soy-bean flours were adjusted to obtain the same protein content in bread (16%). Both fermented and unfermented faba bean flour resulted in larger bread volume (2.1 mL/g and 2.4 mL/g, respectively) compared to bread made with soybean flour (1.5 mL/g). Breads made with unfermented and fermented faba flour had higher porosity (82% and 72%, respectively) than bread with soy flour (61%). The faba breads also were softer than the soy bread. Fermentation of faba flour prior to bread making significantly increased crumb hardness (584 vs. 817 g). Fermentation increased in vitro protein digestibility (72.3% vs. 64.8%). Essential Amino Acid and Biological Value indexes were significantly higher for breads containing fermented faba flour compared to breads made with unfermented faba and soy flour. The Protein Efficiency Ratio and Nutritional Index increased by fermentation from 33 to 36 and 1.6 to 2.7, respectively. Pre-fermentation of faba bean flour improved the nutritional properties of high-protein, gluten-free faba bread. A sensory panel indicated that fermentation did not affect the crumbliness, evenness of pore size and springiness of breadcrumb.

Keywords: gluten-free; legumes; faba beans; fermentation; textural properties; nutritional properties

1. Introduction

Faba bean (*Vicia faba*) is an ancient crop, and together with pea, is one of the most important legume crops in Europe. Faba bean seeds, used in both human and animal nutrition, are rich in essential amino acids: isoleucine, leucine, lysine, phenylalanine, threonine and valine [1,2], and when combined with cereal ingredients (rich in sulphur-containing amino acids) make a plant protein source with well-balanced amino acid composition [3]. Faba bean is also rich in dietary fibre, minerals, phenolics, and non-nutrient secondary metabolites recognized as beneficial in human health [4]. Faba bean storage proteins have been found to lower serum glucose, insulin, total and low-density lipoprotein cholesterol in young men with hypercholesterolemia, whereas L-3,4-dihydroxyphenylalanine (L-DOPA), contained at high levels in seeds, has been reported as effective in the treatment of Parkinson's disease [5,6]. Despite the nutritional and functional properties, faba bean use is limited by the presence of antinutritional factors (ANF), which can interfere with digestion, sometimes causing pathologic conditions. The main ANF are α-galactosides (e.g., raffinose, verbascose, stachyose), condensed tannins, protease inhibitors, phytic acid and vicine and convicine [4,6,7]. The pyrimidine glycosides, vicine and convicine, responsible for causing favism in susceptible individuals (suffering from glucose 6 phosphate dehydrogenase deficiency) [6], are synthesized by the plant as a defence mechanism against fungi and insects [8]. In addition, lectins, found in most legumes, can cause gastrointestinal tract distress [9].

Some of these ANF are heat-labile (e.g., protease inhibitors and lectins) and thus thermal treatments would remove the negative effect during consumption. Others (e.g., phytic acid, raffinose, tannins, saponins) are heat stable [4]. Processing, such as dehulling [10,11], boiling and roasting [12], and germination [13], have been suggested to reduce the ANF content of legumes. Nonetheless, fermentation through lactic acid bacteria (LAB) has proven to be effective in ANF reduction [6,14–17]. Overall, LAB fermentation is a traditional food processing method with many applications, including baked products [18]. It can provide improved sensory, technological and nutritional properties as well as longer shelf life [18]. Recently, fermentation has been reported as being responsible for the enhancement of the overall nutritional and functional quality of faba bean through the increase of the protein digestibility, the release of bioactive peptides and free aminoacids, the decrease of the glycemic index, the increase of mineral bioavailability and antioxidant activity, and by producing γ-aminobutyric acid (GABA), a non-protein amino acid with health promoting effects [6,14–17]. Moreover, the degradation of the pyrimidine glycosides, vicine and convicine, was complete after 48 h of fermentation with a LAB strain selected for the high β-glucosidase activity, and the aglycone derivatives (also toxic) were not detectable. Ex-vivo assays on human blood also confirmed the lack of toxicity of fermented faba bean flour [18].

Like other legume crops, faba bean is gluten-free and could thus be utilized for gluten-sensitive and celiac people. The prevalence of celiac disease has become more and more common in recent years, therefore, the need for tasty and nutritious gluten-free foods is increasing. The structure and texture of gluten-free foods pose challenges for baking technology. Pulses such as chickpea, pea or soybeans have been used in gluten-free bread baking either in the form of flour or protein concentrate supplemented to a starch base [19,20]. Chickpea protein has good emulsifying properties and its supplementation to corn starch (8% of flour mix) improved bread volume, crust/crumb colour and crumb hardness without the need of shortening or emulsifier compared to breads made with cornstarch [19]. Chickpea flour-based gluten-free bread had the best physico-chemical and sensory characteristics compared to bread made with soy or pea protein isolate with equal protein content [20].

We have previously shown that fermentation of faba bean flour by *Lactobacillus plantarum* decreased ANFs, increased the amount of free essential amino acids, improved in vitro protein digestibility and lowered the starch hydrolysis index [13]. After fermentation of faba flour, vicine and convicine contents reduced from 11.46 ± 0.22 to 0.67 ± 0.01 (mg/g dm) and 6.24 ± 0.14 to 0.56 ± 0.01 (mg/g dm), respectively [13]. Trypsin inhibitor (TI) activity dropped from 2.09 ± 0.09 to 0.91 ± 0.04 (TI unit/mg dm), whereas the concentration of condensed tannins halved from 27.10 ± 1.30 to 13.71 ± 0.79 (eq catechin/100 g) by fermentation [13]. The objective of the current work was to evaluate the applicability of fermented faba bean flour for protein enrichment of gluten-free bread especially with respect to structural, textural and nutritional properties.

2. Materials and Methods

2.1. Materials

Faba beans (*Vicia faba* cv Kontu), which were used for the production of faba bean flour, were provided by the University of Helsinki, Finland. The beans were dehulled by using an ultra-fine friction grinder (Supermasscolloider MKZA10-15, Masuko Sangyo Co. Ltd., Kawaguchi, Japan) operated with a 0.5 mm gap, and the hulls were aspirated in a spray drier (Mobile minor, GEA Niro A/S, Soeborg Denmark). Dehulled beans were milled as described in Reference [13] by using a cutting mill (Retsch GmbH, Haan, Germany) to obtain the faba bean flour. Commercial soy flour (native, not heat-treated) (Soyappetit, Helsinki, Finland) (protein 40.3%, fat 21.6%, total dietary fibre 14.2%, total carbohydrates including sugar, 13.3%), corn starch (Maizena, Unilever, Denmark) (starch 86%, total dietary fibre 1%, protein and sugar <0.5%), sugar (Dan Sucker, Suomen sokeri Oy, Kantvik, Finland), baking powder (Meira, Helsinki, Finland), shortening (Sunnuntai, Raisio, Finland), baker's yeast (Suomen hiiva Oy, Lahti, Finland), and salt (JOZO salt, Mariager, Denmark) were obtained from a local store. In addition, emulsifier (Panodan A2020, Dupont, Wilmington, DE, USA) and xanthan gum (Carl Roth GmbH, Karlsruhe, Germany) were purchased.

2.2. *Fermentation of Faba Bean Flour with Lactic Acid Bacteria*

2.2.1. Microorganisms and Culture Condition

Lactobacillus plantarum VTT E-133328, at VTT Culture Collection (Espoo, Finland) was routinely propagated at 30 °C in MRS broth (Oxoid, Basingstoke, Hampshire, England). When used for fermentation, lactic acid bacterial cells were cultivated until the late exponential phase of growth was reached (ca. 10 h), washed twice in 50 mM phosphate buffer, pH 7.0, and suspended in the tap water used for fermentation.

2.2.2. Fermentation Procedure

Faba bean flour was mixed with water to prepare dough in a ratio of 50:50. Fermentation with the lactic acid bacteria (initial cell density of ca. 10^7 CFU/g of dough) was carried out at 30 °C for 48 h as described in Reference [13]. After fermentation, the pH of faba flour was 4.1 and adjusted to higher pH levels around 5.5 by using sodiumbicarbonate to avoid excessive sour taste in the breads. The dough was later freeze-dried (Christ Alpha 2-4 B. Braun, Bioteck International, Osterode am Harz, Germany) and milled in an ultra-centrifugal mill ZM 200 (Retsch GmbH, Haan, Germany) using a 0.5 mm screen and 8000 rpm speed.

The pH drop during fermentation was measured by a TitroLine autotitrator (Alpha 471217, Schott, Mainz, Germany) suspending and aliquot of 10 g of samples in 100 mL distilled water.

Total titratable acidity (TTA) was determined with the TitroLine Alpha autotitrator on 10 g dough homogenized with 90 mL of distilled water and expressed as the amount (ml) of NaOH 0.1 M to reach pH of 8.5.

2.2.3. Pasting Properties by Rapid Visco Analyzer

The Rapid ViscoTM Analyser (RVA) (Newport Scientific Pty Ltd., Warriewood, Australia) was used to determine the pasting properties of the faba bean flours. Pasting properties were determined following the standard Newport Scientific Method 1 (STD1). The heating cycle was 50–95 °C in 282 s, holding at 95 °C for 150 s, and then cooling down to 50 °C in 228 s and holding at 50 °C for 120 s. Each cycle was initiated by a 10 s mixing at 960 rpm paddle speed and 160 rpm paddle speed was used for the rest of the test. The RVA studies were carried out using 3.5 g of dried sample and 25 mL deionized water in an aluminium canister. The parameters recorded were peak, final, breakdown and setback viscosities together with pasting temperature. Flour samples were run in triplicate.

2.2.4. Baking of Gluten-Free Breads

Faba bean flour and fermented faba bean flour (freeze-dried) were applied in gluten-free bread baking and commercial soy flour was used for control breads. The amounts of bean flours were adjusted to obtain the same final protein content (16%) which would enable the "protein-rich" nutritional claim. Bread S37 was soy bread with 37% soy flour and 63% corn starch, bread F50 was faba bread with 50% faba flour and 50% corn starch, and bread FF50 was faba bread with 50% fermented faba flour and 50% corn starch. The proportion of the rest of the ingredients were fixed for each bread recipe and were as follows: sugar 5.7%, baking powder 2.4%, shortening 4.9%, baker's compressed fresh yeast 4.9%, salt 1.7%, xanthan gum 2%, emulsifier 0.6%, water 102.7%. All the percentages were based on baker's percentages. The recipe, mixing, proofing time and baking conditions were optimized in a preliminary work which is not shown here. The dry ingredients were first mixed together and then shortening, yeast and water were added. The temperature of the water was adjusted so that the final dough temperature after mixing was 26 ± 1 °C. Kneading was done for 4 minutes at high speed (level 2/3) with a planetary mixer (Ningbo Sybo Machinery Co., B30CT, Ningbo, China). Small tins (9 × 5 × 6 cm) were filled with 160 g of batter and then proofed for 45 min at 28 °C and 85% relative humidity The breads were baked in a rack oven (Sveba Dahlen, Fristad, Sweden) at 180 °C for 17 min. Baking was conducted twice for each bread recipe.

2.3. Texture Profile Analysis

The texture profile analysis of the crumb was performed after 1 h and 24 h of baking. Two slices were taken from the centre of each loaf. Uniform slices of 2.5 cm thickness were obtained with a cutter and the analysis was carried out with a TA-XT Plus texture analyser (Stable micro systems, Godalming, UK), equipped with a 25 kg load cell and a 25 mm diameter aluminium cylindrical probe. The pre-test and test speeds were identical with a value of 1.7 mm/sand post-test speed was 10 mm/s. A strain value of 40% strain with 5 s pause time between compression cycles was applied. The results are an average of ten replicates.

2.4. Volume Analysis

The specific volume of the bread loaves was measured based on the infrared light scanning method (Bread Vol Scan, Pregesbauer, Germany). Six loaves of each bread type were measured, and the average value was calculated.

2.5. Sensory Analysis

A trained descriptive sensory panel (n = 5) evaluated the characteristics of the breads based on a previously published protocol [21]. The panel consisted of two males and three females with an age range of 27–50 years. Attribute intensities were rated on 5-unit, verbally anchored intensity scales. Altogether, three attribute categories were selected to describe the texture and flavour of the breads. These were: appearance (Evenness of bread on a scale in which 1 = not at all even and 5 = very even, Shape of bread on a scale in which 1 = very asymmetric and 5 = very symmetric, Intensity of bread colour on a scale in which 1 = very light and 5 = very dark), Texture of the crumb (Crumbliness on a scale in which 1 = not at all crumbly and 5 = very crumbly, Evenness of the pore size on a scale in which 1 = very uneven (big pore >20 mm) and 5 = very even, Softness of the crumb on a scale in which 1 = firm and 5 = very soft, Springiness of the crumb on a scale in which 1 = not at all springy (the slice is broken when bended slightly) and 5 = very springy, Toughness on a scale in which 1 = not at all tough (needs only 2 bites before broken down) and 5 = very tough), Crumb flavour and colour (Intensity of the colour on a scale in which 1 = very light and 5 = very dark, Intensity of the flavour on a scale in which 1 = flavourless and 5 = very intense flavour).

2.6. Microstructural Analyses

2.6.1. X-ray Microtomography

Samples for X-ray microtomography (XMT) were made by cutting 1 cubic cm bread samples at 3 different locations from the centre of each bread recipe. Samples were scanned using a desktop XMT system (Model 1172, SkyScan, Aartselaar, Belgium) consisting of an X-ray tube, an X-ray detector and a charge-coupled devices (CCD) camera. The X-ray tube was operated at a voltage of 40 kV/250 μA to obtain optimum contrast between void (air cells) and matter (cell walls) according to a modified method [22]. A 12-bit cooled CCD camera (524 × 1984 pixels) was used to collect the X-ray data. Samples were rotated by a total of 180° during the scanning process with a pixel size of 12.85 μm to obtain optimum resolution, resulting in a total scanning time of 33 min. The initial X-ray radiographs or raw images were obtained at every 0.7° of rotation. Samples were scanned in triplicate. After scanning, radiographs were loaded into NRecon reconstruction software (v. 1.6.6, Micro Photonics, Allentown, PA, USA). The software combines the images graphically into a three-dimensional (3D) object from which 2D cross-sectional images can be taken. Before the reconstruction, the cross-sectionrotation feature was used to rotate the sample cross-sections, making them parallel to the view window. Beam hardening correction and ring artefact corrections were set to 10% and 40% respectively, in order to reduce the number of artefacts. Cell walls of the solid matrix appear grey, whereas air cells appear black. The reconstructed 2D slices were then loaded into CTAn software (v. 1.12, Skyscan, Kontich, Belgium) to obtain the parameters of porosity (%), cell wall thickness (t), and cell diameter (D).

2.6.2. Light Microscopy

Bread samples were first embedded in 2% (*w*/*v*) agar and fixed in 1% (*v*/*v*) glutaraldehyde in 0.1 M Na-K phosphate buffer (pH 7.0), dehydrated in a graded ethanol series, and embedded in hydroxyethyl methylacrylate resin (Leica Historesin embedding kit, Leica Microsystems, Heidelberg, Germany). Polymerized samples were sectioned (2 μm sections) in a rotary microtome HM 355S (Microm Laborgeräte GmbH, Walldorf, Germany) using a tungsten carbon knife and stained after transferring onto glass slides. Staining of protein and starch in 2 mm thick sections were done with Light Green and Lugol's iodine respectively, as described in Reference [13]. The stained sections were examined with an Olympus BX-50 microscope (Olympus Corp., Tokyo, Japan). Micrographs were obtained using a PCO SensiCam CCD colour camera (PCO AG, Kelheim, Germany) and the Cell^P imaging software (Olympus, Tokyo, Japan).

2.7. Nutritional Properties

In vitro starch digestibility was measured based on Reference [23] and represented as hydrolysis index (HI) which refers to the amount of maltose per 1 g of soluble starch as compared to wheat bread. Triplicate sample suspensions containing 1 g of starch in 0.0 5M sodium potassium phosphate buffer (pH 6.9) were placed in water bath (37 °C) and the pH was adjusted to 6.9 with 1M NaOH and 1N HCl. Pancreatic amylase (110 U) was added to the suspensions. Sample aliquots were removed before the enzyme addition and after 30, 60, 120 and 180 min and placed in a boiling water bath for 5 min followed by cooling. Samples were later on analysed for reducing sugar content and absorbance was measured at 540 nm against the reagent blank. The calibration curve was made with maltose. White wheat bread (Isopaahto, Vaasan and Vaasan, Espoo, Finland) was used as control sample.

In vitro protein digestibility was measured based on Akeson and Stahman [24]. One gram of each sample was incubated with 1.5 mg of pepsin in 15 mL of 0.1M HCl at 37 °C for 3 h. After neutralization with 2 M NaOH and addition of 4 mg of pancreatinin 7.5 mL of phosphate buffer (pH 8.0), 1 mL of toluene was added to prevent microbial growth, and the solution was incubated for 24 h at 37 °C. After 24 h, the enzyme was inactivated by the addition of 10 mL of trichloroacetic acid (20%, *w*/*v*), and the undigested protein was precipitated. The volume was made up to 100 mL with distilled water and the mixture centrifuged at 5000 rpm for 20 min. The precipitate was subjected to protein extraction according to Weiss et al. [25]. The concentration of protein in the supernatant and the precipitate was determined by the Bradford method [26]. The in vitro protein digestibility was expressed as the percentage of the total protein, which was solubilized after enzyme hydrolysis.

The supernatant, which contained the digested protein, was freeze-dried and used for further analyses. The method of AOAC 982.30a [27], modified as previously reported by Curiel et al. [28], was used to determine the amino acid profile of the digested protein fraction. Amino acids were analysed by a Biochrom 30 series Amino Acid Analyser [28], with the exception of tryptophan, estimated by the method of Pintér-Szakács and Molnán-Perl [29].

Chemical Score (CS) estimates the amount of protein required to provide the minimal essential amino acid (EAA) pattern, which is present in the reference protein (hen's egg). It was calculated using the equation of Block and Mitchel [30]. The sequence of limiting essential amino acids corresponds to the list of EAA, having the lowest chemical score [30]. The protein score indicates the chemical score of the most limiting EAA that is present in the test protein [30].

Essential Amino Acids Index (EAAI) estimates the quality of the test protein, using its EAA content as the criterion. EAAI was calculated according to the procedure of Oser [31], according to the following equation:

$$EAAI = \sqrt[n]{\frac{(EAA_1 \times 100)(EAA_2 \times 100)(\ldots)(EAA_n \times 100)[sample]}{(EAA_1 \times 100)(EAA_2 \times 100)(\ldots)(EAA_n \times 100)[reference]}}. \quad (1)$$

The Biological Value (BV) indicates the utilizable fraction of the test protein. BV was calculated using the equation of Oser [31]: BV = ((1.09 × $EAAI$) − 11.70). The Protein Efficiency Ratio (PER) estimates the protein nutritional quality based on the amino acid profile after hydrolysis. PER was determined using the model developed by Ihekoronye [32]: PER = −0.468 + (0.454 × (Leucine)) − (0.105 × (Tyrosine)). The Nutritional Index (NI) normalizes the qualitative and quantitative variations of the test protein compared to its nutritional status. NI was calculated using the equation of Crisan and Sands [33], which considers all the factors with an equal importance:

$$\text{NI} = (EAAI \times \text{Protein } (\%)/100). \quad (2)$$

2.8. Statistics

Data were subjected to analysis of variance using IBM SPSS Statistics 21 (IBM Corporation, New York, NY, USA), and significant differences ($p < 0.05$) between individual means were identified by Tukey's test.

3. Results and Discussion

3.1. Chemical and Physical Characterisation of Fermented and Native Faba Flour

Fermentation did not have a significant impact on the protein (35% dm) and soluble starch (42% dm) contents of the faba flour, whereas total dietary fibre (DF) (7% versus 6.7% dm) and fat contents (1.5% versus 0.9% dm) were slightly but significantly reduced. Reduction in total DF and fat content by fermentation of various pulses such as pigeon pea, gram, kidney bean and cowpea has been observed earlier [34,35]. DF solubilisation is known to take place in fermentation of cereals largely due to the action of endogenous enzymes present in the raw material [36]. The dietary fibre content of faba bean is approximately 10% and consists mainly of insoluble fibre such as hemicelluloses [37]. The reduction in DF content might be also associated with the extensive depolymerisation resulting in hydrolysis products, which will not be analysed as DF, or on the ability of fermentative microorganisms to consume hydrolysis products [38].

Fermentation of faba bean flour reduced setback (native flour: 959 ± 13 cP, fermented faba bean flour: 169 ± 8 cP) and final viscosities (native flour: 1642 ± 8 cP, fermented faba bean flour: 864 ± 11 cP) in the RVA analysis of flour slurry, as reported earlier for various starch matrices including, for example, mung bean [39]. Setback viscosity is an indicator of starch retrogradation and can be reduced by decreasing the pH of the flour matrix as the starch granules become fragile and break down easily at low pH by heating [40]. The setback value for faba bean flour reduced by 83% after lactic acid fermentation, where the pH dropped from the initial value of 6.6 to 4.1. This might have lowered the degree of starch retrogradation after heating. Acidic treatment affects the pasting properties of starches by partial hydrolysis of glycosidic linkages of starch molecules [40]. The impact of fermentation on the other RVA parameters of faba flour such as peak and breakdown viscosity and pasting temperature were not significant.

3.2. Macro and Micro Structure of Bread

The impact of flour pre-fermentation on the structure of breads was evaluated both at macro- and micro-scales. Faba breads had higher volume than soy bread, whereas fermentation reduced the volume yield of bread. The volume yields were significantly different among the breads: 2.4 ± 0.1 ^{a} and 2.1 ± 0.1 ^{b} (mL/g) for unfermented and fermented faba bread respectively, and 1.5 ± 0.05 ^{c} mL/g for soy bread. GF bread dough generally resembles a cake batter and added proteins act as emulsifiers by forming a film or skin around oil droplets, which prevents structural failures due to coalescence. The good emulsifying capacity of faba bean proteins was thus reduced by fermentation [41]. Fermentation has been shown to reduce the foaming stability of faba bean flour by 66% [42], which at least partly explains to reduced functionality in gluten-free bread as well. The lowest volume, however, was for the soy bread (S37) used as a benchmark. Earlier, supplementation of gluten-free corn and potato starch-based breads with 10% soy protein concentrate (72% protein content, 7% protein in bread) decreased the specific volume [43]. In addition, supplementation of gluten-free rice flour-based bread with 13% to 25% (flour weight basis) of soybean protein isolate (81% protein content, 10.5%–20% protein in bread) had a detrimental effect on crumb structure [44]. Faba flour thus could be a better option for protein enrichment of GF bread.

Faba breads had better loaf shape, smooth crust and better crumb colour and structure compared to soy bread (Figure 1a). The microstructural characterization results were in line with the macrostructural data as faba breads also had a more porous, open structure (Figure 1b, Table 1). Native and fermented faba breads (F50 and FF50) with the same 16% protein content as the soy bread (S37) had 82% and 72% total porosity respectively (Table 1), flour fermentation thus decreasing porosity. Faba bread had an air cell diameter of 690 μm, fermented faba bread of 561 μm, and soy bread of 230 μm, having the lowest value. The reduction in air cell diameter by fermentation of faba flour was not significant. On the other hand, fermented faba bread had the thickest cell wall (109 μm). We hypothesize that the cell wall thickness was mainly affected by the surface and functional properties of the faba bean protein. Furthermore, the hydrolysis of protein and reduction in foam formation and emulsification capacity of faba bean proteins by fermentation might have adversely affected the air cell formation during bread processing and resulted in smaller but thicker cell wall formation.

It has earlier been shown that fermentation improves the structure, texture and shelf life of wheat bread [45]. The literature is not, however, consistent for gluten-free baking. For example, in another plant protein-rich GF-bread matrix, replacing 20% of oat flour with oat sourdough (produced with *L. plantarum*) did not improve the loaf volume, aroma or microbiological shelf life of the gluten-free oat bread [46]. Lactic acid fermentation of soy-, rice- and buckwheat-based gluten-free mixture hydrolysed soy proteins after 24 h fermentation, but the impact of fermentation in improving texture and structure was minor compared to wheat baking [47]. The microstructures of bread samples as analysed using light microscopy are presented in Figure 2. As expected, no orientated or organized protein network typical for wheat breads was observed. In addition, starch was the continuous phase and protein appeared as the discontinuous phase. However, it is known that gelling properties of faba bean protein is better than soy protein, and the minimum concentration required for a protein dispersion to form a self-supporting network is 14% and 16% for faba and soy proteins, respectively [48]. This might have further contributed to the superior structural properties of faba breads compared to the control soy bread. Large soy protein aggregates (green) surrounded by starch (dark blue) was visible for soy bread (Figure 2). The protein aggregates in both faba breads were clearly smaller than in soy bread. For unfermented faba bread (F50), larger protein particles were dispersed throughout the continuous starch phase. After fermentation of faba flour, these protein particles are degraded to finer particles, as seen in fermented faba bread (FF50) (Figure 2).

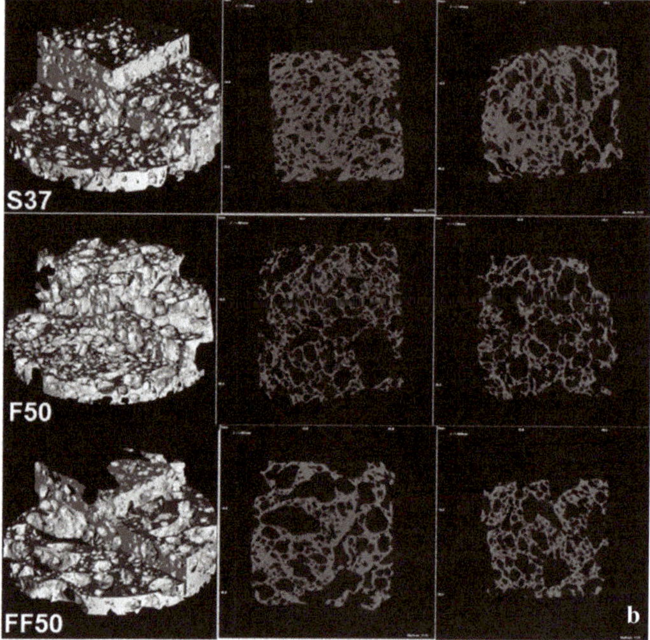

Figure 1. (**a**) Digital images of soy bread (S37), faba bread (F50), fermented faba bread (FF50). (**b**) Rendered three-dimensional (3D) images and selected 2D reconstructed X-ray microtomography (XMT) images of bread samples.

Table 1. Microstructural parameters of image analysis data by XMT.

Bread Samples	Average Cell Wall Thickness (μm)	Average Cell Diameter (μm)	Total Porosity (%)
S37	63 ± 9 [d]	230 ± 54 [a]	61 ± 2 [a]
F50	87 ± 11 [e]	690 ± 8 [bc]	82 ± 2 [c]
FF50	109 ± 29 [f]	561 ± 61 [b]	72 ± 4 [b]

Means in the same column with different letters are significantly different ($p < 0.05$).

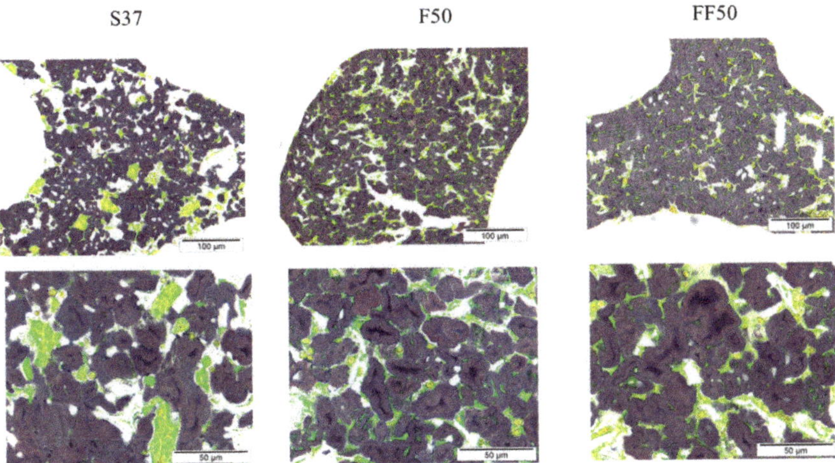

Figure 2. Light green and lugol's iodine staining images of soy bread with 16% protein content (**S37**), faba bread with 16% protein content (**F50**), fermented faba bread with 16% protein content (**FF50**). Green represents protein and dark blue is starch (mainly amylose).

3.3. Textural Properties

Hardness, chewiness and cohesiveness of the breadcrumb as a function of fermentation-induced changes and storage time are presented in Figure 3. Fermentation of faba flour prior to bread making significantly increased crumb hardness (F50 (584 g) versus FF50 (817 g). Air cell diameter, air cell wall thickness and porosity influence the textural properties of baked products [22]. Hardness was negatively related to bread volume, total porosity and pore size. Negative correlation between crumb hardness and volume has also been reported by others [20]. In the current study, the changes in protein structure were proposed to be the key factor responsible for the structural and textural differences as the amount of other structuring agents such as gums, emulsifiers and yeast was constant. Beyond 15% (flour basis) soy flour addition resulted in a hard, compact and low loaf volume for GF bread based on rice and cassava flour [49], whereas soy flour substitution levels lower than 12.5% gave structural failure (e.g., internal cracks). Faba bean protein has water holding capacity similar to soy protein but superior whippability and foam stability [50] which might explain better structural and textural properties of faba breads over soy bread. However, solubility of faba bean protein is reduced at pH levels 4–5. After 48 h of fermentation, the pH of faba flour was 4.1 but before freeze drying it was adjusted to higher pH levels around 5.5 by using sodiumbicarbonate.

Earlier, freeze drying of faba bean protein isolates was shown to reduce the solubility and emulsification properties in protein/water suspension with a maximum 4% protein content compared to spray-dried counter parts [51]. The effect was stated to be lower when protein content increased [51]. As the protein content in this study is much higher (16% of flour mix), the effect of freeze-drying on protein functionality can be expected to be minor compared to the effects caused by fermentation. The TTA values of bread doughs after proofing was 4.66 ± 0.16, 5.53 ± 0.05, 11.04 ± 0.15 mL of NaOH 0.1 M/10 g of dough for S37, F50 and FF50 respectively, whereas the pH values were 6.5, 6.6 and 5.6.

Degradation of the protein matrix by acidification and weakening of the structure reduced the carbon dioxide holding capacity of dough, lowering both the bread volume and porosity that further increased hardness. For all breads, hardness value significantly increased the next day. Chewiness of all faba breads was lower than soy bread (Figure 3). Chewiness reduced after 24 h storage for all breads despite the increase in hardness, which could be linked to the pronounced reduction in cohesiveness. Faba breads were less cohesive than soy bread (Figure 3). Soy crumb was wet and had a more plastic structure (Figure 1a), which could explain the higher cohesiveness value. Fermentation slightly but significantly reduced the cohesiveness of faba breads when they were fresh, but no significant difference was observed after 24 h storage.

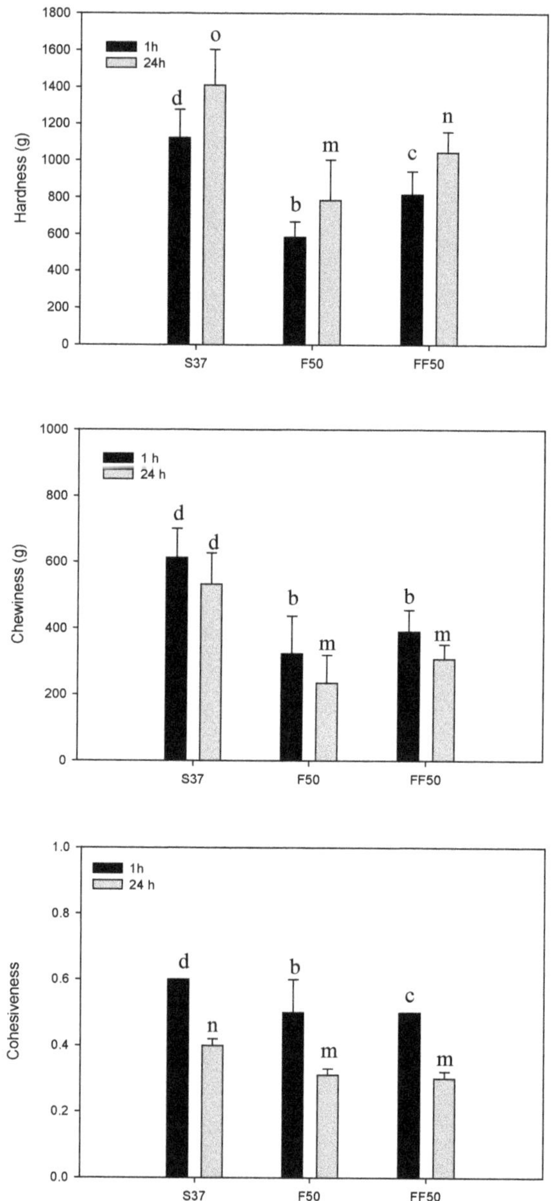

Figure 3. Textural parameters of bread samples after 1 h and 24 h of baking.

3.4. Sensory Properties

The general appearance of faba breads was better than soy bread in terms of evenness, shape and colour intensity of bread (Table 2). No significant difference was caused by fermentation. Crumbliness, evenness of pore size and springiness of crumb were similar for all breads. However, F50 bread made with unfermented faba flour had the highest score for crumb softness. Intensity of colour was similar for all breads (Table 2), whereas fermentation of the faba flour prior to bread making increased the intensity of crumb flavour. During lactic acid fermentation proteolysis, acidification and formation of volatile compounds influenced the flavour profile [21].

Table 2. Sensory properties of bread samples (1 h after baking).

	S37	F50	FF50
Appearance			
Evenness of bread	2.2 ± 0.7 [a]	3.6 ± 0.6 [b]	3.5 ± 0.3 [b]
Shape of bread	2.4 ± 0.9 [a]	3.7 ± 0.5 [b]	3.8 ± 0.5 [b]
Intensity of bread colour	2.5 ± 0.4 [a]	3.5 ± 0.2 [b]	4.0 ± 0.3 [b]
Texture of the crumb			
Crumbliness	2.0 ± 0.1 [a]	2.3 ± 0.4 [a]	1.9 ± 0.4 [a]
Evenness of the pore size	3.5 ± 0.2 [a]	3.3 ±0.3 [a]	3.3 ± 0.2 [a]
Softness of the crumb	3.3 ± 0.3 [ab]	4.2 ± 0.2 [c]	3.0 ± 0.8 [a]
Springiness of the crumb	3.7 ± 0.7 [a]	3.9 ± 0.4 [a]	3.7 ± 0.5 [a]
Toughness	1.7 ± 0.3 [a]	1.3 ± 0.3 [a]	1.7 ± 0.7 [a]
Crumb flavour and colour			
Intensity of the colour	2.8 ± 0.5 [ab]	2.5 ± 0.7 [a]	3.5 ± 0.7 [ab]
Intensity of the flavour	2.8 ± 0.5 [ab]	2.5 ± 0.8 [a]	3.5 ± 0.7 [ab]

Means in the same row with different letters are significantly different ($p < 0.05$).

3.5. Nutritional Characteristics

Bread with fermented faba bean flour (FF50) had a significantly lower hydrolysis index (76 ± 1 [c]) compared to S37 soy bread (93 ± 4 [b]). However, there was no significant difference of HI values between FF50 and F50 (83 ± 2 [bc]). Processing can induce macromolecular interactions between starch and components of the food matrix reducing the availability of starch for digestive enzymes. The chemical changes and interactions between starch-protein that occur in the presence of lactic acid have been suggested to lower the rate of starch hydrolysis [13]. Lactic acid fermentation reduced the in vitro starch hydrolysis rate for sorghum and teff sourdough bread whereas it increased for quinoa and buckwheat sourdough breads [46,52]. The authors attributed the varying effect of fermentation to the structural factors such as increased loaf volume and higher porosity in quinoa and buckwheat sourdough breads [46,52].

Proteins are key components, which contribute to the nutritional value of foods. The quality of proteins is estimated through their amino acid composition, which, combined with protein digestibility, may predict the nutritional value. In vitro digestibility gives information about the stability of protein hydrolyzates, and resistance to digestion. Fermentation increased the in vitro protein digestibility of breads from 53.9% to 72.3% (Table 3) to values even higher than that of soy bread (64.8%) (Table 3). This might be attributed to proteolysis during LAB fermentation [28]. Furthermore, fermentation increased almost all the chemical scores of faba bread to values higher than those of soy bread. All samples contained essential amino acids and the chemical score indicated that Methionine and Cysteine were, respectively, the first and second limiting amino acids. EAAI indicates the ratio of essential amino acids of the sample compared to the reference. BV estimates the nitrogen potentially retained by the human body after consumption. Both EAA and BV indexes were significantly higher for fermented faba bread (FF50) compared to unfermented faba bread F50 and soy bread (S37). The Protein Efficiency Ratio (PER), which reflects the capacity of a protein to support the body weight gain, increased by fermentation from 32.7 to 35.7 but was still lower than soy bread (37.1). Among the indexes that are used to evaluate the nutritional value of foods, only the NI combines qualitative and quantitative factors. Indeed, NI is considered a global predictor of the protein quality. The Nutritional Index (NI) of faba bread (F50, 1.57) increased by fermentation (FF50, 2.47) to levels even higher than the soy bread (S37, 2.1). Fermentation markedly increased protein bioavailability.

Table 3. Nutritional indexes of breads.

	S37	F50	FF50
In vitro protein digestibility (%)	64.8 ± 0.1 [c]	53.9 ± 0.2 [d]	72.3 ± 0.2 [b]
Chemical score (%)			
Histidine	88 ± 2 [b]	85 ± 1 [c]	92 ± 2 [a]
Isoleucine	72 ± 2 [a]	65 ± 2 [b]	64 ± 2 [b]
Leucine	96 ± 1 [a]	88 ± 2 [b]	96 ± 2 [a]
Lysine	105 ± 2 [b]	113 ± 3 [a]	114 ± 1 [a]
Cysteine	35 ± 2 [c]	33 ± 1 [c]	51 ± 2 [a]
Methionine	27 ± 1 [c]	29 ± 1 [c]	32 ± 1 [b]
Phenylalanine + Tyrosine	58 ± 1 [b]	49 ± 1 [c]	63 ± 2 [a]
Threonine	74 ± 3 [b]	78 ± 2 [a]	78 ± 1 [a]
Valine	70 ± 2 [a]	69 ± 2 [a]	70 ± 2 [a]
Tryptophan	35 ± 1 [c]	44 ± 1 [b]	62 ± 2 [a]
Sequence of limiting EAA			
	Methionine	Methionine	Methionine
	Cysteine	Cysteine	Cysteine
	Tryptophan	Tryptophan	Tryptophan
Protein score (%)	27 ± 2 [c]	29 ± 1 [c]	32 ± 2 [b]
Essential Amino Acid Index (EAAI)	61.5 ± 0.4 [b]	58.4 ± 0.3 [c]	63.4 ± 0.4 [a]
Biological Value (BV)	55.4 ± 0.3 [b]	51.9 ± 0.1 [c]	57.4 ± 0.4 [a]
Protein Efficiency Ratio (PER)	37.1 ± 0.4 [a]	32.7 ± 0.3 [c]	35.7 ± 0.4 [b]
Nutritional Index (NI)	2.1 ± 0.07 [c]	1.57 ± 0.12 [d]	2.47 ± 0.08 [b]

Data are expressed as the mean of the results collected in two independent baking tests. [a–d] Values in the same row with different superscript letters differ significantly ($p < 0.05$).

4. Conclusions

The present study showed that fermentation can be used as a modification tool particularly for improving the nutritional properties of faba bean flour. Fermentation of faba bean flour increased in vitro protein digestibility (72.3% versus 53.9%) of gluten-free faba breads. EAA and BV indexes as well as PER were significantly increased for breads made with fermented faba bean flour. Furthermore, fermentation increased the NI of faba bread (2.47) to levels even higher than the soy bread (2.1). No significant difference was caused by fermentation on the general appearance of faba breads (evenness, shape and colour intensity). A sensory panel indicated that crumbliness, evenness of pore size and springiness of crumb were similar for all breads.

Generally, gluten-free products are inferior in their nutritional properties as they are mainly high in carbohydrates and low in protein and dietary fibre. The incorporation of fermented faba bean flour will not only improve nutritional quality, as stated above, but also enable a source of protein (a minimum of 12% of the energy value of the food is provided by protein) or high-protein (a minimum of 20% of the energy value of the food is provided by protein) claims in gluten-free products if used as a protein ingredient. The gluten-free faba bean bread of this study would be qualified for "source of protein" and "source of fibre (at least 3 g of fibre per 100 g of product)" claims.

Author Contributions: Data collection, L.M., S.S. and C.G.R., Planning of the work and supervision: N.S., L.F. and K.P., Writing–original draft paper, N.S., C.G.R., L.M.; Writing—review and editing; all authors.

Funding: No external funds were received for this work.

Acknowledgments: Ulla Holopainen-Manttila and Leila Kostamo (VTT) are acknowledged for their excellent assistance in microscopic and baking analyses, respectively. Fred Stoddard (University of Helsinki) and Chibuzor Eneh (University of Eastern Finland) are acknowledged for providing the faba beans and assistance in XRT imaging of breads, respectively.

Conflicts of Interest: The authors declare no conflict of interest.

References

1. Khalil, A.H.; Mansour, E.H. The effect of cooking, autoclaving and germination on the nutritional quality of faba beans. *Food Chem.* **1995**, *54*, 177–182. [CrossRef]
2. Perez-Maldonado, R.A.; Mannion, P.F.; Farrell, D.J. Optimum inclusion of field pies, faba beans, chick peas and sweet lupins in poultry diets. I. Chemical composition and layer experiments. *Brit. Poult. Sci.* **1999**, *4040*, 667–673. [CrossRef] [PubMed]
3. Derbyshire, E. The nutritional value of whole pulses and pulse fractions. In *Pulse Foods-Processing, Quality and Nutraceutical Applications*; Tiwari, B.K., Gowen, A., McKenna, B., Eds.; Elsevier: London, UK, 2011; pp. 363–383.
4. Gobbetti, M.; De Angelis, M.; Di Cagno, R.; Calasso, M.; Archetti, G.; Rizzello, C.G. Novel insights on the functional/nutritional features of the sourdough fermentation. *Int. J. Food Microbiol.* **2019**, *302*, 103–113. [CrossRef] [PubMed]
5. Duranti, M. Grain legume proteins and nutraceutical properties. *Fitoterapia* **2006**, *77*, 67–82. [CrossRef]
6. Verni, M.; De Mastro, G.; De Cillis, F.; Gobbetti, M.; Rizzello, C.G. Lactic acid bacteria fermentation to exploit the nutritional potential of Mediterranean faba bean local biotypes. *Food Res. Int.* **2019**, *125*, 108571. [CrossRef]
7. Coda, R.; Kianjam, M.; Pontonio, E.; Verni, M.; Di Cagno, R.; Katina, K.; Rizzello, C.G.; Gobbetti, M. Sourdough-type propagation of faba bean flour: Dynamics of microbial consortia and biochemical implications. *Int. J. Food Microbiol.* **2017**, *248*, 10–21. [CrossRef] [PubMed]
8. Multari, S.; Stewart, D.; Russell, W.R. Potential of fava bean as future protein supply to partially replace meat intake in the human diet. *Compr. Rev. Food Sci. Food* **2015**, *14*, 511–522. [CrossRef]
9. Miyake, K.; Tanaka, T.; McNeil, P.L. Lectin-based food poisoning: A new mechanism of protein toxicity. *PLoS ONE* **2007**, *22*, e687. [CrossRef]
10. El-Shemy, H.; Abdel-Rahim, E.; Shaban, O.; Ragab, A.; Carnovale, E.; Fujita, K. Comparison of nutritional and antinutritional factors in soybean and faba bean seeds with or without cortex. *Soil Sci. Plant Nutr.* **2000**, *4646*, 515–524.
11. Kosińska, A.; Karamać, M.; Penkacik, K.; Urbalewicz, A.; Amarowicz, R. Interactions between tannins and proteins isolated from broad bean seeds (*Vicia faba Major*) yield soluble and non-soluble complexes. *Eur. Food Res. Technol.* **2011**, *233*, 213–222. [CrossRef]
12. Cardador-Martínez, A.; Maya-Ocaña, K.; Ortiz-Moreno, A.; Herrera-Cabrera, B.E.; Dávila-Ortiz, G.; Múzquiz, M.; Martín-Pedrosa, M.; Burbano, C.; Cuadrado, C.; Jiménez-Martínez, C. Effect of roasting and boiling on the content of vicine, convicine and L-3, 4-dihydroxyphenylalanine in vicia faba L. *J. Food Qual.* **2012**, *35*, 419–428. [CrossRef]
13. Coda, R.; Melama, L.; Rizzello, C.G.; Curiel, J.A.; Sibakov, J.; Holopainen, U.; Pulkkinen, M.; Sozer, N. Effect of air classification and fermentation by Lactobacillus plantarum VTT E-133328 on faba bean (*Vicia faba* L.) flour nutritional properties. *Int. J. Food Microbiol.* **2015**, *193*, 34–42. [CrossRef] [PubMed]
14. Coda, R.; Varis, J.; Verni, M.; Rizzello, C.G.; Katina, K. Improvement of the protein quality of wheat bread through faba bean sourdough addition. *LWT-Food Sci. Technol.* **2017**, *82*, 296–302. [CrossRef]
15. Verni, M.; Wang, C.; Montemurro, M.; De Angelis, M.; Katina, K.; Rizzello, C.G.; Coda, R. Exploring the microbiota of faba bean: Functional characterization of lactic acid bacteria. *Front. Microbiol.* **2017**, *8*, 2461. [CrossRef] [PubMed]
16. Hammes, W.P.; Brandt, M.J.; Francis, K.L.; Rosenheim, J.; Seitter, M.F.H.; Vogelmann, A. Microbiological ecology of cereal fermentations. *Trends Food Sci. Technol.* **2005**, *16*, 4–11. [CrossRef]
17. Liao, W.; Wang, C.; Shyu, Y.; Yu, R.; Ho, K. Influence of preprocessing methods and fermentation of adzuki beans on γ-aminobutyric acid (GABA) accumulation by lactic acid bacteria. *J. Funct. Foods* **2013**, *5*, 1108–1115. [CrossRef]
18. Rizzello, C.G.; Losito, I.; Facchini, L.; Katina, K.; Palmisano, F.; Gobbetti, M.; Coda, R. Degradation of vicine, convicine and their aglycones during fermentation of faba bean flour. *Sci. Rep.* **2016**, *6*, 32452. [CrossRef]
19. Aguilar, N.; Albanell, E.; Miñarro, B.; Capellas, M. Chickpea and tiger nut flours as alternatives to emulsifier and shortening in gluten-free bread. *LWT-Food Sci. Technol.* **2015**, *62*, 225–232. [CrossRef]
20. Miñarro, B.; Albanell, E.; Aguilar, N.; Guamis, B.; Capellas, M. Effect of legume flours on baking characteristics of gluten free bread. *J. Cereal Sci.* **2012**, *56*, 476–481. [CrossRef]

21. Flander, L.; Suortti, T.; Katina, K.; Poutanen, K. Effects of wheat sourdough process on the quality of mixed oat-wheat bread. *LWT-Food Sci. Technol.* **2011**, *44*, 656–664. [CrossRef]
22. Sozer, N.; Dogan, H.; Kokini, J.L. Textural properties and their correlation to cell structure in porous food materials. *J. Agric. Food Chem.* **2011**, *59*, 1498–1507. [CrossRef] [PubMed]
23. Sozer, N.; Cicerelli, L.; Heiniö, R.L.; Poutanen, K. Effect of wheat bran addition on in vitro starch digestibility, physico-mechanical and sensory properties of biscuits. *J. Cereal Sci.* **2014**, *60*, 105–113. [CrossRef]
24. Akeson, W.R.; Stahmann, M.A. A pepsin pancreatin digest index of protein quality evaluation. *J. Nutr.* **1964**, *83*, 257–261. [CrossRef]
25. Weiss, W.; Vogelmeier, C.; Gorg, A. Electrophoretic characterization of wheat grain allergens from different cultivars involved in bakers' asthma. *Electrophoresis* **1993**, *14*, 805–816. [CrossRef] [PubMed]
26. Bradford, M.M. A rapid and sensitive method for the quantitation of microgram quantities of protein utilizing the principle of protein-dye binding. *Anal. Biochem.* **1976**, *72*, 248–254. [CrossRef]
27. (AOAC) Association of Official Analytical Chemists. *Official Methods of Analysis*, 6th ed.; AOAC: Gaithersburg, MD, USA, 1999.
28. Curiel, J.A.; Coda, R.; Limitone, A.; Katina, K.; Raulio, M.; Rizzello, C.G.; Gobbetti, M. Manufacture and characterization of pasta made with wheat flour rendered gluten-free using fungal proteases and selected sourdough lactic acid bacteria. *J. Cereal Sci.* **2014**, *59*, 79–87. [CrossRef]
29. Pintér-Szakács, M.; Molnán-Perl, I. Determination of tryptophan in unhydrolysed food and feed stuff by the acid ninhydrin method. *J. Agric. Food Chem.* **1990**, *38*, 720–726. [CrossRef]
30. Block, R.J.; Mitchel, H.H. The correlation of the amino acid composition of protein with their nutritive value. *Nutr. Abstr. Rev.* **1946**, *16*, 249–278.
31. Oser, B.L. *Protein and Amino Acid Nutrition*; Albanese Academic Press: New York, NY, USA, 1946; pp. 281–291.
32. Ihekoronye, A.I. A Rapid Enzymatic and Chromatographic Predictive Model for the in-vivo Rat-Based Protein Efficiency Ratio. Ph.D. Thesis, University of Missouri, Columbia, MO, USA, 1981.
33. Crisan, E.V.; Sands, A. *Biology and Cultivation of Edible Mushrooms*; Academic Press: New York, NY, USA, 1978; pp. 137–142.
34. Veena, A.; Urooj, A.; Puttaraj, S. Effect of processing on the composition of dietary fibre and starch in some legumes. *Nahrung* **1995**, *39*, 132–138. [CrossRef]
35. Torres, A.; Frías, J.; Granito, M.; Vidal-Valverde, C. Fermented pigeon pea (*Cajanus cajan*) ingredients in pasta products. *J. Agric. Food Chem.* **2006**, *54*, 6685–6691. [CrossRef]
36. Salmenkallio-Marttila, M.; Katina, K.; Autio, K. Effects of bran fermentation on quality and microstructure of high-fiber wheat bread. *Cereal Chem.* **2001**, *78*, 429–435. [CrossRef]
37. Crépon, K.; Marget, P.; Peyronnet, C.; Carrouée, B.; Arese, P.; Duc, G. Nutritional value of faba bean (*Vicia faba* L.) seeds for feed and food. *Field Crop. Res* **2010**, *115*, 329–339. [CrossRef]
38. Gobbetti, M.; Rizzello, C.G.; Di Cagno, R.; De Angelis, M. How the sourdough may affect the functional features of leavened baked goods. *Food Microbiol.* **2014**, *37*, 30–40. [CrossRef] [PubMed]
39. Chang, Y.H.; Lin, C.L.; Chen, J.C. Characteristics of mung bean starch isolated by using lactic acid fermentation solution as the steeping liquor. *Food Chem.* **2006**, *99*, 794–802. [CrossRef]
40. Chuenkamol, B.; Puttanlek, C.; Rungsardthong, V.; Uttapap, D. Characterization of low-substituted hydroxypropylated canna starch. *Food Hydrocol.* **2007**, *21*, 1123–1132. [CrossRef]
41. Makri, E.; Papalamprou, E.; Doxastakis, G. Study of functional properties of seed storage proteins from indigenous European legume crops (*lupin, pea, broad bean*) in admixture with polysaccharides. *Food Hydrocol.* **2005**, *19*, 583–594. [CrossRef]
42. Chandra-Hioe, M.V.; Wong, C.H.; Arcot, J. The Potential Use of Fermented Chickpea and Faba Bean Flour as Food Ingredients. *Plant Foods Hum. Nutr.* **2016**, *71*, 90–95. [CrossRef]
43. Ziobro, R.; Juszczak, L.; Witczak, M.; Korus, J. Non-gluten proteins as structure forming agents in gluten free bread. *J. Food Sci. Technol.* **2016**, *53*, 571–580. [CrossRef]
44. Marco, C.; Rosell, C.M. Functional and rheological properties of protein enriched gluten free composite flours. *J. Food Eng.* **2008**, *88*, 94–103. [CrossRef]
45. Hartikainen, K.; Poutanen, K.; Katina, K. Influence of bioprocessed wheat bran on the physical and chemical properties of dough and on wheat bread texture. *Cereal Chem.* **2014**, *91*, 115–123. [CrossRef]

46. Wolter, A.; Hager, A.S.; Zannini, E.; Arendt, E.K. In vitro starch digestibility and predicted glycaemic indexes of buckwheat, oat, quinoa, sorghum, teff and commercial gluten-free bread. *J. Cereal Sci.* **2013**, *58*, 431–436. [CrossRef]
47. Moore, M.M.; Juga, B.; Schober, T.J.; Arendt, E.K. Effect of lactic acid bacteria on properties of gluten-free sourdoughs, batters, and quality and ultrastructure of gluten-free bread. *Cereal Chem.* **2007**, *84*, 357–364. [CrossRef]
48. Fernández-Quintela, A.; Macarulla, M.T.; Del Barrio, A.S.; Martínez, J.A. Composition and functional properties of protein isolates obtained from commercial legumes grown in northern Spain. *Plant Foods Hum. Nutr.* **1997**, *51*, 331–341. [CrossRef] [PubMed]
49. Ribotta, P.D.; Ausar, S.F.; Morcillo, M.H.; Pérez, G.T.; Beltramo, D.M.; León, A.E. Production of gluten-free bread using soybean flour. *J. Sci. Food Agric.* **2004**, *84*, 1969–1974. [CrossRef]
50. Sosulski, F.W.; McCurdy, A.R. Functionality of flours, protein fractions and isolates from field peas and faba bean. *J. Food Sci.* **1987**, *52*, 1010–1014. [CrossRef]
51. Cepeda, E.; Villaran, M.C.; Aranguiz, N. Functional properties of faba bean (*Vicia faba*) protein flour dried by spray drying and freeze drying. *J. Food Eng.* **1998**, *36*, 303–310. [CrossRef]
52. Shumoy, H.; Van Bockstaele, F.; Devecioglu, D.; Raes, K. Effect of sourdough addition and storage time on in vitro starch digestibility and estimated glycemic index of tef bread. *Food Chem.* **2018**, *264*, 34–40. [CrossRef] [PubMed]

MDPI
St. Alban-Anlage 66
4052 Basel
Switzerland
Tel. +41 61 683 77 34
Fax +41 61 302 89 18
www.mdpi.com

Foods Editorial Office
E-mail: foods@mdpi.com
www.mdpi.com/journal/foods

www.ingramcontent.com/pod-product-compliance
Lightning Source LLC
LaVergne TN
LVHW070926170726
843515LV00005B/877

* 9 7 8 3 0 3 6 5 0 6 1 4 2 *